Topical Nail Products and Ungual Drug Delivery

Topical Nail Products and Ungual Drug Delivery

EDITED BY

S. Narasimha Murthy
Howard I. Maibach

CRC Press is an imprint of the
Taylor & Francis Group, an **informa** business

CRC Press
Taylor & Francis Group
6000 Broken Sound Parkway NW, Suite 300
Boca Raton, FL 33487-2742

First issued in paperback 2019

ISBN-13: 978-1-4398-1129-0 (hbk)
ISBN-13: 978-0-367-38082-3 (pbk)

Library of Congress Cataloging-in-Publication Data

Topical nail products and ungual drug delivery / editors, S. Narasimha Murthy and Howard I. Maibach.
p. ; cm.
Includes bibliographical references and index.
ISBN 978-1-4398-1129-0 (alk. paper)
I. Murthy, S. Narasimha (Sathyanarayana Narasimha), 1971- II. Maibach, Howard I.
[DNLM: 1. Nail Diseases--drug therapy. 2. Administration, Topical. 3. Nails--drug effects. WR 475]

616.5'47061--dc23 2012031355

Visit the Taylor & Francis Web site at
http://www.taylorandfrancis.com

and the CRC Press Web site at
http://www.crcpress.com

Contents

Preface

Nail diseases, which include infections and inflammatory disorders, could affect human beings at any age. The diseases can result in feelings of inadequacy and depression in patients and affect their quality of life significantly.

Oral and topical delivery of drugs have been the most common approaches to treat nail diseases. Topical therapy is the most preferred mode of drug delivery, due to patient compliance and convenience. Development of topical formulations to deliver effective amounts of drugs into the nail apparatus is highly challenging.

This book provides a comprehensive review regarding the various nail diseases, topical formulations, drug delivery approaches, and unguokinetics of drugs. The chapters are contributed by pioneers in ungual drug delivery.

Editors

S. Narasimha Murthy, PhD, is an associate professor at the University of Mississippi School of Pharmacy, University, Mississippi. After obtaining his PhD in pharmaceutics from Bangalore University, India, Dr. Murthy worked as a research associate at the Roswell Park Cancer Institute, Buffalo, New York, and as an assistant professor at Ohio Northern University, Ada, Ohio. He has published more than 60 research articles in the field of dermal and ungual drug delivery. Dr. Murthy also edited the textbook *Dermatokinetics of Therapeutic Agents*. He has been on the editorial boards of several pharmaceutical journals.

Howard I. Maibach, PhD, is currently serving as a professor of dermatology at the University of California, San Francisco, California. He obtained his MD from Tulane University, New Orleans, Louisiana, in 1955. Later, he served in faculty positions in various levels at the University of California, San Francisco. Dr. Maibach has published more than 2520 articles. He has been on the editorial board of more than 30 scientific journals and is a member of 19 professional societies including the American Academy of Dermatology, San Francisco Dermatological Society, and the Internal Commission on Occupation Health.

Contributors' Biographies

Ali Alikhan, MD, is a resident in dermatology at the Mayo Clinic Department of Dermatology. He received his Bachelors in Science from Kent State Univseríty and Medical Degree from University of California Davis School of Medicine, in addition to completing a transitional residency at MacNeal Hospital in Berwyn, Illinois. He has published well over 60 journal articles and book chapters, and has presented at several national meetings. He has published on melanoma, vitiligo, hidradenitis suppurativa, onychomycosis, medical education, psoriasis, acne, rosacea, allergic contact dermatitis, Fox-Fordyce disease, and dermatopharmacology, among other topics. He holds several local and national leadership positions, and is a reviewer for numerous dermatology and non-dermatology journals.

Corona M. Cassidy, PhD, is an honorary research fellow in the School of Pharmacy at Queen's University Belfast. She obtained an MPharm (First Class) in 2005 and a PhD in pharmaceutical microbiology in 2010, and has since held positions in community pharmacy, academia, and the pharmaceutical industry, where she is based at present. She has research interests in clinical pharmacy and pharmaceutical manufacturing/production technologies.

Antoine Coquerel, PhD, is a neurologist and pediatrician who holds a doctorate degree in neuropsychopharmacology. He has been the head of the Pharmacology Department and laboratory at the University Hospital of Caen, France, since 1999. Dr. Coquerel is also qualified to manage research in the fields of cell biology and neuropsychopharmacology. Since 1999, he has been serving as the head of the Regional Pharmacovigilance Centre of Lower Normandy and the Centre for Evaluation and Information on Drug Dependence for the northwest of France. He is also the director of teaching of the pharmacology of Caen faculty of medicine. Dr. Coquerel does clinical research at the University of Caen as well as the University hospital. He leads a mixed research team at the University of Caen, Basse-Normandie, France. Her clinical research at the University of Caen is focused on "drug and driving," "aging and drugs," and chronobiology in healthy humans. He also performs experimental research on drug dependence in rats and mice. He has 57 publications and is also a coauthor of a French handbook of pharmacology (Masson, Paris, 2002). He is also an active member of French Society of Pharmacology and Therapeutics and the National College of Medical Pharmacology. He is an expert pharmacologist in the scientific committee of the regional clinical research management as well as an expert in the French drug safety agency (AFSSaPS).

Danièle Debruyne, PhD, is a senior scientist in the Department of Pharmacology at the University Hospital Centre of Caen, France. Dr. Debruyne manages the therapeutic monitoring of drugs mainly used in infectious and neurological diseases and in the prevention of graft rejection, with the objective being better efficacy and minor toxicity of the administered treatments owing to an optimal management; according to an agreement with the French Office of Drugs and Drug Addiction, Dr. Debruyne assumed the scientific responsibility of the analysis of illicit and licit recreational drugs that circulate in France to evaluate their dangerousness. Within the research team entitled "Laboratory of Methodological Developments for Positron Emission Tomography (PET)," Dr. Debruyne coordinates the biological characterization (including blood and tissue distribution, metabolism, evaluation of the *in vitro* and *in vivo* specific binding) of the new PET radiotracers developed by radiochemists. Dr. Debruyne has been recognized as an expert by the French Agency for the Safety of Health Products in the field of experimental pharmacokinetics. Dr. Debruyne has published five original research or review articles (on a total of 76 papers indexed in PubMed) concerning antifungal drugs.

Ryan F. Donnelly, PhD, is reader in pharmaceutics in the School of Pharmacy at Queen's University Belfast, UK. Dr. Donnelly's research is centered on design and physicochemical characterization of advanced polymeric drug delivery systems for transdermal and topical drug delivery, with a strong emphasis on improving therapeutic outcomes for patients. His bioadhesive patch design was used in successful photodynamic therapy of over 100 patients with neoplastic and dysplastic gynecological conditions. This technology has now been licensed to Swedish Pharma AB, for whom Dr. Donnelly acts as a technical director. His microneedle technology is currently undergoing commercial development by two leading pharmaceutical companies. Still at a relatively early stage of his career, he has authored over 200 peer-reviewed publications, including three patent applications, three textbooks, seven book chapters, and approximately 90 full papers. He has been an invited speaker at several national and international conferences. Dr. Donnelly is the associate editor of *Recent Patents on Drug Delivery & Formulation* and a member of the editorial advisory boards of *Pharmaceutical Technology Europe* and *Journal of Pharmacy and Bioallied Sciences* and is a visiting scientist at the Norwegian Institute for Cancer Research, where he is an associate member of the Radiation Biology Group. Dr Donnelly is the current holder of the Royal Pharmaceutical Society's prestigious Science Award and is a previous winner of an Innovation Leader Award from the NHS Research & Development Office, a research scholarship from the Research Council of Norway, and the Pharmaceutical Society of Northern Ireland's Gold Medal.

Laila Elkeeb, MD, is a fellow in dermatopathology at the University of Cincinnati Dermatology department. She completed her dermatology residency and a clinical trials fellowship from the University of California, Irvine. She also completed a Melanoma and Cutaneous Oncology fellowship from the University of California, San Francisco. Dr. Elkeeb has several years of research experience in the field of dermatology. She was a co-investigator on several studies that involved; skin cancer biology, laser and light therapy of dermatological diseases, psoriasis, rosacea, acne, and dermatopharmacology. She has published several journal articles, and has presented at several national meetings. Dr. Elkeeb has published in the field of skin cancer biology, laser and light therapy of dermatological diseases, phototoxicity, photoallergic dermatitis, dermatopharmacology, rosacea, onychomycosis, and dermatopathology.

Rania Elkeeb, PhD, is a researcher at Surge Lab, Department of Dermatology, University of California, San Francisco, California. She received her bachelor of pharmacy from Petra University, formerly Jordan University for Women, Amman, Jordan, in 1997. Dr. Elkeeb obtained her PhD in pharmaceutics and industrial pharmacy from the Massachusetts College of Pharmacy and Health Sciences, Boston, Massachusetts, in May 2005. She has done her postdoctoral training in dermatopharmacokinetics and onychopharmacokinetics as relates to their biologic/clinical effects in man and animals in the Department of Dermatology at the University of California, San Francisco, California, in 2008. She served as an adjunct assistant professor of pharmaceutical sciences at the Massachusetts College of Pharmacy and Health Sciences, Worcester, Massachusetts. Her current research interests are trans-ungual and transdermal drug delivery and their absorption efficiency; dermatopharmacokinetic and onychopharmacokinetics analysis; and transfollicular drug delivery.

Jinsong Hao, PhD, obtained her PhD in pharmaceutics from Shenyang Pharmaceutical University, China. After her graduation, she worked at Shenyang Pharmaceutical University (China), National University of Singapore (Singapore), and Nova Southeastern University (United States). She is currently a research assistant professor in the College of Pharmacy at the University of Cincinnati (United States). Her research is in the field of drug delivery. She has published more than 40 research articles in transdermal, transscleral, transcorneal, and trans-ungual drug delivery.

Xiaoying Hui, MD, is an associate research dermatologist in the Department of Dermatology at the University of California, San Francisco, California. He has worked as a principal investigator in Dr. Howard Maibach's laboratory for 20 years. His major interests are dermatotoxicokinetics and risk assessment modeling of environmental chemicals following human skin absorption; antifungal drugs trans-ungual delivery and absorption efficiency; dermal absorption and transdermal delivery; and dermatopharmacokinetic analysis. He has published more than 60 peer-reviewed research articles and is the author of 17 book chapters.

Abhishek Juluri, is a graduate student in the Department of Pharmaceutics at the University of Mississippi, University, Mississippi. He is an NIH predoctoral fellow and a member of the American Association of Pharmaceutical Scientists and Rho Chi. Mr. Juluri received his bachelor's degree in 2009 from Kakatiya University, India.

Majella E. Lane, PhD, is a senior lecturer of pharmaceutics in the School of Pharmacy at University College London. To date, she has been involved in the supervision of over 20 PhD students. She is a visiting professor in the Department of Pharmaceutical Sciences at the University of Michigan, Ann Arbor, Michigan. Over the years, she has contributed to more than 60 peer-reviewed articles and 10 book chapters and serves on the editorial boards of several pharmaceutical science journals. Her major research interests are in the application of physical chemistry to tissue characterization and modulation with special reference to the skin and nail. Her research group collaborates worldwide and uses a range of biophysical techniques (attenuated total reflectance-Fourier transform infra-red spectroscopy, confocal Raman spectroscopy, high-speed differential scanning calorimetry) to probe the mechanisms of skin penetration and its modulation. She also coordinates the Skin Forum, which evolved from an Engineering and Physical Sciences Research Council funded network on skin permeability.

S. Kevin Li, PhD, is an associate professor of pharmaceutics in the College of Pharmacy at the University of Cincinnati, Cincinnati, Ohio. He is also an adjunct associate professor in the College of Pharmacy at the University of Utah, Salt Lake City, Utah. Dr. Li graduated summa cum laude from Brigham Young University, Provo, Utah, where he obtained his bachelor's degree in chemistry. Subsequently, he earned his PhD in pharmaceutics and pharmaceutical chemistry from the University of Utah under the supervision of Dr. William Higuchi and with the support of a predoctoral fellowship from Pharmaceutical Research and Manufacturers of America (PhRMA) Foundation. Dr. Li has published more than 80 articles, patents, and book chapters on transdermal, ocular, and trans-ungual drug delivery and noninvasive pharmacokinetic study using MRI. He is a principal investigator and coinvestigator of research grants funded by the National Institutes of Health (NIH) in the United States. He has frequently served as a reviewer for scientific journals in pharmaceutical sciences, ophthalmology, and MRI research and a member in grant review panels.

Katarzyna Madej, PhD, graduated from the Faculty of Chemistry at the Jagiellonian University, Krakow, Poland, where she defended her master thesis. Her postgraduate study took place at the Faculty of Chemistry in 1987–1991. In 1993, she defended her doctoral thesis on "Development of computerized-aid potentiometric multicomponent titration methods." Then, she was employed in the Institute of Forensic Research, Krakow, Poland, for 7 years. In 1999, she began her career in the Faculty of Chemistry at the Jagiellonian University as an assistant and from 2002 as an assistant professor. Dr. Madej's scientific work mainly concerns development and optimization of analytical procedures for medicaments, especially psychotropic drugs, in biological samples. Her research interests include clinical and forensic toxicology, biological sample preparation techniques, and development of chromatographic and capillary electrophoretic methods. She also participated actively in many national and international conferences and symposia.

Howard I. Maibach, PhD, is currently serving as a professor of dermatology at the University of California, San Francisco, California. He obtained his MD from Tulane University, New Orleans, Louisiana, in 1955. Later, he served in faculty positions in various levels at the University of California, San Francisco. Dr. Maibach has published more than 2520 articles. He has been on the editorial board of more than 30 scientific journals and is a member of 19 professional societies including the American Academy of Dermatology, San Francisco Dermatological Society, and the Internal Commission on Occupation Health.

Sudaxshina Murdan, PhD, is a senior lecturer in Pharmaceutics at the University College London School of Pharmacy. She studied pharmacy at The University of Nottingham and gained her PhD from The School of Pharmacy, University of London (now UCL School of Pharmacy). Her research is in the fields of ungual drug delivery, vaccine delivery, and in pharmacy and development education, and she has authored over 40 peer-reviewed papers, over 70 conference papers, a number of book chapters and articles in industry newsletters and student newspaper. She teaches on the MPharm, MSc and PhD programs and is a fellow of the Higher Education Academy.

S. Narasimha Murthy, PhD, is an associate professor at the University of Mississippi School of Pharmacy, University, Mississippi. After obtaining his PhD in pharmaceutics from Bangalore University, India, Dr. Murthy worked as a research associate at the Roswell Park Cancer Institute, Buffalo, New York, and as an assistant professor at Ohio Northern University, Ada, Ohio. He has published more than 60 research articles in the field of dermal and ungual drug delivery. Dr. Murthy also edited the textbook *Dermatokinetics of Therapeutic Agents*. He has been on the editorial boards of several pharmaceutical journals.

Anroop B. Nair, PhD, is an assistant professor in the College of Clinical Pharmacy at King Faisal University, Al Ahasa, Saudi Arabia. He received his PhD in pharmaceutics from Jadavpur University, Kolkata, India. He was a postdoctoral fellow in S. N. Murthy Research group at the University of Mississippi, University, Mississippi, during which he worked extensively on the iontophoretic drug delivery in nail. He is an active member of several pharmaceutical councils/forums and a reviewer for several peer-reviewed journals in the field of pharmaceutics. He has authored more than 50 peer-reviewed articles.

Michael A. Repka, PhD, is chair and professor of the Department of Pharmaceutics at the University of Mississippi, University, Mississippi, as well as the director of the Center for Pharmaceutical Technology. Dr. Repka joined the faculty at Ole Miss after receiving his PhD from the University of Texas College of Pharmacy and founded a pharmaceutical research/development company that specializes in drug delivery. His research interests include oral transmucosal and transdermal/transnail delivery systems, as well as other novel dosage forms. Many of these systems are directed toward the solubilization and delivery of poorly soluble bioactives via hot-melt extrusion technology, which is a primary focus of his research. In the nail drug delivery area, Dr. Repka worked on development of bioadhesive nail patches and nail etching technology. Her publications include more than 70 peer-reviewed journal articles and book chapters and well over 250 presentations at national/international scientific meetings. She serves on the editorial advisory boards of six prestigious journals, is an associate editor for *AAPS PharmSciTech*, and has been credentialed as a member of the U.S. Pharmacopeial Convention.

H. N. Shivakumar, PhD, is currently working as professor and head in the Department of Pharmaceutics at the KLE University's College of Pharmacy, Bangalore, India. Dr. Shivakumar received his doctoral degree in pharmacy from Rajiv Gandhi

University of Health Sciences, Bangalore, India. He completed his postdoctoral research in S. N. Murthy Research Group at the University of Mississippi, University, Mississippi. His postdoctoral research was focused on passive targeting of microparticulate systems to the lymphatics for various therapeutic interventions. He was also involved in developing innovative drug delivery strategies for transdermal and transungual applications. He holds an appointment as a principal scientist with DermPerm Research Inc., Bangalore, India. Dr. Shivakumar has around 25 research articles in peer-reviewed journals and two book chapters to his credit.

Michael M. Tunney, PhD, is a reader of clinical pharmacy in the School of Pharmacy at Queen's University Belfast. His research interests are centered on clinical pharmacy and pharmaceutical microbiology. His current work focuses on the detection and treatment of polymicrobial infection in a range of respiratory diseases including cystic fibrosis and chronic obstructive pulmonary disease. Other key areas of interest include treatment of biofilm infection of indwelling implants and determination of the factors associated with success or failure of methicillin-resistant *Staphylococcus aureus* decolonization in hospital inpatients. Dr. Tunney has published more than 50 research papers and has contributed to a number of microbiology and biomaterials textbooks.

1 The Nail

Anatomy, Physiology, Diseases, and Treatment

Sudaxshina Murdan

CONTENTS

1.1 THE NAIL UNIT

The nail plate lies on the nail bed, is produced by the nail matrix, and is framed and ensheathed by the nail folds and the hyponychium (Figure 1.1). All these components, that is, nail plate, nail bed, nail folds, matrix, and hyponychium, make up the nail unit (Gonzalez-Serva 1997) and are described below. The nail unit starts to develop in the tenth week of embryogenesis and is almost completely formed by the 17th week, after which changes in the nail unit are mainly associated with growth; it is well perfused by blood and lymphatic vessels, has a rich nerve supply, and is anchored in place by attachment to the distal phalanx (Dawber et al. 2001; Fleckman 2005; de Berker et al. 2007; de Berker and Forslind 2004; Zaias 1990).

The functions of the nail unit are manifold. Equivalent to claws and hooves in other mammals, the nail allows one to manipulate objects, enhances the sensation of fine touch, protects the delicate tips of fingers and toes against trauma, and is used for scratching and grooming (Barron 1970; Dawber and Baran 1984; Chapman 1986; Gonzalez-Serva 1997). The multibillion dollar industry devoted to nail cosmetics and nail salons attests to our use of the nail as a cosmetic organ, even to the extent of sacrificing function for beauty.

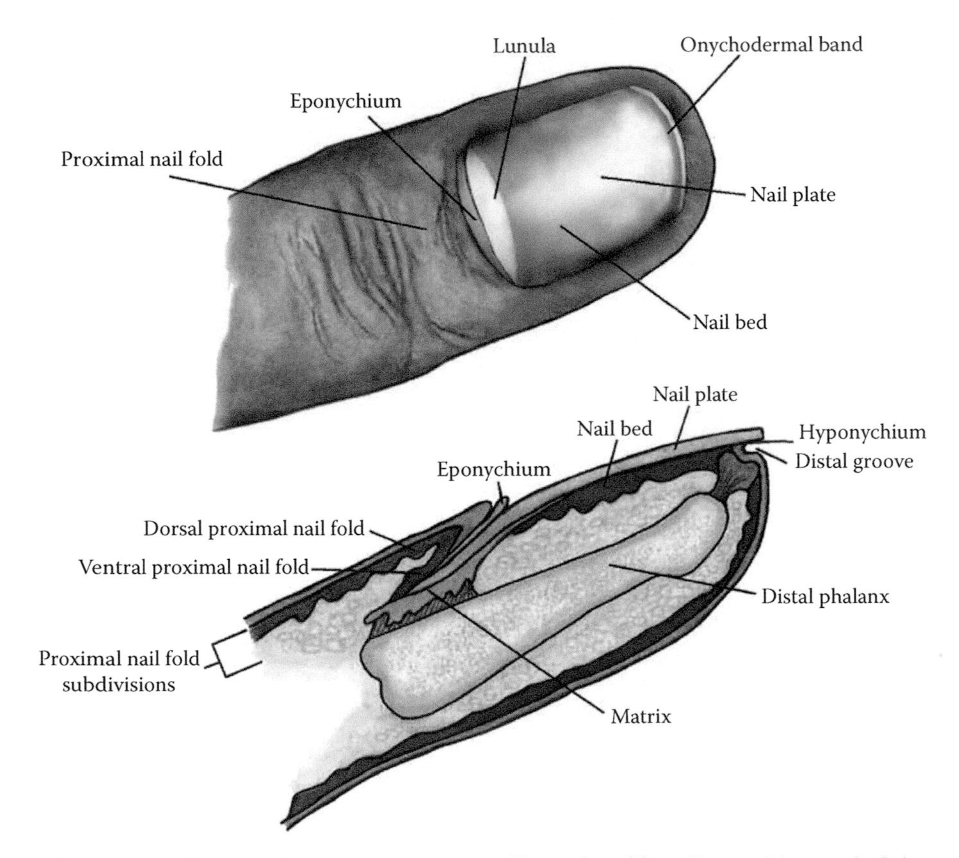

FIGURE 1.1 **(See color insert.)** Nail structure. (Reproduced from Jiaravuthisan et al., *J. Am. Acad. Dermatol.* 57, 1, 2007. With kind permission from the American Academy of Dermatology, Inc.)

For example, the use of the computer keyboard is hampered by long nails, and yet, many of us are happy to cope with such difficulties. Such extremes of a prerogative of beauty over function have even been actively cultivated to communicate social status, exemplified by Chinese mandarin culture. At the other extreme of beauty versus function, nail biting, which severely compromises the beauty of nails, is known to serve several needs, including providing a sense of relief as well as alleviating boredom (Williams et al. 2007).

1.1.1 Nail Matrix, Nail Bed, Hyponychium, Nail Folds, and Cuticle

The nail matrix—also called the root of the nail—is the germinative epithelial tissue whose cell division gives rise to the nail plate. Its distal portion is sometimes (especially in the thumbs and great toes) visible through the transparent nail plate as a white, semilunar area, called the lunula. The nail bed is a very thin epithelium on which the nail plate rests (and strongly adheres to), and slides over during its growth. The nail bed's dorsal surface is characterized by longitudinal ridges, and a complementary set of ridges is also found on the underside of the nail plate (though not at its free edge), which has led to the nail plate being described as led on rails as it grows out (Dawber et al. 1994), and to the suggestion that the ridges contribute to adhesion between the nail plate and the nail bed (Rand and Baden 1984). It has, however, been pointed out that it is unclear whether the ridges seen on the nail plate actually belong to the nail bed epithelium and remain attached to the undersurface of the nail plate or whether they are etchings on the nail plate (Fleckman 2005). There has also been vigorous debate for more than a century about whether or not the nail bed contributes to production of the nail plate. The interested reader should refer to Branca (1910) and Johnson et al. (1991), and references within.

The hyponychium is the region underneath the free edge of the nail plate where the latter starts to separate from the nail bed. It includes the space, the epithelium bordered by the nail bed and the distal groove, and the keratinous products of that area. The distal groove (Figure 1.2) marks the most distal boundary between the nail

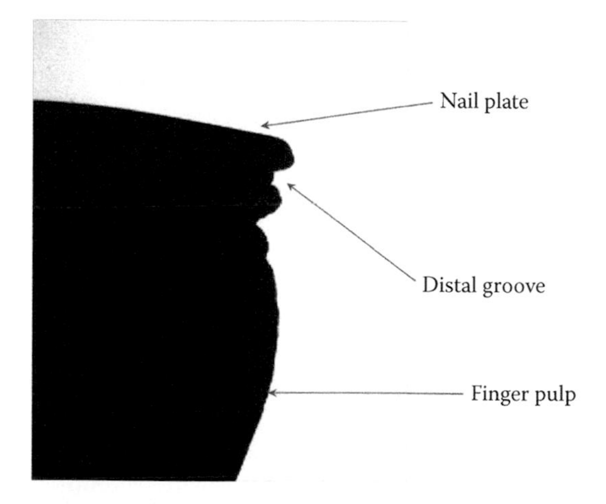

FIGURE 1.2 A side view photograph of the distal end of a fingertip, showing the distal groove of the nail unit.

unit and the finger pulp. The onychodermal band—observed as a deeper pink band than the adjacent distal nail bed, especially when the finger is extended—marks the site of the "seal" between the nail plate and the nail bed before these two separate at the distal end of the nail unit.

The lateral and proximal nail folds are folded skin structures that enclose the nail plate at its lateral and proximal edges, respectively. The dorsal surface of the proximal nail fold covers part of the nail matrix and is continuous with the cuticle. The latter, also called the eponychium, extends from the proximal nail fold and adheres strongly to the nail plate surface, creating a physical seal against the entry of exogenous materials.

1.1.2 The Nail Plate

The nail plate is transparent, hard yet slightly elastic, and curved in both longitudinal (in the direction of growth) and transverse (perpendicular to the direction of growth) directions. Its size, shape, thickness, surface ridging, curvature, and mechanical properties such as flexibility vary within and among individuals, with site (finger/toe), age, and other endogenous and exogenous factors such as disease states and seasons. The nail plate grows throughout life, and it has been estimated that on average, 3 g nail plate is produced every year (Vellar 1970). Much effort has been devoted to the factors that influence the rate of nail growth. Though variable intra- and interindividually, average figures of 0.1 mm/day for fingernails and 0.03–0.05 mm/day for toenails are often quoted (Fleckman 2005). Thus, on average, fingernails grow out completely in 6 months and toenails in 12–18 months. The rate of nail plate growth is influenced by gender (faster in males), age (slower with age), pregnancy (faster), local and systemic disease (can increase or decrease), nutrition (slower in malnutrition), trauma/nail biting (faster), hand dominance (faster in the dominant hand), the environment such as the weather (slower in cold climate), and chemicals, for example, drugs (can increase or decrease) (Hamilton et al. 1955; Bean 1980; Geoghegan et al. 1958; Hewitt and Hillman 1966; Landherr et al. 1982; Sibinga 1959; Dawber et al. 1994; Le Gros Clark and Dudley Buxton 1938; Gilchrist and Dudley Buxton 1939).

Fingernail plates are thinner (~0.5 mm at the distal edge) than the big toe (up to 1 mm at the distal edge), in children (~0.25 mm), and their thickness is in the order of thumb > index > middle > ring > little finger (Hamilton et al. 1955). In males, the fingernail plate is longer, broader, thicker (Hamilton et al. 1955), and less curved in the transverse direction (Murdan 2010) compared to females although the rate of nail plate growth does not seem to differ greatly in males and females (Hamilton et al. 1955). Age has a significant influence on the nail plate, with the greatest changes occurring during the first two decades of life, when the nail plate length, breadth, and thickness show marked increases. Interestingly, the ratio of nail plate breadth to length decreases rapidly in the first 10 years (Hamilton et al. 1955) such that the nail plate's shape changes from a landscape to a portrait orientation. In adults, aging is accompanied by thicker, broader, and less curved (in the transverse direction) nail plates that have a reduced rate of linear growth (Hamilton et al. 1955; Lewis and Montgomery 1955; Murdan 2010). In addition to gender and age, the transverse fingernail curvature is dependent on digit (in the order of thumb > index = middle >

ring > little finger), hand breadth, and dominance, with nail plates being flatter in the dominant hand (especially evident in right-handed individuals) and in individuals with wider hands (Murdan 2010). The nail plate's convex shape in both longitudinal and transverse directions is thought to contribute to its mechanical rigidity.

Differences in light refraction at the solid–solid interface between the nail plate and the nail bed and at the solid–gas interface at the nail plate's distal free edge is responsible for the nail plate's transparency where it overlies the nail bed and its opacity at its free edge (Gonzalez-Serva 1997). Thus, abnormal separation of the nail plate from the nail bed, as occurs in certain diseases, is easily observed. The transparency of the nail plate enables visualization of the supporting nail bed whose capillary network imparts a pink color to the nail plate.

1.1.2.1 Nail Plate Surfaces

Longitudinal ridges are often seen on the nail plate's dorsal surface, and these ridges have been suggested as a means of personal identification (Diaz et al. 1990). Ridging increases with aging and in certain diseases, such as rheumatoid arthritis (Lewis and Montgomery 1955; Tosti and Piraccini 2007; Michel et al. 1997).

The dorsal and ventral surfaces of the nail plate have distinctly different natures and functions. The ventral surface ensures the nail plate's attachment to the nail bed and hence to the body, while the dorsal surface—the barrier between the environment and the body at the digit tips—ensures that the "outsides remain outside and the insides remain inside." Scanning electron micrographs of the nail plate surfaces show a relatively smooth dorsal surface whose cells' edges seem to overlap, which should aid the barrier function (Figure 1.3). In contrast, the ventral surface has an irregular appearance (Figure 1.4). Indeed, Kobayashi et al. showed that the dorsal nail surface had lower permeability to topically applied drugs compared to the ventral surface (Kobayashi et al. 1999).

The surface free energy of the nail plates is expected to influence the adhesion of topical drug delivery vehicles as well as that of microorganisms, which

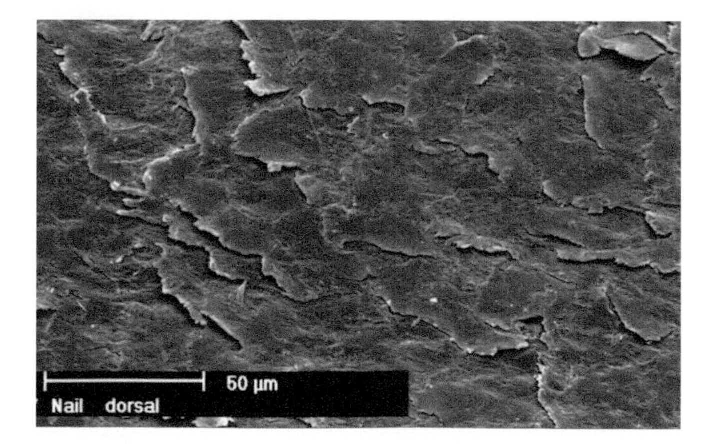

FIGURE 1.3 Scanning electron micrograph of the dorsal nail surface. (Reproduced from Murdan *Expet. Opin. Drug Deliv*, 4, 453, 2007. With kind permission from Elsevier.)

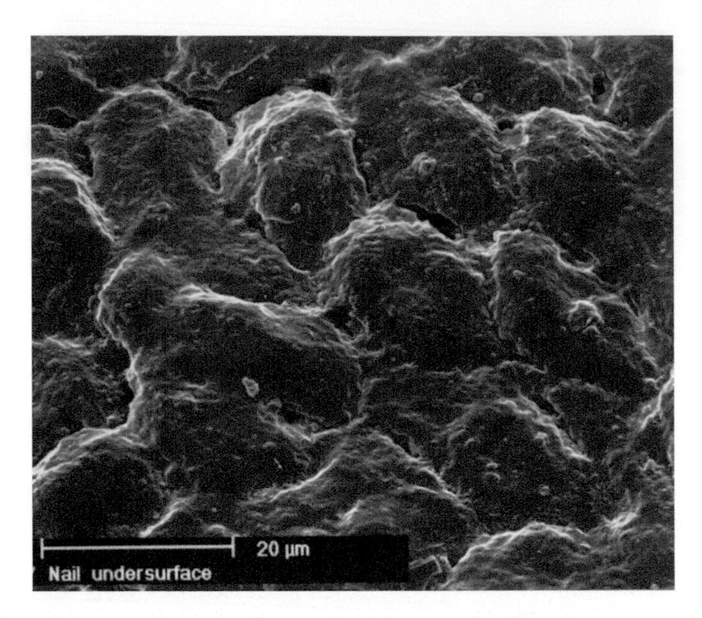

FIGURE 1.4 Scanning electron micrograph of the ventral nail surface.

would in turn influence drug permeation into the nail and microbial colonization, respectively. We therefore measured the surface free energy of the dorsal fingernail plates in 11 adult volunteers using a goniometer and the theory of van Oss as explained by Mavon et al. (1998). The nail plate's surface free energy was found to be 33.8 ± 5.6 mJ/m^2 (n = 59) (Murdan et al. 2010). This is slightly lower than that of the skin, whose surface free energy was reported to be 38 ± 6 and 43 ± 4 mJ/m^2 for the volar forearm and forehead, respectively (Mavon et al. 1997). Filing the dorsal surface of the nail plate using the files provided with a pharmaceutical lacquer (Curanail)—as patients would do prior to application of topical formulations—did not lead to any significant change in surface energy, which was 33.7 ± 4.7 mJ/m^2 (n = 40) Similarly, hydrating the nail plate by incubating the nail in a beaker of water for 10 min—to mimic possible hydration when occlusive topical formulations are applied on the nail plate—had no significant influence on the nail plate surface energy, which became 31.1 ± 6.6 mJ/m^2 (n = 39).

We also quantified the sebum found on the surface of the nail plate using a Sebumeter (Courage + Khazaka Electronic GmbH, Koln, Germany) in adult volunteers following washing, drying, and equilibrating for 20 min in a temperature-controlled room (temperature was 22.8 ± 1.1°C; RH was 28.0 ± 3.4%). Sebum—an oily substance produced by the skin's sebaceous glands—was not expected to be present at any significant level on the nail plate surface although its spreadability could lead to it flowing onto the nail plate surface from the surrounding skin. The sebum values measured by the Sebumeter on 234 fingernails and (great) toenails in 22 individuals were indeed low, ranging from 0 to 12 μg/cm^2; they had a mode of 1 μg/cm^2 and a mean ± standard deviation (sd) of 2.1 ± 1.7 μg/cm^2. The large sd reflects the high inter- and intraindividual variability for both fingernails and toenails. The sebum levels for fingernails and toenails were not significantly different from

TABLE 1.1
Sebum Levels Measured by Sebumeter

	Sebum levels (μg/cm²)				
	N	Minimum	Maximum	Mode	Mean (sd)
Fingernail	196	0	7	1	2.1 (1.5)
Toenail	38	0	12	1	2.3 (2.5)
Inner forearm	44	0	8	0	0.7 (1.9)
Foot	40	0	3	0	0.5 (0.8)

each other (*t* test, $p > 0.05$), and the separate values for fingernails and toenails can be seen in Table 1.1. Interestingly, although still low, the dorsal nail surface sebum values were statistically higher (paired *t* test using the nail/skin sebum values for each volunteer, $p < 0.05$) than the skin sebum levels on the forearms and midfeet of the same volunteers, which ranged from 0 to 8 μg/cm^2, had a mode of 0 μg/cm^2, and had a mean ± sd of 0.6 ± 1.5 μg/cm^2. The forearm and midfeet sebum levels did not differ from each other (*t* test, $p > 0.05$), and the separate values can be seen in Table 1.1. The low forearm sebum levels measured in our volunteers correlate with previous reports (Agache 2004); however, no literature on foot sebum levels was found.

1.1.2.2 Nail Plate Layers

The nail plate is not a homogeneous structure. Lewis (1954) proposed that the nail plate consists of three distinct parallel strata: dorsal, intermediate, and ventral. The intermediate layer is said to be almost twice as thick as the dorsal layer and forms the main bulk of the nail plate, with the ventral layer being only one or two cells thick. Jarrett and Spearman (1966) showed that the three layers stain differently and contain different proportions of disulfides, thiols, phospholipids, calcium, and acid phosphatase. The three nail layers could also be tentatively discerned by infrared spectroscopy (Sowa et al. 1995). The thin ventral layer is, however, not always easy to observe, and other researchers have divided the nail plate into two layers only: a thin hard dorsal layer and a thicker plastic ventral one (Forslind and Thyresson 1975). These two distinct layers are easily visible by ultrasound imaging (Jemec and Serup 1989; Wortsman and Jemec 2006), and the border between the two constitutes a natural cleaving plane, as can be observed when dissecting nails (Forslind 1970). The different nail layers have different organizations of cells as discussed in Section 1.1.2.3.

1.1.2.3 Nail Plate Cells

The nail plate is made up of 80–90 layers (about 30–40 in the dorsal and about 50 in the ventral nail plate of the big toenail, the number varying with age and site) of dead, keratinized cells, whose size, shape, attachment to their neighbors, and architectural organization change as one moves from the dorsal to the ventral surface (Achten et al. 1991). At the dorsal surface end, the onychocytes (nail plate cells) are on average 34-μm long × 60-μm wide × 2.2-μm high and become thicker toward the ventral surface (the cell size is approximately 40-μm long × 53-μm wide × 5.5-μm

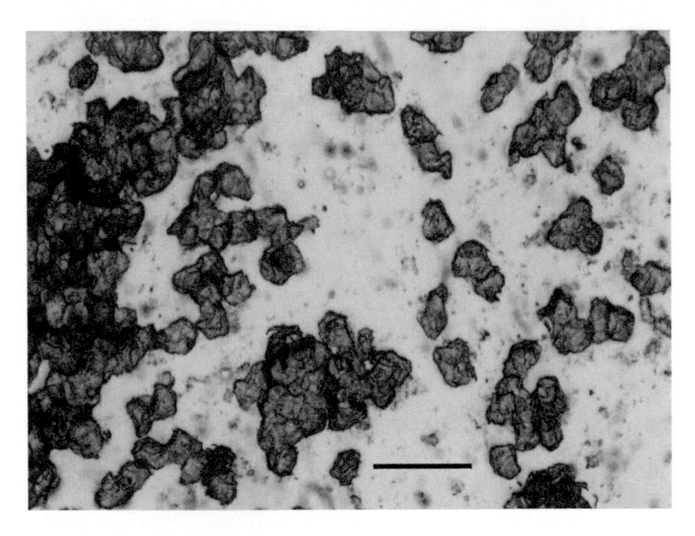

FIGURE 1.5 Light micrograph of onychocytes, obtained after tape stripping the dorsal nail surface. Bar represents 100 μm.

high), with a gradation in between. The cell contour also changes, becoming more meandering, and interdigitations between adjacent cells become more frequent, but shallower, toward the ventral side. In addition, complex "anchoring knots" are observed in the deeper nail layers (Dawber et al. 2001).

Individual onychocytes can be visualized microscopically following tape stripping (Figure 1.5). Germann et al. reported healthy dorsal nail onychocytes to have a fairly regular polyhedrical shape with rounded corners and average areas of 920 ± 176 μm^2 in adults. While the corneocyte area was unaffected by site (fingernail versus toenail), it did increase with age and was greatly affected by disease states. In nail psoriasis, where cell growth is increased, onychocytes were smaller, while in patients with lichen planus where cell growth is slowed, onychocytes were larger (Germann et al. 1980; Germann and Plewig 1977).

A 25–35-nm-thick space between onychocytes can be visualized under an electron microscope (de Berker and Forslind 2004). Not much is known about the intercellular "cement" that is present in this space, though it is thought to consist of proteins and/or mucopolysaccharides (de Berker and Forslind 2004). The "cement" completely fills the intercellular spaces in the dorsal nail plate but is discontinuous in the intermediate nail layer. Such differential content of the intercellular cement could explain the lower permeability of the dorsal nail surface to topically applied drugs compared to the ventral nail surface. The enzyme keratinase seems to act on this cement, such that onychocytes separate from one another (Figure 1.6) when nail plates are exposed to the enzyme (Mohorcic et al. 2007). A greater knowledge of the composition of this intercellular "cement" will assist investigations into chemicals to destabilize this component of the nail plate and thereby enhance the ungual (pertaining to the nail; used interchangeably with "onycheal") permeation of topically applied drugs.

FIGURE 1.6 Onychocytes separate from one another when the nail plate has been exposed to keratinase enzyme, which seems to act on the intercellular "cement." (Reproduced from Murdan *Int. J. Pharm.*, 332, 196–201, 2007. With kind permission from Elsevier, Inc.)

1.1.2.4 Nail Plate Composition

Keratins—intermediate-filament-forming proteins that provide mechanical support in epithelial cells and tissues—are the major component of the nail onychocytes. Human keratins can be classified into hair keratins or hair-follicle-specific epithelial keratins, or epithelial keratins (Schweizer et al. 2006; Ramot et al. 2009). All of these groups can be further subdivided into type I (i.e., acidic) and type II (i.e., basic to neutral) keratins. A keratin molecule consists of an α-helical central rod composed of about 310 amino acids flanked by nonhelical head and tail domains. Association of the central rod domains of one type I and of one type II keratin molecules in a coiled–coiled fashion results in the formation of a heterodimer. Pairs of heterodimers align to form tetramers, which polymerize into long chains and pack laterally to form keratin filaments (Moll et al. 2008; Arin 2009). The keratin filaments are embedded in a nonfilamentous matrix of proteins called keratin-associated proteins. Hair keratins differ from epithelial ones by their much higher content of cysteine, which forms numerous chemical cross-links between the filament and the nonfilamentous matrix, thereby increasing the protein's stability. Both hair and epithelial keratins are present in the nail plate, though the majority (80%–90%) of the nail plate keratin is of the more stable hair type (Lynch et al. 1986). Raman spectroscopy confirmed the stability of the nail plate keratin: nail proteins are highly folded, interacting minimally with their surroundings, and a large part of the disulphide bonds are in the most energetically favorable conformation (Gniadecka et al. 1998). Interestingly, the location and orientation of the hair and epithelial keratins differ. Considering the nail plate to consist of three (dorsal, intermediate, and ventral) strata, Garson et al. (2000) reported that hair keratin is present only in the intermediate nail layer and that the keratin filaments are oriented parallel to the nail plate surface, but perpendicular to the growth axis. In contrast, the epithelial keratin filaments are found in the dorsal and the ventral layers and are oriented both parallel and perpendicular to the growth axis, though some are randomly dispersed. Such sandwich orientation of the keratin fibers in the nail plate has been suggested to contribute to the hardness and mechanical rigidity of the nail plate. The amino acid content of the nail plate has been determined, and the proportions of the different amino acids (number of residues per 100) are as follows: lysine (3.1%), histidine (1.0%), arginine (6.4%), aspartic acid (7.0%), threonine (6.1%), serine (11.3%), glutamic acid (13.6%), proline (5.9%), glycine (7.9%), alanine (5.5%), valine (4.2%), methionine (0.7%), isoleucine (2.7%), leucine (8.3%), tyrosine (3.2%), phenylalanine (2.5%), and halfcystine (10.6%), while the sulfur content was 3.2% of the nail plates's dry weight (Baden et al. 1973).

The nail plate also contains significant amounts of water, though the exact amount in situ has been difficult to measure so far. Stern et al. (2007) found the water content of cut nail clippings to be between 8% and 22%, although these were underestimates due to water loss prior to measurement, while Egawa et al. (2006) measured (using near infrared spectroscopy) the *in vivo* water content of the distal free edge of the nail plate to be between about 5% and 30% w/w in volunteers. The distal free edge of the nail plate is likely to have less water than the plate overlying the nail bed, as it is not replenished with water from below, though this has not been confirmed.

In vitro, nail plates readily absorb water vapor from the surrounding atmosphere (Baden et al. 1973; Martinsen et al. 2008; Gunt and Kasting 2006). This was reflected *in vivo*, where the water content of the distal free edge of nail plates was lower in (the drier) winter than in summer (Egawa et al. 2006). The nail plate also rapidly absorbs water and becomes soft and malleable when in contact with liquid water, for example, during hand washing. In the laboratory, periodic weighing of nail clippings incubated in water to determine the change in mass has shown that water is taken up within minutes of the nail clipping being placed in water (Figure 1.7). Subsequent removal of the

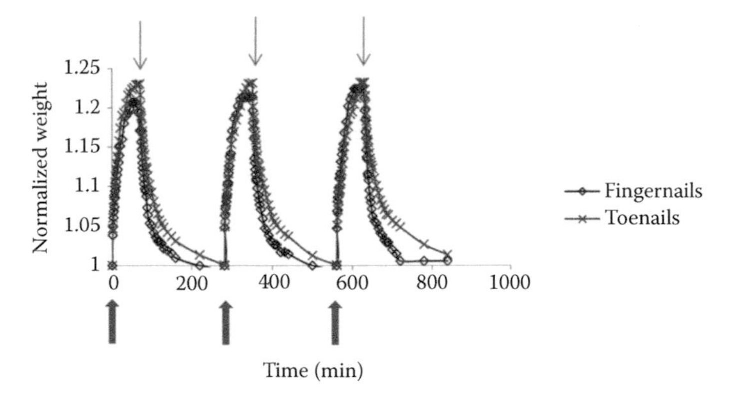

FIGURE 1.7 **(See color insert.)** Rapid uptake and loss of water when a nail clipping is sequentially placed in (up arrows) and removed from water (down arrow) to air-dry. The curves show averages (±sd) of three clippings each for fingernail and toenail. Fingernail and toenail clippings absorb similar amounts of water, with consequent increase in mass of between 20% and 23%. When placed in water, nail clippings' mass increases rapidly and saturation is reached after about 70 min. Once removed from water, most of the absorbed water is lost rapidly although the initial nail mass is not regained for 3 h. Toenail clippings seem to lose the last of the absorbed water at a slower rate compared to fingernail clippings.

nail clipping from the water bath results in rapid loss of the absorbed water (Figure 1.7). Repeated uptake by, and loss of water from, the nail plates *in vivo*, for example, after repeated washings, leading to frequent changes in the nail plate hydration, has been suggested as one of the causes of brittle nails (Scher and Bodian 1991; Scher 1989), although no statistical difference was later found between the water content of normal and brittle nails when cut nail *clippings* were compared (Stern et al. 2007), and the role of water in nail plate cohesion has been challenged (Duarte et al. 2009). Increased nail plate hydration enhances drug diffusivity within the nail plate and thereby enhances topical ungual drug delivery (Gunt and Kasting 2007). Increased nail plate hydration has also been linked to greater enhancement of ungual drug permeation by chemical enhancers (Khengar et al. 2007). Hydration of the nail plate is obviously an important parameter in nail plate health and for ungual drug delivery.

In order to find a simple, commercially available, and portable method of measuring the *in vivo* nail plate hydration, we investigated the ability of the Moisture Checker MY-808S (Scalar, Tokyo, Japan) to give an indication of nail plate hydration. The Moisture Checker uses electrical impedance to measure water content and, like the corneometer, is often used to measure skin hydration (in arbitrary units). While the hard and curved nail plate surface renders measurements challenging, it was possible to get comparative values, though these were much lower than typical values for healthy skin, and had greater variability (coefficients of variation of 7%–14% for nail plate compared to 1%–3% for the forearm for repeated measurements of the same site). The fingernail hydration measured in 26 adult volunteers ranged from 3.8% to 11.1% and had a mean ± sd of 5.5 ± 1.4% (n = 190), while the (big) toenail hydration in the same volunteers ranged from 3.8% to 13.1% and had a mean ± sd of 4.8 ± 1.8% (n = 24). It was not possible to measure the hydration of the other toenails due to their small surface area. The toenail hydration was significantly lower than the fingernail hydration (independent samples t test, $p < 0.05$). In contrast, gender, side (right or left), and digit (thumb, index, middle, ring, and little) did not have any influence on the nail plate hydration. In the same volunteers, the forearm and midfoot hydrations were found to be 35.2 ± 3.1% (n = 43) and 34.8 ± 3.9% (n = 42), respectively, which correlate closely with the manufacturer's measurements of 35.5 ± 1% for healthy skin (Scalar 2010).

To ascertain the sensitivity of the Moisture Checker at the low hydration levels found for the nail plates, the change in water content of a wetted filter paper with time (due to evaporation of the water) was followed by gravimetric analysis as well as by the Moisture Checker. A plot of the water content of the filter paper (in grams) against Moisture Checker hydration values shows a fairly good linear relationship (Figure 1.8) and shows the Moisture Checker to be sensitive at the low hydration values found in the nail plate. To determine whether such a direct relationship would exist *in vivo*, a volunteer's fingernail was immersed in water for increasing durations (to differentially increase the nail plate hydration) and immediately afterward, the fingernail was blotted dry, and the hydration was measured by the Moisture Checker. The latter is able to differentiate between different nail plate hydrations as shown in Figure 1.9, where nail plate hydration is plotted against duration of nail immersion in water. These showed that the Moisture Checker could give an indication of nail plate hydration, although not absolute values, and ideally, the Moisture Checker would be designed to fit the nail plate.

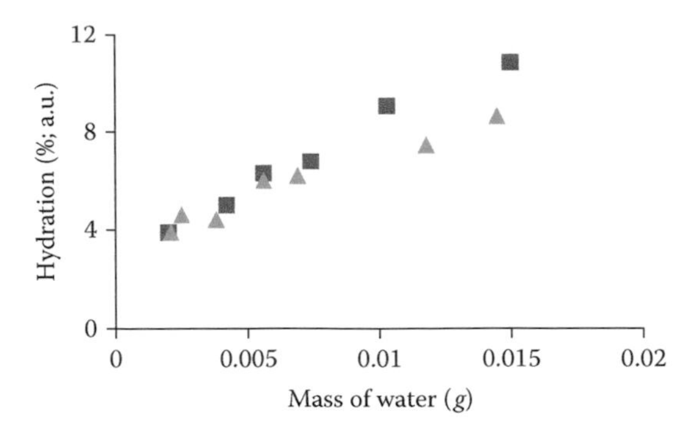

FIGURE 1.8 Linear relationship between mass of water and Moisture Checker hydration values. R2 = 0.98 (■) and 0.96 (▲) in two separate experiments.

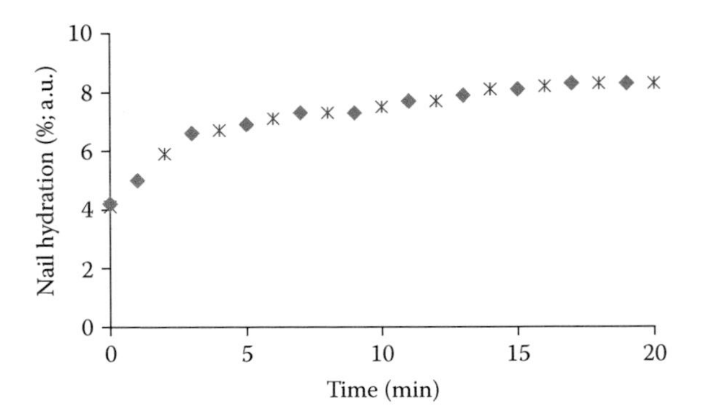

FIGURE 1.9 Moisture Checker hydration values of two fingernails (represented by the different symbols) that had been incubated in water for increasing durations.

Water is constantly being lost from the body to the outside environment through the nail plate. This transonychial water loss (TOWL) has been measured in a number of laboratories, and *in vivo* TOWL values of 16 $g/m^2 \cdot h$ (Spruit 1972), 19–31 $g/m^2 \cdot h$ (Spruit 1971), 12–34 $g/m^2 \cdot h$ (Jemec et al. 1989), mean 38 $g/m^2 \cdot h$ (Nuutinen et al. 2003), mean 13 $g/m^2 \cdot h$ (Kronauer et al. 2001), 28–75 $g/m^2 \cdot h$ for fingernails and 26–48 $g/m^2 \cdot h$ for toenails (Murdan et al. 2008), 25–51 $g/m^2 \cdot h$ (Dutet and Delgado-Charro 2009), and 7 $g/m^2 \cdot h$ (Tudela et al. 2008) for healthy, normal nails have been reported. *In vitro* values of 12–24 $g/m^2 \cdot h$ (Burch and Winsor 1946) and 0–2.8 $g/m^2 \cdot h$ (Vejnovic et al. 2010) for normal nails have also been reported. The wide range of reported values can partly be explained by the different devices and techniques used to measure TOWL. Within studies (and hence the same measuring methodology), *in vivo* TOWL has been found to be variable inter- and intraindividually, to be inversely related to the nail plate thickness, and to increase when the dorsal nail surface is filed or removed by tape stripping and is lowered in old age and in patients with atopic eczema, psoriasis, and onychomycosis (Spruit 1971; Spruit 1972; Jemec et al. 1989; Kronauer et al. 2001; Nuutinen et al. 2003; Murdan et al. 2008; Tudela et al. 2008). TOWL has also been used to investigate the influence of iontophoresis (Dutet and Delgado-Charro 2009) and of topical ungual formulations (Vejnovic et al. 2010) on the nail plate and to assess nail quality

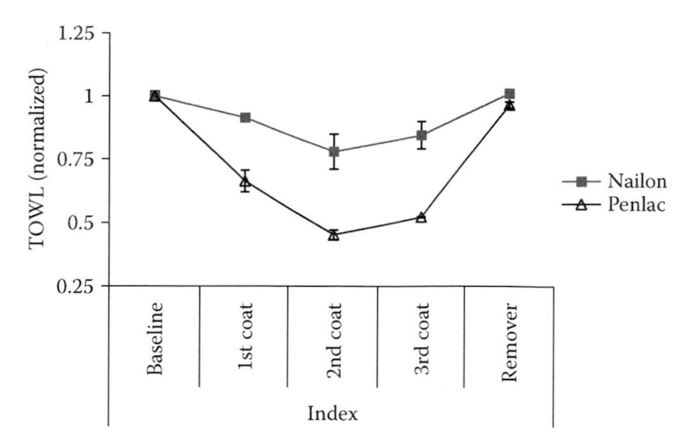

FIGURE 1.10 Comparison of two nail lacquer formulations of 8% Penlac® (ciclopirox) and Nailon® using TOWL.

(Kruger et al. 2006; Zaun 1997). Topical application of ungual formulations, such as nail lacquers, leads to decreased TOWL (Spruit 1972; Murdan et al. 2008), which is expected to increase the nail plate hydration (Spruit 1972) and consequently ungual drug permeation (Marty 1995). TOWL could thus be used to compare topical ungual formulations, for example, generic and proprietary nail lacquers, as shown in Figure 1.10.

Nail plate lipids include cholesterol sulfate, ceramides, free sterols, free fatty acids, triglycerides, sterol and wax esters, and squalene (Helmdach et al. 2000). The lipid content is influenced by gender (which indicated an influence of sex hormones) and age and decreases sharply from infancy (when it is up to 16% w/w) into childhood and adulthood, when it is less than 5% w/w. Most of the lipid content is found in the ventral and dorsal (rather than the intermediate) nail layers (Kobayashi et al. 1999) and is in a highly ordered, lamellar crystalline state (Gniadecka et al. 1998).

Analysis for elements in the nail plate revealed the presence of calcium, magnesium, sodium, potassium, iron, copper, zinc, aluminum, and chlorine, amongst others (Vellar 1970; Sirota et al. 1988). The growing nail plate incorporates substances from its surroundings through the systemic circulation, and drugs (therapeutic and recreational), isotopes (e.g., 13C, 15N), hormones, fluoride, and environmental pollutants (e.g., arsenic) have been detected and quantified from the nail plate (for example, see Wilhelm et al. 2005; Nardoto et al. 2006).

1.1.2.5 Nail Plate pH

While there is much information on skin pH, there is almost none on nail plate pH. It is possible that the latter influences nail plate colonization by pathogens, as happens in skin (Agache and Humbert 2004). It is also known that fungal spore formation is influenced by the pH of the medium (Yazdanparast and Barton 2006). In this context, we measured the pH of healthy nail plates *in vivo* in an attempt to explore whether the skin pH meter could be used to measure nail plate pH and to establish baseline values (Murdan et al. 2010). Nail plate pH is defined here as the pH measured by a flat glass electrode at the nail plate surface with a hydrated nail–electrode interface, in an analogous manner to the definition of skin pH (Fluhr et al. 2006).

Following washing, drying, and equilibrating for 20 min in a temperature-controlled room, the pH of fingernails (digits 1–5) and toenails (big toenails) in 22 volunteers was measured using a skin pH meter PH-905® (Courage + Khazaka Electronic GmbH). pH was found to be 5.1 ± 0.6 ($n = 204$) for fingernails and 5.4 ± 0.8 ($n = 33$) for toenails. Multiple regression, used to investigate the influence of gender, age, site (i.e., fingernail or toenail), and side (i.e., right or left) on the nail plate pH, showed that only the site, that is, finger or toe, was significant ($p = 0.002$). Toenail plate pH was statistically higher than fingernail plate pH, while the subject's gender, age, and hand or foot side (right or left) had no influence on pH. The nails had been washed prior to pH measurement to ensure that the pH values were not influenced by extraneous factors such as dirt, sweat, hand creams, etc. While washing significantly increased (paired t test, $p < 0.0005$) nail plate pH from 5.1 ± 0.4 to 5.3 ± 0.5 ($n = 140$ fingernails, from 14 individuals), the increase in pH was not sustained with time and the pH dropped to baseline (before washing) values within ~20 min, as shown in Figure 1.11. This indicates that equilibrating for about 20 min following washing allowed the pH to return to baseline values.

In order to determine the pH inside the nail plates, fingernail plate surface layers were removed by tape stripping using D-Squame (Cuderm, Dallas, USA) adhesive discs in six volunteers. The number of tape strips used depended on the volunteers, and stripping was stopped when the volunteer felt tingling and discomfort at the nail. Thus, the different volunteers received a total of either 15 (1 volunteer), 20 (1 volunteer), 30 (2 volunteers), 40 (1 volunteer), or 50 (1 volunteer) strippings. Comparison of the initial prestripping pH and the final pH for each volunteer showed a statistically significant difference (paired t test, $p < 0.05$), with the mean (±sd) prestripping and final pH being 4.7 ± 0.7 and 4.1 ± 0.7, respectively ($n = 12$, two thumbnails in six volunteers). It appears that the pH of the nail plate surface is higher than that in its bulk. In this experiment, nails had not been washed prior to any measurement.

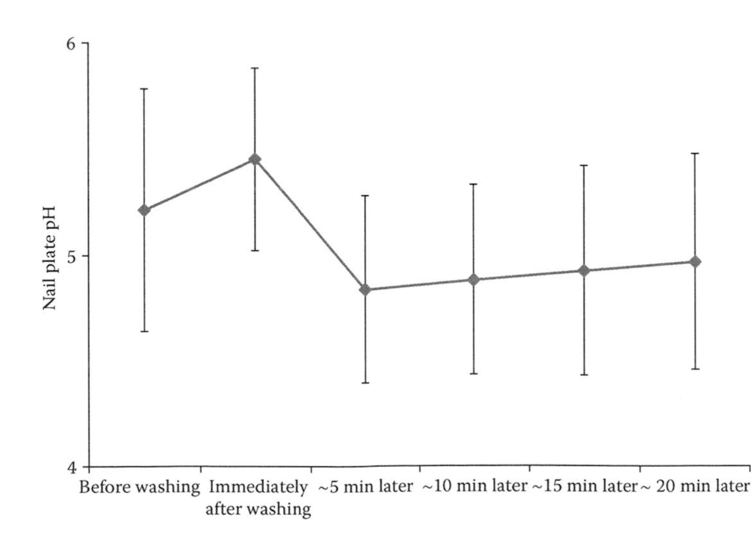

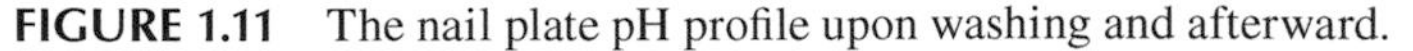

FIGURE 1.11 The nail plate pH profile upon washing and afterward.

1.2 NAIL DISEASES AND CURRENT TREATMENT APPROACHES

The nail apparatus can suffer from a very wide range of disorders, from benign, for example, nail yellowing in smokers, to extremely painful and serious, for example, infections and malignant tumors of the nail apparatus. Not all the parts of the nail apparatus are affected; for example, in some diseases, only the perionychial tissues (the nail folds) are affected while the nail plate is normal. In other diseases, the nail plate's shape, surface features, color, mechanical properties, and attachment to underlying soft tissues are altered as described in more detail below, and the symptoms are used in disease diagnosis.

1.2.1 Disorders of the Nail Plate Size and Shape

The nail plate may be totally or partially absent in newborns and in children (Figure 1.12) as a result of genetic disorders (e.g., epidermolysis bullosa) or teratogen intake during pregnancy (Tosti and Piraccini 2005). In other conditions, the nail plate may be excessively long or short compared to its normal length-to-width ratio, which is 1 ± 0.1, as well as excessively large or small, thickened, or hypertrophied. Nail plate curvature can also be affected. Nail clubbing, where the nail plates are overcurved in both the longitudinal and transverse axes and the underlying soft tissues are enlarged, has long been recognized as a cause of bronchopulmonary and cardiovascular diseases. Nail overcurvature confined to the transverse axis can also occur, remaining the same along the whole nail length or occurring sharply at one or both of the lateral margins of the nail plate, or increasing along the length of the nail and becoming maximal at the distal fingertip. Nail transverse overcurvature may be hereditary or acquired due to trauma or dermatoses. The nail plate can also be

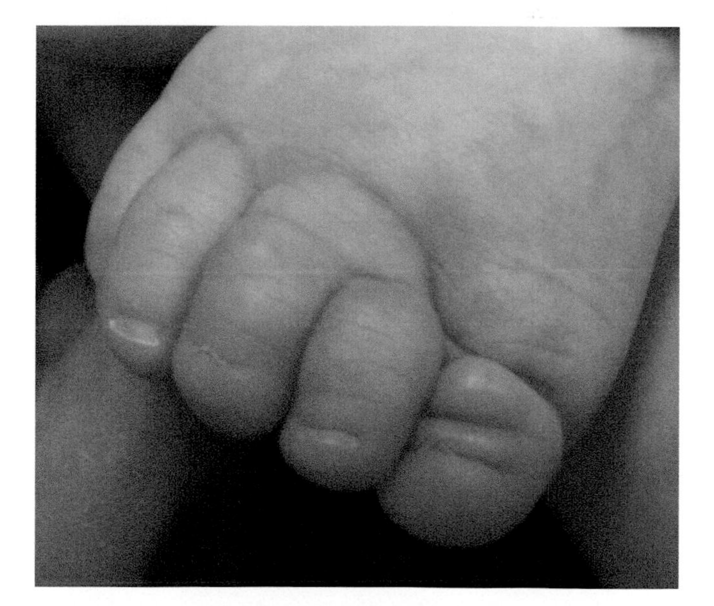

FIGURE 1.12 **(See color insert.)** Partially and totally absent toenails in a newborn on digits 3 and 1, respectively.

"undercurved" in the longitudinal direction, where it is concave with everted edges, the so-called "spoon nail." Common in healthy children's toenails (where it can disappear spontaneously) and a possible racial feature in Tibetans, the "spoon nail" can be a symptom of systemic diseases, such as iron deficiency, and a result of trauma, for example, nail biting or occupational hazard, such as the handling of petrol, solvents, and engine oils (Dawber et al. 2001; Rich and Scher 2003).

1.2.1.1 Disorders of the Nail Plate Surface

The nail plate surface may be rough, excessively ridged with a dull or shiny appearance or covered in pits (shallow depressions in the nail plate) whose number, size, shape, and depth vary and whose distribution on the nail plate may be uniform or random. A pit is formed when a group of nail plate surface cells is shed during routine activity, shedding being due to the fact that those cells were lined by loosely adherent parakeratotic cells. In some diseases, part of the nail plate may be missing, appearing to be "punched out." Longitudinal or transverse lines may appear on the nail plate surface (Figures 1.13 through 1.15). The longitudinal lines may be indented

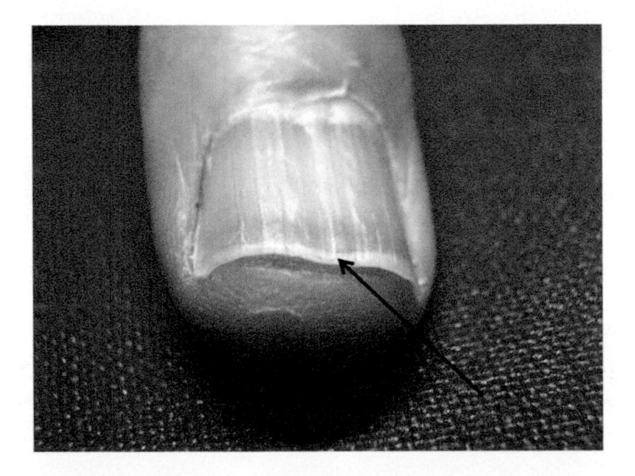

FIGURE 1.13 **(See color insert.)** Longitudinal grooves in a fingernail.

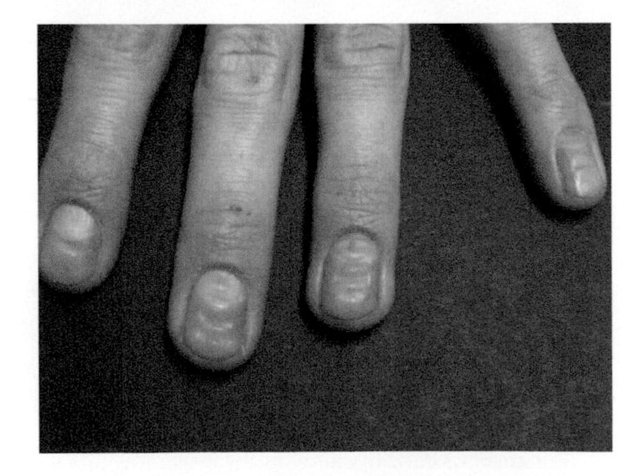

FIGURE 1.14 **(See color insert.)** Transverse ridges and grooves on fingernails.

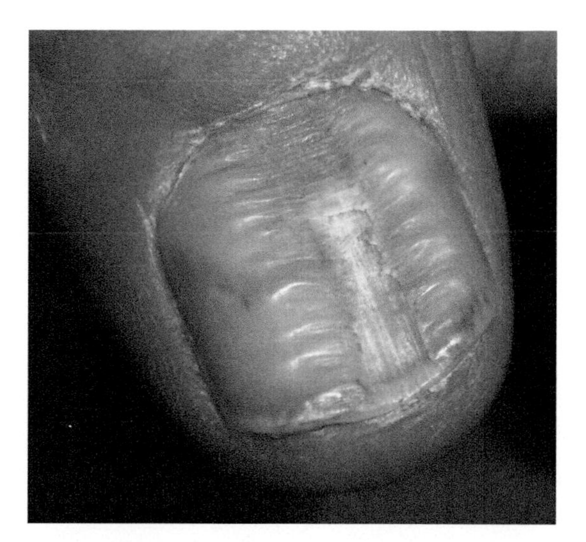

FIGURE 1.15 **(See color insert.)** Transverse grooves in the nail plate; part of the nail plate has been lost due to the damage.

grooves (depressions in the nail plate) or projected ridges, and may appear on part or all of the nail plate from the proximal nail fold to the nail's free edge. Longitudinal grooves along the whole length of the nail plate can result in nail splitting. Oblique ridges that converge distally toward the center are seen in early childhood and may coexist on the same nail plate with longitudinal ridges in teenagers. Transverse grooves on the nail plate are caused by a change in the activity of the nail matrix, for example, in disease, trauma, and antimitotic drug therapy. The transverse lines may occur singly or in a series, affect all or some of the nail plates, be shallow or deep, and have an abrupt or a sloping nature of the depression, depending on the cause of the line. A fairly common condition in healthy female adults, especially those whose nails are repeatedly exposed to water, is the horizontal splitting of the distal part of the nail (Dawber et al. 2001; Rich and Scher 2003).

1.2.1.2 Disorders of the Nail Plate Color

Changes in the color of the nail (chromonychia) are due to changes in the bulk or at the surface of the nail plate, or in the underlying tissues, which are then seen through the translucent nail plate. White nails are the most common abnormality although a range of colors, such as brown, blue, yellow, orange, green, grey, purple, red, black, or even a spectrum of colors, can be seen, depending on the cause of the chromonychia. Exogenous (e.g., topical cosmetics, tobacco smoking, fungal and bacterial infections) as well as endogenous (e.g. trauma to the nail matrix, systemic drugs, chemicals, infections, dermatological and other diseases) factors can affect all or part of the nail plate. For example, single or multiple (in regular or otherwise arrangements) white spots (Figure 1.16) are commonly seen, especially in women, as a result of excessive manicuring, as the nail matrix is repeatedly subjected to minor trauma. The pattern of discoloration can also indicate the cause of the chromonychia. For example, discoloration following the shape of the proximal nail fold

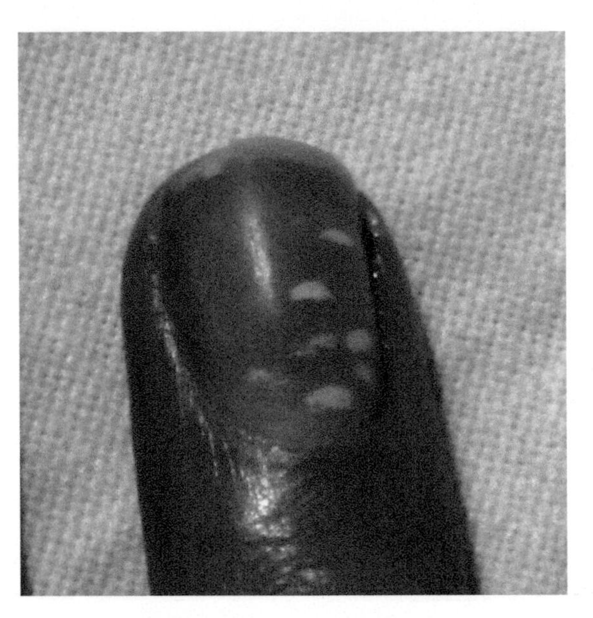

FIGURE 1.16 **(See color insert.)** Multiple white spots on the nail plate that are commonly seen.

and growing out with the nail plate is obviously caused by the topical application of agents. In contrast, discoloration that corresponds to the shape of the lunula has an obvious systemic origin (Dawber et al. 2001).

1.2.1.3 Disorders of the Nail Plate's Mechanical Properties

Diseased nail plates are extremely hard and difficult to trim in pachyonychia congenita (an autosomal disease where the nails are very thick). Conversely, other congenital and noncongenital diseases, trauma, and occupational exposure to water, detergents, and chemicals cause the nail plates to become excessively soft, flexible, friable, fragile, and prone to breaking or splitting at the free edge (Dawber et al. 2001).

1.2.1.4 Disorders of the Nail Plate's Attachment to Underlying Tissues

In some diseases, or after trauma (e.g., sporting activities), the nail plate is shed following its separation from the nail matrix and nail bed. A less extreme detachment of the nail plate from the rest of the nail unit is onycholysis when the nail plate becomes detached from the nail bed at its distal and/or lateral edges. The open space created between the nail plate and the nail bed can subsequently house dirt, keratin debris, and water and become infected. Onycholysis can also occur in the middle of the nail plate, for example, in nail psoriasis. Other causes of onycholysis include other skin diseases, congenital and/or hereditary conditions, systemic drugs/states, e.g., pregnancy, and local insults such as trauma, infections, and exposure to certain chemicals and cosmetics. Hyperkeratosis of the nail bed is another condition that moves the nail plate from its "normal" position within the nail unit (Dawber et al. 2001; Rich and Scher 2003).

1.2.1.5 Disorders of the Perionychium

The perionychium (nail folds surrounding the nail plate) can be affected without any overt changes in the nail plate (Figure 1.17). It can become swollen and inflamed, and pus may be present. Trauma, ingrown toenails, certain chemicals, and bacterial/fungal/viral/parasitic infections can all cause paronychia (inflammation and/or infection of the nail folds). Paronychia can become chronic, especially when caused by an occupational hazard, and a wide range of workers, such as cleaners, bakers, bricklayers, cosmetic workers, pianists, and fishmongers, have been reported to be affected (Dawber et al. 2001).

1.2.1.6 Treatment of Nail Disorders

It can be seen from the previous examples that nail disorders can have numerous origins, including chemicals, infections, trauma, and congenital, hereditary, systemic, and local diseases. Treatment of these disorders will, of course, depend on the originating cause. Some of the disorders, such as transverse lines and white spots, grow out with the growing nail plate (although new ones can reappear). Changes in the nail due to exposure to local/systemic chemicals and trauma can be treated by removal of the insult; for example, rubber gloves can be worn to protect brittle nails from excessive exposure to water, detergents, and other occupational hazards. Manicuring can be performed less aggressively, and systemic drugs that cause nail changes can be replaced. Other symptoms of nail diseases are resolved following the treatment of the underlying condition, such as drug treatment of infections, surgical removal of tumors or of foreign bodies, and surgical treatment of abnormalities, such as ingrowing toenails.

For the purposes of this book on trans-ungual drug delivery, the two nail diseases—onychomycosis and nail psoriasis—which make up the majority of nail

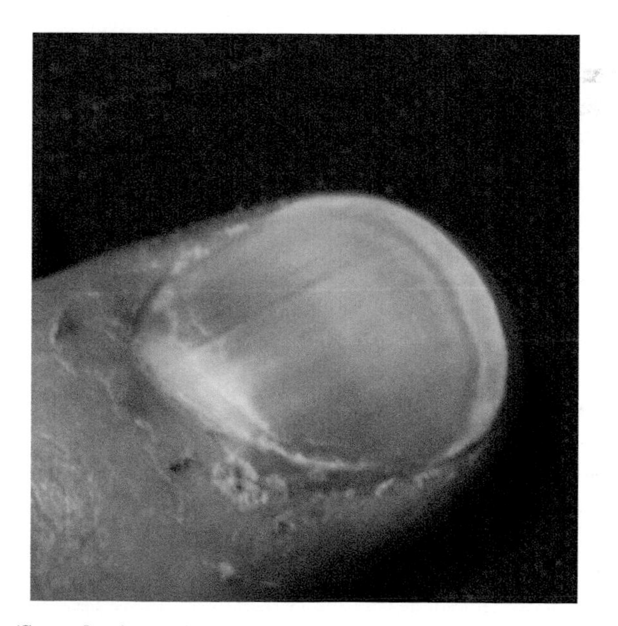

FIGURE 1.17 **(See color insert.)** Damaged to the nail folds without any overt abnormality on the nail plate.

disorders and which are the most common targets for local treatment with topically applied preparations, are focused on and are described in Sections 1.2.2 and 1.2.3.

1.2.2 Onychomycosis

Onychomycosis (fungal infections of the nail plate and/or nail bed), responsible for between 18% and 40% of all nail disorders (Achten and Wanet-Rouard 1978; Pardo-Castello 1960), is a very common problem, affecting 10%–40% of the population (Rich and Scher 2003), prevalence being higher in certain groups such as the elderly, diabetics, miners, and sports-active individuals (especially their toes) (Roberts 1999; Chabasse et al. 2000; Pierard 2001; Hay et al. 2001; Caputo et al. 2001). Occurrence seems to be on the increase due to a growing elderly population, the spread of HIV infection and AIDS, a higher frequency of iatrogenic immunosuppression due to the use of immunosuppressant drugs, lifestyle factors such as the wearing of tight-fitting shoes, and the use of communal recreational facilities and health clubs, as well as improved detection and higher public awareness (Daniel 1991; Cohen and Scher 1994; Scher 1996; Gupta and Shear 1997). Most (90%–95%) of the infections are caused by dermatophytes (especially *Trichophyton rubrum* and *Trichophyton mentagrophytes*), the rest being caused by yeasts (*Candida albicans* mainly) and nondermatophytes, such as *Scytalidium dimidiatum, Scytalidium hyalinum*, and *Fusarium* species. Toenails are affected more than fingernails (Midgley et al. 1994) and are also more recalcitrant to treatment.

Clinically, onychomycosis can be divided into categories depending on where the infection begins (Hay et al. 2001):

1. Distal and lateral subungual onychomycosis (DLSO; Figure 1.18). The fungal (mainly *T. rubrum*) infection starts at the hyponychium and the distal or lateral nail bed. This is the most common route of the fungal infection of the nail. The fungus then progresses toward the nail matrix, infecting the nail plate from its undersurface, and the nail plate becomes opaque. Histopathology shows the fungal hyphae to have a parallel, longitudinal arrangement. Tunnels in the nail plate (whose diameter is larger than those of the fungi) produced by the fungi can be seen by high-power microscopy. Sometimes, whitish-yellow longitudinal streaks extending from the free edge of the nail plate toward the proximal nail fold are seen. These streaks—loci of fungal infection containing large amounts of fungal hyphae and spores and keratin debris—are very difficult to treat successfully.
2. Superficial onychomycosis (Figure 1.19). The nail plate is invaded directly by the causative organism and patches appear on the nail plate surface. The patches may coalesce to cover the whole plate, whose surface may crumble. The patches are chalky white (in superficial white onychomycosis (SWO), mainly caused by *Trichophyton mentagrophytes*), but may be black (when caused by dematiaceous or black fungi in superficial black onychomycosis). Histopathology reveals chains of round fungal spores in the nail plate between splits in the surface of the nail plate. Superficial onychomycosis is normally confined to the toenails and is a fairly rare pattern of fungal infection, except in HIV-infected patients where the superficial onychomycosis

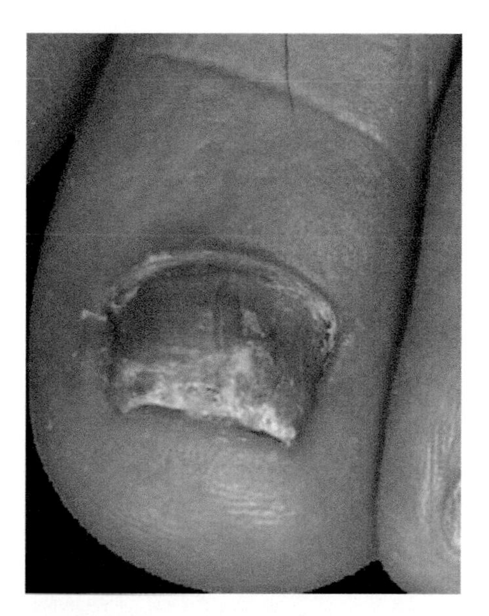

FIGURE 1.18 **(See color insert.)** Distal and lateral subungual onychomycosis. (Reproduced from Welsh et al., *Clin. Dermatol.*, 28, 151, 2010. With kind permission from Elsevier Inc.)

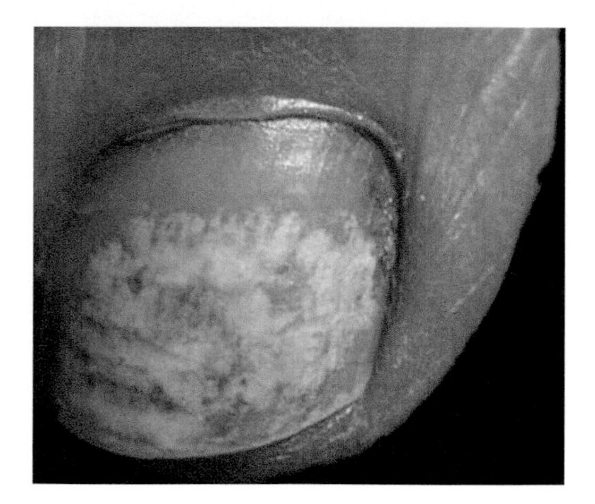

FIGURE 1.19 **(See color insert.)** Superficial white onychomycosis. (Reproduced from Welsh et al., *Clin. Dermatol.*, 28, 151, 2010. With kind permission from Elsevier Inc.)

can occur in fingernails or toenails and can occur concurrently with proximal subungual infection.

3. Proximal white subungual onychomycosis (PWSO; Figure 1.20). The fungus (mainly *T. rubrum*) penetrates through the proximal nail fold and invades the undersurface of the newly formed nail plate, producing a white discoloration in the area of the lunula. When the nail plate grows, the white discoloration moves distally. Histopathology reveals fungal hyphae and spores in different layers of the nail plate. Microscopic slits can result, which can in turn give

rise to onycholysis. PWSO is a very rare pattern of fungal infection although in patients with AIDS, a rapidly spreading PWSO infection has been noted. PWSO can also occur secondary to paronychia, due to the loss of the mechanical barrier to invading organisms when the proximal nail folds are inflamed.

4. Endonyx onychomycosis (EO). This new form of onychomycosis, caused by *T. soudanense*, has recently been added to the classification of onychomycosis by Hay et al. (2001). In EO, there is less thickening and opacification of the nail plate although the latter is pitted and lamellar splits occur at the plate's distal end, causing the latter to be friable and split. Over the nail plate, areas of superficial and deep fungal invasion coexist.
5. Total dystrophic onychomycosis (Figure 1.21). This is the potential endpoint of all forms of onychomycosis, especially of DLSO, and the entire nail plate and bed are invaded by the fungus. The nail plate can crumble off to expose a thickened and abnormal nail bed.

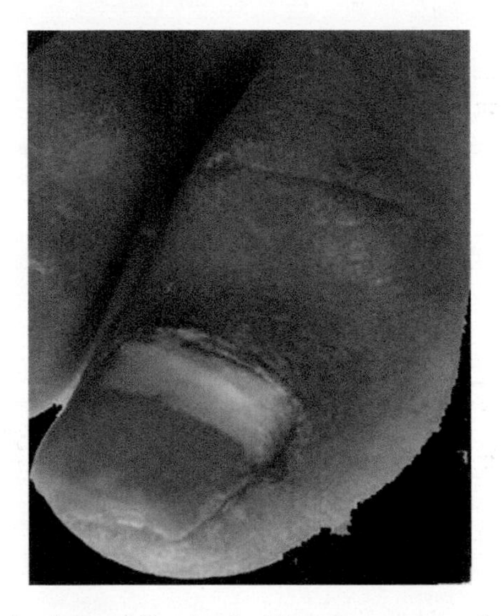

FIGURE 1.20 **(See color insert.)** Proximal white subungual onychomycosis. (Reproduced from Welsh et al., *Clin. Dermatol.*, 28, 151, 2010. With kind permission from Elsevier Inc.)

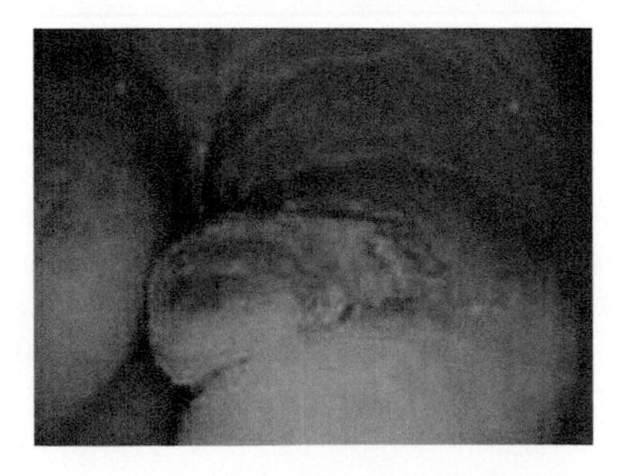

FIGURE 1.21 **(See color insert.)** Total dystrophic onychomycosis. (Reproduced from Welsh et al., *Clin. Dermatol.*, 28, 151, 2010. With kind permission from Elsevier Inc.)

1.2.2.1 Treatment of Onychomycosis

Onychomycosis can be treated using systemic and/or topical antifungal agents, depending on the severity of the disease, patient population and choice, and cost-effectiveness, among other factors (Baran et al. 2005; Gupta and Lynch 2004; Welsh et al. 2010). Baran et al. (2008) have suggested a stepwise approach to therapy, including the options for special populations, such as the elderly, children, pregnant women, and immunocompromised patients. It should be noted that even after successful treatment, a normal-looking nail is not always achieved (Scher et al. 2007).

1.2.2.1.1 Systemic Therapy of Onychomycosis

Systemic therapy is the mainstay of onychomycosis treatment, due to the low cure rate of topical therapy. Currently, terbinafine, itraconazole, and fluconazole are the most commonly used oral antifungal agents. Following oral administration and absorption into the systemic circulation, the drugs diffuse from the blood vessels into the nail plate via the nail bed. Terbinafine, itraconazole, and fluconazole concentrate and persist in the nail plate even after cessation of therapy, with the longest duration for itraconazole (6–9 months) compared to 3–6 months for fluconazole and terbinafine (Baran et al. 2008a,b). The long persistence of itraconazole in the nail plate led to the development of its intermittent dosing regimen (200 mg twice daily for 1 week per month, for a total of 2 months for fingernails and 3 months for toenails) instead of the daily (200 mg) dose for 3 months. The pulse treatment reduces the systemic exposure of the drug and hence the occurrence of adverse effects. Fluconazole is also administered in pulses—it is taken once weekly (100–300 mg) for 6 months. Following oral administration, it rapidly penetrates into the nail. Both itraconazole and fluconazole inhibit the cytochrome P450 3A4 isoenzyme system. Fluconazole also inhibits CYP2C9. These two drugs thus have the potential to increase plasma concentrations of other concomitantly administered drugs that are metabolized by these enzyme systems, and are therefore not ideal systemic therapies. Terbinafine is the drug of choice in the United Kingdom due to its greater efficacy and safety profile (Sigurgeirsson et al. 2002); it interacts with few drugs, unlike itraconazole and fluconazole, and is dosed at 250 mg daily for 6–12 weeks.

Unfortunately, treatment failure is common; around 20% of patients do not respond to treatment (Roberts 1999) and onychomycosis has been called a "stubborn clinical problem" (Arrese and Pierard 2003). Numerous reasons for treatment failure have been identified and include poor patient compliance, misdiagnosis, clinical and mycological variants [e.g., dermatophytoma (a hyperkeratotic mass containing densely packed thick-walled dermatophyte hyphae), thickened nails, extensive onycholysis (which is a barrier to drug movement within the nail unit), development of resistant organisms], local predisposing factors (e.g., wearing of improperly fitted shoes), systemic factors (e.g., diabetes), host response (e.g., immunological factors), and a large load of dormant, drug-resistant fungal spores (Baran et al. 2008; Pierard et al. 2000a). Relapse is also common, and recurrence rates of 10%–53% have been reported (Scher et al. 2007). Systemic therapy also has inherent disadvantages such as adverse events (e.g., headache and gastrointestinal symptoms, taste disturbance, and more rarely, but seriously, liver and kidney disturbances) and drug interactions.

1.2.2.1.2 Topical Therapy of Onychomycosis

Topical therapy avoids the problems associated with the adverse events and drug interactions of systemic drugs and may have greater patient compliance. However, because of the low efficacy of topical therapy (Pierard et al. 2000a,b), it is generally recommended for mild and distal infection in up to two nails, for SWO, or when systemic therapy is contraindicated, for example, in pregnancy (BNF 2008). Topical therapy is sometimes used in combination with systemic therapy as this improves cure rate (Baran et al. 2000) and costs less. The most convenient topical preparations are the nail lacquers (nail varnish) containing the antifungal agents amorolfine (Loceryl®) and ciclopirox (Penlac®). Following application to nail plates, the lacquers dry within a few minutes to leave a water-insoluble film. Loceryl is applied using an applicator one to two times weekly to filed nail plates for up to 6 months for fingernails and for 9–12 months for toenails. Penlac is applied with a brush once daily, preferably at bedtime, for up to 48 weeks. Every 7 days the existing Penlac film is removed with alcohol before reapplication of the lacquer. Penlac and Loceryl are generally well tolerated though slight burning with Loceryl and periungual erythema with Penlac have been reported (Patient Information Leaflets for Penlac and Loceryl). The lacquers are cost-effective compared to oral therapy (Einarson et al. 1997) and are discussed in more detail in Section 3.3. Other topical antifungal formulations for onychomycoses include tioconazole nail solution (Trosyl®), an undecenoate solution (Monphytol® discontinued in the United Kingdom), and salicylic acid paint (Phytex®), described in more detail in Section 3.1. Trosyl is applied to infected nails and surrounding skin twice daily for up to 6 months or in some cases 12 months. Phytex and Monphytol are also applied twice daily but are not the first line of treatment. Following topical application, the drug partitions out of the formulation and into the nail plate, then diffuses through the nail plate and hopefully into the nail bed. Drug concentration in the nail plate is routinely measured in the investigations of topical ungual formulations, and often the ungual drug concentration far exceeds the minimum inhibitory concentrations (MICs) of the target fungi. Achievement of high ungual drug concentrations is not, however, equivalent to treatment cure, as evidenced by the low efficacy of existing topical formulations. A steep gradient in drug concentrations within the nail plates has been found from the dorsal nail surface (that was previously in contact with the drug formulation) toward the inner nail (Hui et al. 2002). Low (or no) drug presence in the deeper parts of the nail prevents fungal eradication and cure. It should also be noted that the MICs of drugs is most often measured in standard media—a very different environment to the nail plate. The presence of keratin is known to increase the MICs (Osborne et al. 2004; Tatsumi et al. 2001); the “real” MIC in the nail plate may be much higher than those measured *in vitro*. The low efficacy of topical ungual formulations is further limited by the factors that limit systemic therapy, such as the presence of fungal spores, dermatophytoma, and others discussed previously. Topical therapy can be improved by removal of as much diseased nail as possible or by abrasion of the nail plate by using a high-speed handpiece or sandpaper nail files. Removal of the whole (not recommended) or part of the nail plate can be achieved using a urea ointment or by surgery. Following avulsion of the diseased part of the nail plate, topical antifungal agents, such as imidazoles, are applied under occlusion for several months.

1.2.2.1.3 Boosted Antifungal Therapy

As mentioned above, the existence of dormant fungal spores within the nail plate leads to incomplete cure rates for both topical and systemic therapies for onychomycosis. Months after cessation of therapy and an apparent onychomycotic cure, these spores can germinate and reinfect the nail, and the patient experiences a relapse. To address the existence of the dormant spores which form a fungal reservoir in the nail plate, boosted oral antifungal therapy (BOAT) and boosted antifungal topical treatment (BATT) have been developed (Pierard et al. 2000a,b). During systemic or topical therapy, a piece of Sabouraud's agar slide is secured onto affected nails plates. Because the Sabouraud medium converts dormant fungal cells into hyphae *in vitro*, it was hypothesized that this could also occur *in vivo*. Thus, germination of the fungal spores into drug-susceptible hyphae within the nail plate would be induced by the Sabouraud agar slide, and the fungi could be eradicated.

1.2.2.1.4 Keratin Binding to Drugs

Once permeated into the nail plate following topical or systemic administration, antifungal agents that have high affinity to keratin can bind to the latter. For example, terbinafine is known to strongly bind to keratin (Uchida and Yamaguchi 1993). However, to be effective as an antifungal agent, the drug must be in a free form to kill the fungus. Drug binding to keratin can therefore have a deleterious effect on its antifungal efficacy (Tatsumi et al. 2002). On the other hand, drug binding to keratin can result in drug concentration and persistence in the nail, such that a drug reservoir is formed in the nail plate. If the drug is subsequently released from the keratin at sufficient concentration and rate, the drug's antifungal efficacy would not be compromised, and keratin binding would be beneficial. During systemic treatment of onychomycosis, the drug permeates the nail plate through the nail bed—an easier route of drug entry into the nail plate (compared to after topical application)—and high drug concentrations can be achieved in the nail plate. Keratin binding becomes beneficial for therapy by providing a drug reservoir in the nail plate, such that systemic treatment can be given in pulses. On the other hand, during topical therapy, the drug penetrates the nail plate following application to its dorsal surface—a very good barrier to drug penetration. Keratin binding to the small amounts of drug that permeated into the nail plate could result in insufficient free drug available for antifungal activity. Drug binding to keratin in this case would have an adverse effect on antifungal activity. It is possible that the low efficacy of many topical antifungal agents is due to keratin binding and insufficient release of the bound drug. Thus, despite drug concentrations in the nail plate being well above the fungal organisms' MICs, the cure rate can be low.

1.2.3 Psoriasis

Skin psoriasis—characterized by epidermal thickening and scaling—affects between 1% and 3% of most populations but is most common in Europe and North America (Schofield and Hunter 1999). It is thought that over a lifetime, 80%–90% of patients with skin psoriasis will also suffer from psoriasis of the nail (Samman 1978), while 1%–5% of patients with nail psoriasis do not present any overt cutaneous disease (Del Rosso 1997). The nail matrix, nail bed, and nail folds may all be affected by psoriasis and influence the symptoms of the nail disease, which include pitting, discoloration, onycholysis, subungual hyperkeratosis (resulting from the deposition and collection of cells under the nail plate that have not undergone desquamation), nail plate abnormalities, splinter hemorrhages, and paronychia (Figures 1.22 through 1.24). Transient and limited dysfunction of the

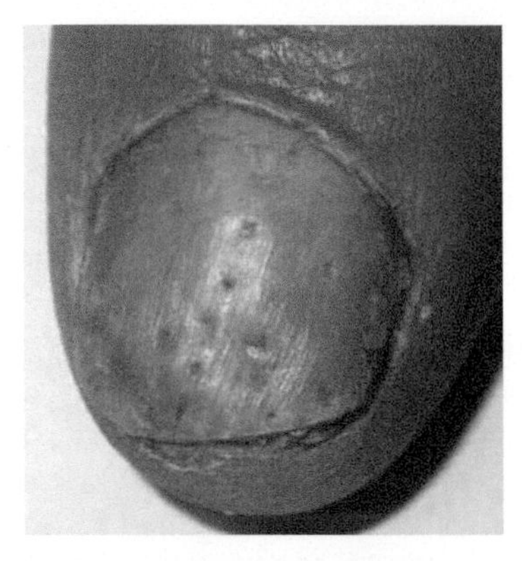

FIGURE 1.22 **(See color insert.)** Nail pitting. (Reproduced from Jiaravuthisan et al., *J. Am. Acad. Dermatol.* 57, 1, 2007. With kind permission from the American Academy of Dermatology, Inc.)

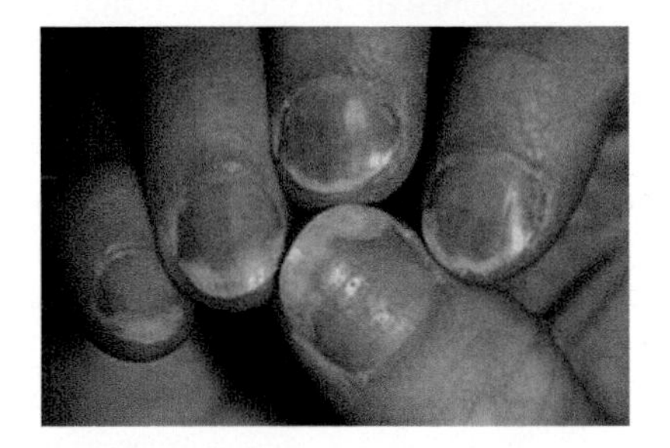

FIGURE 1.23 **(See color insert.)** Onycholysis at the distal nail plate. (Reproduced from Jiaravuthisan et al., *J. Am. Acad. Dermatol.* 57, 1, 2007. With kind permission from the American Academy of Dermatology, Inc.)

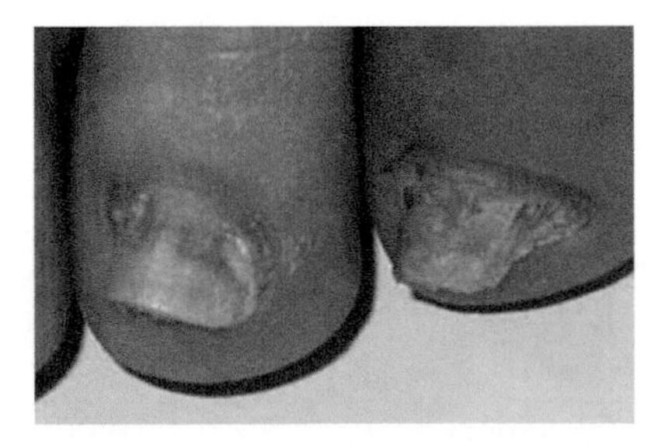

FIGURE 1.24 **(See color insert.)** Subungual hyperkeratosis. (Reproduced from Jiaravuthisan et al., *J. Am. Acad. Dermatol.* 57, 1, 2007. With kind permission from the American Academy of Dermatology, Inc.)

nail matrix results in pitting, white spots and transverse furrows, while a greater/more persistent extent of matrix dysfunction causes nail fragility, crumbling, and nail loss. Nail bed psoriasis produces oval, salmon-colored "oil drop" discoloration of the nail plate, subungual hyperkeratosis, splinter hemorrhages, and onycholysis. The latter is also caused by psoriasis of the hyponychium. Psoriatic nail folds result in paronychia, which leads to ridging of the nail plate. When paronychia is severe, the matrix may be injured with consequent nail abnormalities (Del Rosso 1997).

Like onychomycosis, nail psoriasis is a long-term condition and is difficult to cure, and therapy has to be maintained for very long durations. A range of, rather than a standardized, treatment options exist as certain manifestations of the disorder have been found to respond better to certain therapies, patient preferences for this long-term condition have been taken into account, and few large and well-controlled clinical trials have been undertaken. These treatment options include topical, intralesional, radiation, and systemic therapies, which have sometimes been used in combination, for example, oral and topical (Jiaravuthisan et al. 2007; Edwards and de Berker 2009; and references therein).

1.2.3.1 Topical Treatment of Nail Psoriasis

Creams, ointments, gels, lotions, and lacquers are applied to the nail plate and/or nail fold and/or nail bed (or as close to it as possible) depending on the location of the psoriatic lesion. If onycholysis is present, the nail plate is trimmed to the point where it separates from the nail bed. Sometimes, the nail plate is chemically avulsed, and the formulation is applied onto the nail bed. To enhance the drug movement into the nail unit, patients may be advised to cover the topical formulations with an occlusive dressing, such as by wearing plastic gloves. Glucocorticosteroids (e.g., betamethasone, clobetasol propionate, fluocinolone acetonide, and triamcinolone acetonide) and vitamin D3 analogs (calcipotriol, tacalcitol, calcitriol) are the most popular topical drugs and have been used concomitantly, for example, vitamin D3 analogs during weekdays and the steroid on weekends. Other drugs, 5-fluorouracil, anthralin, tazarotene, and topical cyclosporine have also shown benefits though much more clinical evidence for their efficacy is needed. 5-fluorouracil, whose efficacy has been variable, is recommended only when pitting and hypertrophy are the main symptoms of nail psoriasis, and is avoided when onycholysis is present. Anthralin has been suggested to be more appropriate when psoriatic symptoms are due to nail bed and hyponychium lesions, rather than nail matrix ones. Tazarotene—a third generation retinoid—was found to reduce onycholysis and pitting. Cyclosporine, in an oily vehicle, has shown very promising results in a patient with pitting and onycholysis. Urea and salicylic acid are sometimes included in the topical drug formulations as keratolytic agents that might assist ungual drug permeation. Urea (in a formulation with propylene glycol) has also shown some benefits in the absence of other agents. The adverse events related to these topical agents include erythema, skin atrophy, and possible atrophy of the bone underlying the nail treated with the potent corticosteroids, skin irritation with calcipotriol, pain, swelling, and onycholysis with 5-fluorouracil, staining of the skin with anthralin, and local erythema and skin irritation with tazarotene (Jiaravuthisan et al. 2007; Edwards and de Berker 2009).

1.2.3.2 Intralesional Therapy

Repeated (e.g., monthly, weekly) injection of long-acting corticosteroids (most commonly triamcinolone acetonide) into the nail folds (most commonly the proximal ones) is considered the mainstay for nail psoriasis treatment by many physicians. It appears to be most efficacious for treating the manifestations of nail matrix psoriasis, such as pitting, ridging, nail thickening, and subungual hyperkeratosis. Injection is conducted using a fine gauge needle; a high-pressure needle-less injector used previously has largely been abandoned due to "splash back" of blood. The pain of the injection is obviously a major adverse effect although a local anesthetic may be used. Other adverse events of these intralesional injections include subungual hematomas, reversible atrophy at the injection site, and potential atrophy of the underlying bone (Jiaravuthisan et al. 2007; Edwards and de Berker 2009).

1.2.3.3 Photochemotherapy

Oral or topical psoralen plus ultraviolet A has shown benefits for nail dystrophies arising from the nail bed, but not those from the nail matrix psoriasis. Pitting was not improved while onycholysis, subungual hyperkeratosis, oil drop, nail plate crumbling, and proximal nail fold symptoms were improved. Other types of radiation—superficial radiotherapy, Grenz ray therapy, and electron beam therapy—have demonstrated variable response rates in the few studies conducted. The adverse events and perceived harmfulness of radiation further limit their use (Jiaravuthisan et al. 2007; Edwards and de Berker 2009).

1.2.3.4 Systemic Treatment

Nail psoriasis can be treated with orally administered agents, such as retinoids, cyclosporine, etretinate. In addition, psoriasis being a disorder of the immune system (Mueller and Herrmann 1979), new biological agents such as alefacept and infliximab have been tried (Lawry 2007). However, due to the potential for adverse events and the greater cost of treatment, systemic therapy is only recommended when there is widespread skin or joint involvement in addition to nail psoriasis (Jiaravuthisan et al. 2007; Edwards and de Berker 2009).

1.3 TOPICAL UNGUAL FORMULATIONS

While lotions, paints, lacquers, creams, ointments, and gels, among other formulations, have been applied to the nail plate, there have been relatively few topical preparations developed specifically for the treatment of nail diseases. This could be due to the previous perception that nail diseases did not provide a large enough market for drug development and/or that nail diseases were mainly a cosmetic issue. Fortunately, the situation is now improved with two new products (Loceryl and Penlac nail lacquers) launched fairly recently, more products in the pipeline and much research on topical ungual drug delivery (Gupchup and Zatz 1999; Sun et al. 1999; Murdan 2002; Elkeeb et al. 2010; Murdan 2008). An ideal ungual formulation should be cosmetically acceptable, be easy to apply (and easy to remove when removal is desired), have a long residence time on the nail plate such that frequent application is not required, and most importantly, release the drug that can then permeate into the nail. Pharmaceutical nail lacquers possess these desirable qualities to some extent, and the anti-onychomycosis lacquers have largely replaced the older antifungal formulations, which include tioconazole nail solution, salicylic acid paint, and an undecenoate solution (Monphytol® paint). The topical formulations are discussed in detail in Chapter 3.

1.4 CONCLUSIONS

Research into the basic science of the nail unit, nail diseases and their treatment, and especially into the topical therapy continues at a dynamic pace, as evidenced by the number of books, journal articles, conference abstracts, and even dedicated meetings (Murdan 2007) on the subject. More research is needed to optimize therapy for the recalcitrant nail disorders and to assist in the rational design of topical ungual medicines, such that treatment of nail diseases becomes a routine and successful endeavor.

REFERENCES

Achten, G., J. Andre, and M. Laporte. 1991. "Nails in Light and Electron-Microscopy." *Seminars in Dermatology* 10 (1): 54–64.

Achten, G., and J. Wanet-Rouard. 1978. "Onychomycoses in the Laboratory." *Mykosen Supplement* 1: 125–27.

Agache, P. 2004. "Sebaceous Physiology." In *Measuring the Skin. Non-Invasive Investigations, Physiology, Normal Constants*, edited by P. Agache and P. Humbert. Berlin: Springer.

Agache, P., and P. Humbert. 2004. *Measuring the Skin*. Berlin: Springer

Arin, M. J. 2009. "The Molecular Basis of Human Keratin Disorders." *Human Genetics* 125 (4): 355–73.

Arrese, J. E., and G. E. Pierard. 2003. "Treatment Failures and Relapses in Onychomycosis: A Stubborn Clinical Problem." *Dermatology* 207 (3): 255–60.

Baden, H. P., L. A. Goldsmith, and B. Fleming. 1973. "Comparative Study of Physicochemical Properties of Human Keratinised Tissues." *Biochimica et Biophysica Acta* 322 (2): 269–78.

Baran, R., M. Feuilhade, A. Datry, S. Goettmann, P. Pietrini, C. Viguie, G. Badillet et al. 2000. "A Randomized Trial of Amorolfine 5% Solution Nail Lacquer Combined with Oral Terbinafine Compared with Terbinafine Alone in the Treatment of Dermatophytic Toenail Onychomycoses Affecting the Matrix Region." *British Journal of Dermatology* 142 (6): 1177–83.

Baran, R., A. K. Gupta, and G. E. Pierard. 2005. "Pharmacotherapy of Onychomycosis." *Expert Opinion on Pharmacotherapy* 6 (4): 609–24.

Baran, R., R. J. Hay, and J. I. Garduno. 2008a. "Review of Antifungal Therapy and the Severity Index for Assessing Onychomycosis: Part I." *Journal of Dermatological Treatment* 19 (2):72–81.

Baran, R., R. J. Hay, and J. I. Garduno. 2008b. "Review of Antifungal Therapy, Part II: Treatment Rationale, Including Specific Patient Populations." *Journal of Dermatological Treatment* 19 (3): 168–75.

Barron, J. N. 1970. "The Structure and Function of the Skin of the Hand." *Hand* 2 (2): 93–6.

Bean, W. B. 1980. "Nail Growth—35 years of Observation." *Archives of Internal Medicine* 140 (1): 73–6.

BNF, ed. 2008. *British National Formulary*. 56th ed. London: BMJ Group & RPS Publishing.

Branca, A. 1910. "Notes Sur la Structure de L'ongle" [Notes on the Structure of the Nail]. *Annales de dermatologie et de syphiligraphie* 1: 353–71.

Burch, G. E., and T. Winsor. 1946. "Diffusion of Water through Dead Plantar, Palmar and Torsal Human Skin and through Toe Nails." *Archives of Dermatology and Syphilology* 53 (1): 39–41.

Caputo, R., K. De Boulle, J. Del Rosso, and R. Nowicki. 2001. "Prevalence of Superficial Fungal Infections among Sports-Active Individuals: Results from the Achilles Survey, a Review of the Literature." *JEADV* 15: 312–16.

Chabasse, D., R. Baran, and M. F. De Chauvin. 2000. "Onychomycosis I: Epidemiology and Etiology." *Journal De Mycologie Medicale* 10 (4): 177–90.

Chapman, R. E. 1986. "Hair, Wool, Quill, Nail, Claw, Hoof and Horn." In *Biology of the Integument.Vol 2, Vertebrates*, edited by J. Bereiter-Hahn, A. G. Matoltsy, and K. S. Richards. New York: Springer-Verlag.

Cohen, P. R., and R. K. Scher. 1994. "Topical and Surgical Treatment of Onychomycosis." *Journal of the American Academy of Dermatology* 31 (3): S74–S77.

Daniel, C. R. 1991. "The Diagnosis of Nail Fungal Infection." *Archives of Dermatology* 127 (10): 1566–67.

Dawber, R. P. R., and R. Baran. 1984. "Structure, Embryology, Comparative Anatomy and Physiology of the Nail." In *Diseases of the Nails and Their Management*, edited by R. P. R. Dawber and R. Baran. Oxford: Blackwell Scientific Publications.

Dawber, R. P. R., D. de Berker, and R. Baran. 1994. "Science of the Nail Apparatus." In *Diseases of the Nails and Their Management*, edited by R. P. R. Dawber, D. A. R. de Berker, and R. Baran. London: Blackwell Scientific Publications.

Dawber, R. P. R., D. A. R. de Berker, and R. Baran. 2001. "Science of the Nail Apparatus." In *Baran and Dawber's Diseases of the Nails and Their Management*, edited by D. R. R. Baran, D. A. R. de Berker, E. Haneke, and A. Tosti. Oxford: Blackwell Science Ltd.

de Berker, D. A. R., J. Andre, and R. Baran. 2007. "Nail Biology and Nail Science." *International Journal of Cosmetic Science* 29 (4): 241–75.

de Berker, D., and B. Forslind. 2004. "The Structure and Properties of Nails and Periungual Tissues." In *Skin, Hair, and Nails, Structure and Function*, edited by B. Forslind, M. Linberg, and L. Norlén. New York: Marcel Dekker.

Del Rosso, J. Q., P. J. Basuk, R. K. Scher, and A. R. Ricci. 1997. "Dermatologic Diseases of the Nail Unit." In *Nails: Therapy, Diagnosis, Surgery*, edited by R. K. Scher and C. R. Daniel. Philadelphia: WB Saunders.

Diaz, A. A., A. F. Boehm, and W. F. Rowe. 1990. "Comparison of Fingernail Ridge Patterns of Monozygotic Twins." *Journal of Forensic Sciences* 35 (1): 97–102.

Duarte, A. F., O. Correia, and R. Baran. 2009. "Nail Plate Cohesion Seems to be Water Independent." *International Journal of Dermatology* 48 (2): 193–95.

Dutet, Julie, and M. Begona Delgado-Charro. 2009. "In Vivo Transungual Iontophoresis: Effect of DC Current Application on Ionic Transport and on Transonychial Water Loss." *Journal of Controlled Release* 140 (2): 117–25.

Edwards, F., and D. de Berker. 2009. "Nail Psoriasis: Clinical Presentation and Best Practice Recommendations." *Drugs* 69 (17): 2351–61.

Egawa, M., Y. Ozaki, and M. Takahashi. 2006. "In Vivo Measurement of Water Content of the Fingernail and Its Seasonal Change." *Skin Research and Technology* 12 (2): 126–32.

Einarson, T. R., P. I. Oh, A. K. Gupta, and N. H. Shear. 1997. "Multinational Pharmacoeconomic Analysis of Topical and Oral Therapies for Onychomycosis." *Journal of Dermatological Treatment* 8 (4): 229–35.

Elkeeb, R., A. AliKhan, L. Elkeeb, X. Y. Hui, and H. I. Maibach. 2010. "Transungual Drug Delivery: Current Status." *International Journal of Pharmaceutics* 384 (1–2): 1–8.

Fleckman, P. 2005. "Structure and Function of the Nail Unit." In *Nails. Diagnosis Therapy Surgery*, edited by R. Scher and C. R. Daniel III. Philadelphia: Elsevier Saunders.

Fluhr, J., L. Bankova, and S. Dikstein. 2006. *Handbook of Non-Invasive Methods and the Skin*. 2nd ed. Boca Raton: Taylor & Francis.

Forslind, B. 1970. "Biophysical Studies of Normal Nail." *Acta Dermato-Venereologica* 50 (3): 161–80.
Forslind, B., and N. Thyresson. 1975. "Structure of Normal Nail-Scanning Electron-Microscope Study." *Archiv Fur Dermatologische Forschung* 251 (3): 199–204.
Garson, J. C., F. Baltenneck, F. Leroy, C. Riekel, and M. Muller. 2000. "Histological Structure of Human Nail as Studied by Synchrotron X-Ray Microdiffraction." *Cellular and Molecular Biology* 46 (6): 1025–34.
Geoghegan, B., D. F. Roberts, and M. R. Sampford. 1958. "A Possible Climatic Effect on Nail Growth." *Journal of Applied Physiology* 13 (1): 135–38.
Germann, H., W. Barran, and G. Plewig. 1980. "Morphology of Corneocytes from Human Nail Plates." *Journal of Investigative Dermatology* 74 (3): 115–18.
Germann, H., and G. Plewig. 1977. "Size of Corneocytes from Human Nail Plate—Effects of Age and Sex." *Archives of Dermatological Research* 258 (1): 94.
Gilchrist, M. L., and L. H. Dudley Buxton. 1939. "The Relation of Fingernail Growth to Nutritional Status." *Journal of Anatomy* 73: 575–82.
Gniadecka, M., O. F. Nielsen, D. H. Christensen, and H. C. Wulf. 1998. "Structure of Water, Proteins, and Lipids in Intact Human Skin, Hair, and Nail." *Journal of Investigative Dermatology* 110 (4): 393–98.
Gonzalez-Serva, A. 1997. "Structure and Function." In *Nails: Therapy, Diagnosis, Surgery*, edited by R. K. Scher and C. R. Daniel. Philadelphia: WB Saunders.
Gunt, H., and G. B. Kasting. 2006. "Hydration Effect on Human Nail Permeability." *Journal of Cosmetic Science* 57 (2): 183–84.
Gunt, H. B., and G. B. Kasting. 2007. "Effect of Hydration on the Permeation of Ketoconazole through Human Nail Plate In Vitro." *European Journal of Pharmaceutical Sciences* 32 (4–5): 254–60.
Gupchup, G. V., and J. L. Zatz. 1999. "Structural Characteristics and Permeability Properties of the Human Nail: A Review." *Journal of Cosmetic Science* 50 (6): 363–85.
Gupta, A. K., and L. E. Lynch. 2004. "Management of Onychomycosis: Examining the Role of Monotherapy and Dual, Triple, or Quadruple Therapies." *Cutis* 74 (1): 5–9.
Gupta, A. K., and N. H. Shear. 1997. "Onychomycosis—Going for Cure." *Canadian Family Physician* 43: 299–305.
Hamilton, J. B., H. Terada, and G. E. Mestler. 1955. "Studies of Growth throughout the Lifespan in Japanese: Growth and Size of Nails and Their Relationship to Age, Sex, Heredity, and Other Factors." *Journals of Gerontology* 10 (4): 401–15.
Hay, R. J., R. Baran, and E. Haneke. 2001. "Fungal (Onychomycosis) and Other Infectins Involving the Nail Apparatus." In *Baran and Dawber's Diseases of the Nails and Their Management*, edited by D. R. Baran, D. A. R. De Berker, E. Haneke, and A. Tosti. Oxford: Blackwell Science.
Helmdach, M., A. Thielitz, E. M. Ropke, and H. Gollnick. 2000. "Age and Sex Variation in Lipid Composition of Human Fingernail Plates." *Skin Pharmacology and Applied Skin Physiology* 13 (2):111–9.
Hewitt, D., and R. W. Hillman. 1966. "Relation between Rate of Nail Growth in Pregnant Women and Estimated Previous General Growth Rate." *American Journal of Clinical Nutrition* 19 (6): 436–39.
Hui, X. Y., Z. Shainhouse, H. Tanojo, A. Anigbogu, G. E. Markus, H. I. Maibach, and R. C. Wester. 2002. "Enhanced Human Nail Drug Delivery: Nail Inner Drug Content Assayed by New Unique Method." *Journal of Pharmaceutical Sciences* 91 (1): 189–95.
Jarrett, A., and R. I. Spearman. 1966. "Histochemistry of Human Nail." *Archives of Dermatology* 94 (5): 652–57.
Jemec, G. B. E., T. Agner, and J. Serup. 1989. "Transonychial Water-Loss—Relation to Sex, Age and Nail-Plate Thickness." *British Journal of Dermatology* 121 (4): 443–46.

Jemec, G. B. E., and J. Serup. 1989. "Ultrasound Structure of the Human Nail Plate." *Archives of Dermatology* 125 (5): 643–46.

Jiaravuthisan, M. M., D. Sasseville, R. B. Vender, F. Murphy, and C. Y. Muhn. 2007. "Psoriasis of the Nail: Anatomy, Pathology, Clinical Presentation, and a Review of the Literature on Therapy." *Journal of the American Academy of Dermatology* 57 (1): 1–27.

Johnson, M., J. S. Comaish, and S. Shuster. 1991. "Nail Is Produced by the Normal Nail Bed—A Controversy Resolved." *British Journal of Dermatology* 125 (1): 27–29.

Khengar, R. H., S. A. Jones, R. B. Turner, B. Forbes, and M. B. Brown. 2007. "Nail Swelling as a Pre-Formulation Screen for the Selection and Optimisation of Ungual Penetration Enhancers." *Pharmaceutical Research* 24: 2207–12.

Kobayashi, Y., M. Miyamoto, K. Sugibayashi, and Y. Morimoto. 1999. "Drug Permeation through the Three Layers of the Human Nail Plate." *Journal of Pharmacy and Pharmacology* 51 (3): 271–78.

Kronauer, C., M. Gfesser, J. Ring, and D. Abeck. 2001. "Transonychial Water Loss in Healthy and Diseased Nails." *Acta Dermato-Venereologica* 81 (3):175–7.

Kruger, N., T. Reuther, S. Williams, and M. Kerscher. 2006. "Effect of Urea Nail Lacquer on Nail Quality. Clinical Evaluation and Biophysical Measurements." *Hautarzt* 57 (12): 1089–93.

Landherr, G., O. Braunfalco, C. Hofmann, G. Plewig, and A. Galosi. 1982. "Fingernail Growth in Patients with Psoriasis Receiving PUVA Therapy." *Hautarzt* 33 (4): 210–13.

Lawry, M. 2007. "Biological Therapy and Nail Psoriasis." *Dermatologic Therapy* 20 (1): 60–67.

Le Gros Clark, W. E., and L. H. Dudley Buxton. 1938. "Studies in Nail Growth." *British Journal of Dermatology* 50 (5): 221–35.

Lewis, B. L. 1954. "Microscopic Studies of Fetal and Mature Nail and Surrounding Soft Tissue." *AMA Archives of Dermatology and Syphilology* 70 (6): 733–47.

Lewis, B. L., and H. Montgomery. 1955. "The Senile Nail." *Journal of Investigative Dermatology* 24 (1): 11–18.

Lynch, M. H., W. M. Oguin, C. Hardy, L. Mak, and T. T. Sun. 1986. "Acidic and Basic Hair Nail (Hard) Keratins—Their Colocalization in Upper Cortical and Cuticle Cells of the Human-Hair Follicle and Their Relationship to Soft Keratins." *Journal of Cell Biology* 103 (6): 2593–606.

Martinsen, O. G., S. Grimnes, and S. H. Nilsen. 2008. "Water Sorption and Electrical Properties of a Human Nail." *Skin Research and Technology* 14 (2): 142–46.

Marty, J. P. 1995. "Amorolfine Nail Lacquer: A Novel Formulation." *Journal of the European Academy of Dermatology and Venereology* 4 (Suppl 1): S17–S21.

Mavon, A., D. Redoules, P. Humbert, P. Agache, and Y. Gall. 1998. "Changes in Sebum Levels and Skin Surface Free Energy Components Following Skin Surface Washing." *Colloids and Surfaces B-Biointerfaces* 10 (5): 243–50.

Mavon, A., H. Zahouani, D. Redoules, P. Agache, Y. Gall, and P. Humbert. 1997. "Sebum and Stratum Corneum Lipids Increase Human Skin Surface Free Energy As Determined from Contact Angle Measurements: A Study on Two Anatomical Sites." *Colloids and Surfaces B-Biointerfaces* 8 (3): 147–55.

Michel, C., B. Cribier, J. Sibilia, J. L. Kuntz, and E. Grosshans. 1997. "Nail Abnormalities in Rheumatoid Arthritis." *British Journal of Dermatology* 137 (6): 958–62.

Midgley, G., M. K. Moore, and J. C. Cook. 1994. "Mycology of Nail Disorders." *Journal of the American Academy of Dermatology* 31: S68–S74.

Mohorcic, M., A. Torkar, J. Friedrich, J. Kristl, and S. Murdan. 2007. "An Investigation into Keratinolytic Enzymes to Enhance Ungual Drug Delivery." *International Journal of Pharmaceutics* 332 (1–2): 196–201.

Moll, R., M. Divo, and L. Langbein. 2008. "The Human Keratins: Biology and Pathology." *Histochemistry and Cell Biology* 129 (6): 705–33.

Mueller, W., and B. Herrmann. 1979. "Cyclosporin A for Psoriasis." *New England Journal of Medicine* 301 (10): 555.

Murdan, S. 2002. "Drug Delivery to the Nail Following Topical Application." *International Journal of Pharmaceutics* 236 (1–2): 1–26.

Murdan, S. 2007. "1st Meeting on Topical Drug Delivery to the Nail." *Expert Opinion on Drug Delivery* 4 (4): 453–55.

Murdan, S. 2008. "Enhancing the Nail Permeability of Topically Applied Drugs." *Expert Opinion on Drug Delivery* 5 (11): 1267–82.

Murdan, S. 2010. "Transverse Fingernail Curvature: A Quantitative Evaluation, and An Exploration into the Influence of Gender, Age, Handedness, Height and Hand Size." Presentation at *UK Pharmaceutical Science 2010*, September 2010.

Murdan, S., G. Equizi, and G. Goriparthi. 2010. "An In Vivo Determination of the Human Nail Plate pH." Presentation at *UK Pharmaceutical Science 2010*, September 2010.

Murdan, S., D. Hinsu, and M. Guimier. 2008. "A Few Aspects of Transonychial Water Loss (TOWL): Inter-Individual, and Intra-Individual Inter-Finger, Inter-Hand and Inter-Day Variabilities, and the Influence of Nail Plate Hydration, Filing and Varnish." *European Journal of Pharmaceutics and Biopharmaceutics* 70 (2): 684–89.

Murdan, S., C. Poojary, D. R. Patel, J. Fernandes, A. Haman, and Z. Sheikh. 2010. "An Investigation into the Surface Energy of Human Fingernail Plates." Presentation at *UK Pharmaceutical Science 2010*, September 2010.

Nardoto, G. B., S. Silva, C. Kendall, J. R. Ehleringer, L. A. Chesson, E. S. B. Ferraz, M. Z. Moreira, J. P. H. B. Ometto, and L. A. Martinelli. 2006. "Geographical Patterns of Human Diet Derived from Stable-Isotope Analysis of Fingernails." *American Journal of Physical Anthropology* 131 (1): 137–46.

Nuutinen, J., I. Harvima, M. R. Lahtinen, and T. Lahtinen. 2003. "Water Loss through the Lip, Nail, Eyelid Skin, Scalp Skin and Axillary Skin Measured with a Closed-Chamber Evaporation Principle." *British Journal of Dermatology* 148 (4): 839–41.

Osborne, C. S., I. Leitner, B. Favre, and N. S. Ryder. 2004. "Antifungal Drug Response in an In Vitro Model of Dermatophyte Nail Infection." *Medical Mycology* 42 (2): 159–63.

Pardo-Castello, V. 1960. *Diseases of the Nail*. 3rd ed. Springfield, IL: Charles C Thomas.

Pierard, G. 2001. "Onychomycosis and Other Superficial Fungal Infections of the Foot in the Elderly: A Pan-European Survey." *Dermatology* 202 (3): 220–24.

Pierard, G. E., C. Pierard-Franchimont, and J. E. Arrese. 2000a. "The Boosted Antifungal Topical Treatment (BATT) for Onychomycosis." *Medical Mycology* 38 (5): 391–92.

Pierard, G., C. Pierard-Franchimont, and J. E. Arrese. 2000b. "The Boosted Oral Antifungal Treatment for Onychomycosis Beyond the Regular Itraconazole Pulse Dosing Regimen." *Dermatology* 200 (2): 185–87.

Ramot, Y., R. Paus, S. Tiede, and A. Zlotogorski. 2009. "Endocrine Controls of Keratin Expression." *Bioessays* 31 (4): 389–99.

Rand, R., and H. P. Baden. 1984. "Pathophysiology of Nails—Onychopathophysiology." In *Pathophysiology of Dermatologic Diseases*, edited by H. Baden and N. A. Soter. New York: McGraw-Hill.

Rich, P., and R. K. Scher. 2003. *An Atlas of Diseases of the Nail, The Encyclopedia of Visual Medicine Series*. London: The Parthenon Publishing Group.

Roberts, D. T. 1999. "Onychomycosis: Current Treatment and Future Challenges." *British Journal of Dermatology* 141: 1–4.

Samman, P. 1978. *The Nails in Disease*. 3rd ed. London: Heinemann.

Scalar. 2010 [cited 30 June 2010]. "Evaluation of skin moisture contents for atopic dermatitis using the Scalar Moisture checker". http://www.scalar.co.jp/english/products/my-808s.html.

Scher, R. K. 1989. "Brittle Nails." *International Journal of Dermatology* 28 (8): 515–16.

Scher, R. K. 1996. "Onychomycosis: A Significant Medical Disorder." *Journal of the American Academy of Dermatology* 35 (3): S2–S5.

Scher, R. K., and A. B. Bodian. 1991. "Brittle Nails." *Seminars in Dermatology* 10 (1): 21–25.

Scher, R. K., A. Tavakkol, B. Sigurgeirsson, R. J. Hay, W. S. Joseph, A. Tosti, P. Fleckman et al. "Onychomycosis: Diagnosis and Definition of Cure." *Journal of the American Academy of Dermatology* 56 (6): 939–44.

Schofield, O. M. V., and J. A. A. Hunter. 1999. "Diseases of the Skin." In *Davidson's Principles and Practice of Medicine*, edited by C. Haslett, E. R. Chilvers, J. A. A. Hunter, and N. A. Boon. London: Churchill Livingstone.

Schweizer, J., P. E. Bowden, P. A. Coulombe, L. Langbein, E. B. Lane, T. M. Magin, L. Maltais. et al. 2006. "New Consensus Nomenclature for Mammalian Keratins." *Journal of Cell Biology* 174 (2): 169–74.

Sibinga, M. S. 1959. "Observations on Growth of Fingernails in Health and Disease." *Pediatrics* 24 (2): 225–33.

Sigurgeirsson, B., J. H. Olafsson, J. B. Steinsson, C. Paul, S. Billstein, and E. G. Evans. 2002. "Long-term Effectiveness of Treatment with Terbinafine vs Itraconazole in Onychomycosis: A 5-year Blinded Prospective Follow-Up Study." *Archives of Dermatology* 138 (3): 353–57.

Sirota, L., R. Straussberg, P. Fishman, F. Dulitzky, and M. Djaldetti. 1988. "X-Ray—Microanalysis of the Fingernails in Term and Preterm Infants." *Pediatric Dermatology* 5 (3): 184–86.

Sowa, M. G., J. Wang, C. P. Schultz, M. K. Ahmed, and H. H. Mantsch. 1995. "Infrared Spectroscopic Investigation of In-Vivo and Ex-Vivo Human Nails." *Vibrational Spectroscopy* 10 (1): 49–56.

Spruit, D. 1971. "Measurement of Water Vapor Loss through Human Nail In-Vivo." *Journal of Investigative Dermatology* 56 (5): 359–61.

Spruit, D. 1972. "Effect of Nail Polish on the Hydration of the Fingernail." *Americas Cosmetic Perfils* 87: 57–58.

Stern, D. K., S. Diamantis, E. Smith, H. C. Wei, M. Gordon, W. Muigai, E. Moshier, M. Lebwohl, and P. Spuls. 2007. "Water Content and Other Aspects of Brittle versus Normal Fingernails." *Journal of the American Academy of Dermatology* 57 (1): 31–36.

Sun, Y., J.-C. Liu, and J. C. T. Wang. 1999. "Nail Penetration. Focus on Topical Delivery of Antifungal Drugs for Onychomycosis Treatment." In *Percutaneous Absorption. Drugs Cosmetics Mechanisms Methodology*, edited by R. L. Bronaugh and H. I. Maibach. New York: Marcel Dekker, Inc.

Tatsumi, Y., M. Yokoo, T. Arika, and H. Yamaguchi. 2001. "In Vitro Antifungal Activity of KP-103, a Novel Triazole Derivative, and Its Therapeutic Efficacy against Experimental Plantar Tinea Pedis and Cutaneous Candidiasis in Guinea Pigs." *Antimicrobial Agents and Chemotherapy* 45 (5): 1493–99.

Tatsumi, Y., M. Yokoo, H. Senda, and K. Kakehi. 2002. "Therapeutic Efficacy of Topically Applied KP-103 against Experimental Tinea Unguium in Guinea Pigs in Comparison with Amorolfine and Terbinafine." *Antimicrobial Agents and Chemotherapy* 46 (12): 3797–801.

Tosti, A, and B. M. Piraccini. 2005. "Pediatric Diseases." In *Nails. Diagnosis Therapy Surgery*, edited by R. K. Scher and C. R. I. Daniel. Philadelphia: Elsevier Saunders.

Tosti, A., and B. M. Piraccini. 2007. "Biology of Nails and Nail Disorders." In *Fitzpatrick's Dermatology in General Medicine*, edited by K. Wolff, L. A. Goldsmith, S. I. Katz, B. A. Gilchrest, A. S. Paller, and D. J. Leffell. New York: McGraw-Hill.

Tudela, E., A. Lamberbourg, M. C. Diaz, H. Zhai, and H. I. Maibach. 2008. "Tape Stripping on a Human Nail: Quantification of Removal." *Skin Research and Technology* 14 (4): 472–77.

Uchida, K., and H. Yamaguchi. 1993. "Studies on the Affinity of Terbinafine with Keratin." *Japanese Journal of Medical Mycology* 34: 207–12.

Vejnovic, I., L. Simmler, and G. Betz. 2010. "Investigation of Different Formulations for Drug Delivery through the Nail Plate." *International Journal of Pharmaceutics* 386 (1–2): 185–94.

Vellar, O. D. 1970. "Composition of Human Nail Substance." *American Journal of Clinical Nutrition* 23 (10): 1272–74.

Welsh, O., L. Vera-Cabrera, and E. Welsh. 2010. "Onychomycosis." *Clinics in Dermatology* 28 (2): 151–59.

Wilhelm, M., B. Pesch, R. Wittsiepe, P. Jakubis, P. Miskovic, T. Keegan, M. J. Nieuwenhuijsen, and U. Ranft. 2005. "Comparison of Arsenic Levels Fingernails with Urinary As Species As Biomarkers of Arsenic Exposure in Residents Living Close to a Coal-Burning Power Plant in Prievidza District, Slovakia." *Journal of Exposure Analysis and Environmental Epidemiology* 15 (1): 89–98.

Williams, T. I., R. Rose, and S. Chisholm. 2007. "What Is the Function of Nail Biting: An Analog Assessment Study." *Behaviour Research and Therapy* 45 (5): 989–95.

Wortsman, X., and G. B. E. Jemec. 2006. "Ultrasound Imaging of Nails." *Dermatologic Clinics* 24 (3): 323–28.

Yazdanparast, S. A., and R. C. Barton. 2006. "Arthroconidia Production in *Trichophyton Rubrum* and a New Ex Vivo Model of Onychomycosis." *Journal of Medical Microbiology* 55 (Pt 11): 1577–81.

Zaias, N. 1990. *The Nail in Health and Disease*. 2nd ed. Connecticut: Appleton & Lange.

Zaun, H. 1997. "Brittle nails. Objective Assessment and Therapy Follow-up." *Hautarzt* 48: 455–61.

2 Permeability of the Nail Plate

Jinsong Hao and S. Kevin Li

CONTENTS

The characteristic structure of human nails renders the nail resistant to penetration of permeants. Although nail permeability is generally poor, previous studies have shown the penetration of small to moderate hydrophilic permeants and slightly hydrophobic antifungal agents across the nail. This chapter covers the nail properties critical for nail permeability, such as hydration and electrical properties, and reviews nail permeability and formulation factors influencing nail permeability.

2.1 PHYSICOCHEMICAL NATURE OF THE NAIL

2.1.1 Hydration and Water Content

Water content is not only an indicator of nail physical conditions (Baden 1970) but also an important factor influencing the physical properties of nails (Baden et al. 1973; Finlay et al. 1980). The *in vivo* water content of the nail plate of human fingernails is reported to be between 7% and 25% under physiological conditions, which were measured from the free edge of the nail using portable near infrared spectroscopy (Egawa et al. 2005). The water content changes seasonally and is affected by the relative humidity (RH) in the environment. Dilatation of intercellular space between nail keratinocytes correlated with nail brittleness and dehydration may induce intercellular dilatation (Kitamori et al. 2006). However, recent studies found no relationship between the water content of the nail plate and nail brittleness (Stern et al. 2007); lipid and keratin contents are factors influencing the nail plate cohesion (Duarte et al. 2009).

The *in vitro* measurement of the water content of nails and nail water uptake are mainly based on the studies of water absorption and desorption in liquid phase (Hao and Li 2008a, b; Hao et al. 2008; Smith et al. 2009) and in vapor phase (Gunt and Kasting 2006, 2007a; Gunt et al. 2007) using cadaver fingernails and nail clippings from healthy subjects. Ultrasound was also used to determine the water content of nail plate specimen in postmortem studies (Jemec and Serup 1989). Methods to measure *in vivo* water content of nails include Raman spectroscopy (Gniadecka et al. 1998; Wessel et al. 1999) and near infrared spectroscopy (Egawa et al. 2003, 2005).

Kinetics of nail water uptake shows a two-phase hydration process, that is, a fast phase and then a slow phase. In nail clipping hydration studies (Hao and Li 2008a), the nails approached 90% of complete hydration within half an h in the fast phase upon soaking the nails in phosphate buffered saline (PBS). This was followed by a slow phase in which equilibrium was observed in 1 day. No further change in nail hydration was observed up to 7 days of monitoring. In an earlier *in vitro* transport study (Walters et al. 1981), the permeability of methanol across human nails was not affected by the duration of nail hydration after 24-h hydration pretreatment before the transport study, indicating no further significant hydration of the nail over a period of up to 49 days after the initial 24 h.

There was no significant difference in water content between nail water uptake in deionized water and in PBS (Hao and Li 2008b). The *in vitro* average water content of fully hydrated nail clippings obtained from uptake studies in PBS was 40% (Hao et al. 2008). Other water content results of fully hydrated nails have been reported, such as 143 mg/cm^3 for great toenails and 242–362 mg/cm^3 for fingernails (Malhotra and Zatz 2000) corresponding to 11% for great toenails and 19–28% for fingernails, given the nail density of 1.3 g/cm^3 (Leider and Buncke 1954). The nail water content obtained from the liquid-phase desorption study was as high as approximately 50% (Gunt et al. 2007). These differences may be related to the different measurement methods employed. For example, the water content (24%) measured when the fully hydrated nail clippings were subjected to air drying at room temperature was significantly lower than the water content (36%) when the nail was measured using the oven dry method at 60°C (Hao and Li 2008a). This is in agreement with the report that over 80% of water in the nail under physiological conditions was in bound form and less than 20% of water was in free form (Gniadecka et al. 1998).

The relationship between the water content of the nail and the surrounding RH has been previously established (Baden 1970; Gunt and Kasting 2007a). In *in vitro* vapor-phase water uptake studies, the results showed that the equilibrium water content of nail increased with increasing RH (Gunt and Kasting 2006). At 32°C, the water content of the nail increased from 0.035 g H_2O/g dry nail at 15% RH to 0.214 g H_2O/g dry nail at 100% RH (Gunt et al. 2007). It was also found that the water content increased gradually with an increase in RH in the range of up to 80%, and then the water content drastically increased between 80% and 100% RH (Baden 1970; Gunt and Kasting 2007a). Data analysis using the D'Arcy–Watt model suggests that most of the water in the nail is strongly bound to keratin fibers at low RH (Gunt and Kasting 2007a). Under physiological conditions, the dorsal side of the human nail is exposed to lower RH and the ventral side of the human nail is in contact with the nail bed, which is fully hydrated. Therefore, a water content gradient from the nail bed to the nail surface is expected *in vivo*. This suggests that the water content measured from the free edge of the nail plate *in vivo* could be different from the water content measured from the nail plate above the nail bed.

The uptake of water into nails has also been studied under different pH and ionic strength conditions. In a moderate pH range, *in vitro* water uptake of nails was not affected by pH or ionic strength of the equilibrating solutions (Hao and Li 2008b; Smith et al. 2009). However, extreme pH (e.g., pH <2 and pH >10) increased *in vitro* water uptake of the nails, indicating a change in the integrity of the nails (Smith et al. 2010). When the nail hydrates, the nail keratin network swells. Nail water uptake, expressed as weight gain after hydration with respect to the normal nail weight, is a measurement of nail swelling. The human nail swelling (or water uptake) was about 27% (Khengar et al. 2007) and affected by the ungual penetration enhancer treatment (Kobayashi et al. 1998). It has been demonstrated that nail permeability is sensitive to nail swelling (Khengar et al. 2007). For example, nail swelling has been used to screen for ungual penetration enhancers.

Transonychial water loss (TOWL) is a method used to measure water loss through the nail *in vivo*. A relatively large range of TOWL values have been reported in human nails (Jemec et al. 1989; Spruit 1971). The TOWL value varies highly with the sex, age, and nail plate thickness of the individuals (Jemec et al. 1989). Intraindividual, interfinger, interhand, and interday variabilities were also found (Dutet and Delgado-Charro 2009; Murdan et al. 2008). Nail hydration affected TOWL measurements (Murdan et al. 2008). Diseased nails showed a significantly lower TOWL than healthy nails (Kronauer et al. 2001; Nuutinen et al. 2003). Tape stripping increased the TOWL from 6.9 to 9.3 g/m^2/h (Tudela et al. 2008). In an *in vitro* study, TOWL values of human cadaver nails were found to be affected by ungual penetration enhancer treatments (Vejnovic et al. 2010).

2.1.2 Electrical Properties of the Nail

2.1.2.1 Electrical Resistance

Trace amounts of minerals and electrolytes exist in the human nail plate (Murdan 2002). The main electrolytes in the nail plate are similar to those circulating in the plasma and other body fluids, such as K, Na, Mg, and Ca (Fleckman 2005). The amounts of Na in the nails of adults have been reported to be on average in the order of 1 mg/g nail (Fleckman 2005). Trace metals like Cu, Fe, Zn, Au, Cr, Se, and Ag have also been found in human nails (Fleckman 2005; Gupchup and Zatz 1999). Under physiological conditions, the nail plate is partially hydrated, and the ions in the nail are not highly conductive. Therefore, the nail electrical resistance *in vivo* is generally high. In a human *in vivo* nail resistance study (Hao et al. 2010), the nail resistance was as high as ~5.6 MΩ cm^2 immediately after the application of PBS (0.15 M, pH 7.4) on fingernails. Upon 2-min nail hydration, the nail electrical resistance decreased rapidly to ~0.5 MΩ cm^2. The nail resistance was found to decrease in the first 2 h after PBS application and reached a relatively constant value of ~10 kΩ cm^2 in the 6-h nail hydration study. The decrease in nail resistance was mainly attributed to nail hydration. The concentration of the ions in the solution in contact with the nails also affected nail electrical resistance, but to a much lesser extent than hydration (Hao et al. 2010). The presence of the endogenous ions in the nail plays an important role in the resistance of the nail under hydration. Nail resistance *in vitro* was observed to decrease in a similar manner as that *in vivo* upon hydration (Hao et al. 2010).

2.1.2.2 Nail Isoelectric Point and Charge

Keratins are the main proteins in the nail plate, and have an isoelectric point (pI) between 4.9 and 5.4 (Gupchup and Zatz 1999; Marshall 1983; Murdan 2002). The nail plate is expected to be essentially uncharged at a pH value about the same as its pI, net negatively charged at pH above the pI, and net positively charged at pH below the pI. The charges and charge distribution in the pores of the nail depend on the pH and ionic strength of the solution in contact with the nail. At the physiological pH of 7.4, the effective surface charge density of the pores in the nail was in the order of 10^{-4}–10^{-3} C/m^2, estimated using the partitioning, passive transport, or electroosmosis data (Hao and Li 2008b; Smith et al. 2009). The effective surface charge density of the pore pathways in the nail affects the extent of charge–charge interactions between the permeant and the nail during trans-ungual transport.

2.2 PERMEABILITY OF THE NAIL

2.2.1 Methodology

Side-by-side diffusion cells, Franz diffusion cells, and their modifications are commonly used in nail permeability studies *in vitro* and *in vivo*. Table 2.1 lists the representative methods in trans-ungual drug delivery literature that will be described here. A number of nail studies employed the side-by-side diffusion cells to study *in vitro* transport of ciclopirox and other model permeants through hydrated human cadaver fingernail plate (Hao and Li 2008a, b; Hao et al. 2008, 2009; Smith et al. 2009, 2010). Franz cells were also commonly used. Murthy et al. (2007a, b) used the Franz diffusion cells to study the transport of salicylic acid, glucose, and griseofulvin across human nail plate. Terbinafine penetration through human nail was investigated using the Franz diffusion cells *in vitro* (Murthy et al. 2009;

TABLE 2.1
Methodology Used in Trans-Ungual Transport Studies of Drugs *In Vitro*

Diffusion Cells	Experimental Conditions	Drugs Studied	References
Side-by-side diffusion cells	Human fingernail; effective diffusion area 0.64 cm^2; room temperature	Ciclopirox	Hao et al. (2009)
Franz diffusion cells	Human fingernail and nail clippings; effective diffusion area 0.2 cm^2; room temperature	Griseofulvin; terbinafine	Murthy et al. (2007a, 2009), Nair et al. (2009c, d)
Modified Franz diffusion cells	Human nails; effective diffusion area 0.7 cm^2; 37°C	Ketoconazole	Repka et al. (2004)
One-chamber Franz-type cells	Human fingernails; effective diffusion areas 0.38 and 0.64 cm^2; 32°C	Ketoconazole	Gunt and Kasting (2007b)
ChubTur cell	Human nail clippings; effective diffusion area 0.05 cm^2; 28°C	Terbinafine	Brown et al. (2009)
Teflon one-chamber cell	Human fingernail; normal saline cotton ball as nail bed; effective diffusion area 0.78 cm^2; 25°C	Ketoconazole; econazole; ciclopirox; oxaboroles	Hui et al. (2002, 2003, 2004, 2007a)
Vertical diffusion cells	Bovine hoof membrane; effective diffusion area 1.23 cm^2; 32°C	Ciclopirox	Monti et al. (2005)
Modified horizontal cells	Human nail clippings; 37°C	Miconazole; ketoconazole; itraconazole	Quintanar-Guerrero et al. (1998)
Plastic penetration chamber	Porcine hoof membrane; 32°C	Ciclopirox	Kim et al. (2001); Myoung and Choi (2003)
Side-by-side diffusion cells (small area)	Human nail clippings; effective area 0.05 cm^2; 37°C	5-Fluorouracil; tolnaftate; flurbiprofen	Kobayashi et al. (1998, 1999, 2004)
Stainless steel diffusion cells	Human nail plate; effective area 0.38 cm^2; 37°C	Miconazole	Walters et al. (1985b)

Nair et al. 2009c, d). Franz diffusion cells were also modified and employed to study trans-ungual penetration of ketoconazole (Repka et al. 2004) and 5-aminolevulinic acid from their formulations (Donnelly et al. 2005). A vertical diffusion cell system similar to the Franz diffusion cells was used to study ciclopirox nail lacquer with bovine hoof membrane as a model for human nail (Monti et al. 2005). Another modification was the specially designed one-chamber Franz-type diffusion cell system that had the shape to accommodate the natural curvature of the nail (Gunt and Kasting 2007b). The effect of hydration on the penetration of ketoconazole through human nail was investigated with this device. In another study, the ChubTur® cell was used to examine terbinafine trans-ungual transport through small distal nail clippings (3 mm × 3 mm) (Brown et al. 2009). Hui et al. (2002) developed an experimental system to simulate *in vivo* conditions of therapeutic, nonoccluded application of drug to human nail. In this system, the human nail surface was open to air and the inner surface was kept in contact with a small, normal, saline–wetted cotton ball acting as the nail bed. A unique drilling/removal method was developed to sample the different nail sections and allowed the measurement of drugs in the inner nail. This method has been used in the penetration experiments of ketoconazole (Hui et al. 2002), econazole (Hui et al. 2003), ciclopirox (Hui et al. 2004), oxaboroles (Hui et al. 2007a), and panthenol (Hui et al. 2007b). A poloxamer gel-based method to measure the amount of drug penetrated across the nail was developed (Kim et al. 2001) and used in the development of ciclopirox-loaded pressure-sensitive adhesive patches (Myoung and Choi 2003). Newly designed toenail applicators were used to study terbinafine penetration ex vivo (Nair et al. 2009b). For *in vivo* penetration studies, nail clippings could be collected and then assayed for drug amounts in the whole nail clippings and/or different layers of nail specimens. This method was used in the studies of a number of antifungal agents on repeated topical applications of drug formulations for several weeks *in vivo* (Amichai et al. 2010; Susilo et al. 2006; van Hoogdalem et al. 1997).

Bovine hoof was commonly used as a model for human nail plates in nail studies due to the limited availability of human nail plates, particularly in trans-ungual transport experiments (Donnelly et al. 2005; Kim et al. 2001; Mohorcic et al. 2007; Monti et al. 2005; Myoung and Choi 2003). Similarities between human nail plates and bovine hoof membranes have been reported, in terms of keratinic nature of the tissues, conformation of the nail keratins, effects of keratinolytic enzymes, and immunologic properties (Edwards et al. 1998; Mohorcic et al. 2007). However, the human nail plate has denser keratin networks and lesser water content than bovine hoof membrane (Mertin and Lippold 1997a, b). In addition, the human nail plate generally has lower permeability than bovine hoof membrane. Despite the higher permeability of bovine hoof membrane, bovine hoof was considered as a reliable model and used to investigate trans-ungual transport *in vitro* (Mertin and Lippold 1997a, b, c). The relationship between the permeability of human nail plate (P_N) and bovine hoof membrane (P_H) was established as $\log P_N = 3.723 + 1.751 \log P_H$ (Mertin and Lippold 1997b).

Lateral diffusion in bovine hoof membrane was observed by using photothermal beam deflection spectroscopy (Gotter et al. 2010). In nail transport studies, lateral transport in the nail can introduce errors. In transport of a permeant across a thick membrane (e.g., nail) through a small diffusion area with significantly larger membrane surface compared to the surface area for diffusion, edge effect should be considered. The degree of edge effect on transport under this condition depends on the thickness-to-diffusion area aspect ratio. For a circular aperture of radius (a) and a membrane of thickness (h), the critical aspect ratio (h/a) is 0.2 (Crank 1975). Below this value, the edge effect is negligible.

In nail transport experiments, the steady-state flux (J) can be calculated experimentally from the slope of the linear portion of the plot of the cumulative amount of a permeant across the nail (Q) against time (t), correcting for the effective diffusion area (A):

$$J = \frac{\Delta Q}{A \Delta t} \tag{2.1}$$

The apparent permeability coefficient (P) is defined as the flux divided by the concentration of the permeant in the donor chamber (C_d) in the transport experiment:

$$P = \frac{J}{C_d} \tag{2.2}$$

The transport lag time (t_L) can be estimated by extrapolating the linear portion of the Q versus t plot to the x-axis (i.e., the x-intercept). The effective diffusion coefficient (D_{eff}) can be obtained from the transport lag time by the relationship

$$t_L = \frac{h^2}{6D_{eff}} \tag{2.3}$$

2.2.2 Transport Mechanisms

Table 2.2 summarizes the nail transport studies published in the literature. It is generally accepted that the nail plate behaves like a concentrated hydrogel rather than a lipophilic membrane, as supported by *in vitro* penetration data of homologous *n*-alcohols (Walters et al. 1983), nicotinic acid esters (Mertin and Lippold 1997a), and *p*-hydroxybenzoic acid esters (Kobayashi et al. 2004). The nail keratin networks contain pores filled with aqueous solution when the nail hydrates. The effective size of these pores in the fully hydrated nail has been estimated to be around 0.7 nm (radius of a cylindrical pore model) using passive transport data of mannitol and urea (Hao and Li 2008a), which is in the same order of magnitude of the Stokes–Einstein radii of most drugs. When the dimensions of pores are of the same order as those of permeants, permeant transport through liquid-filled pores can be significantly hindered (Deen 1987). For nail permeation under hydration, transport across the human nail can be modeled as hindered transport across aqueous pore pathways through diffusion and convection (Hao and Li 2008a). This model suggests that the pore size of the nail transport pathways, nail porosity, and the molecular size and shape of a permeant all play important roles in trans-ungual transport. Accordingly, nail permeability can be increased by enlarging the nail effective pore size and/or increasing its porosity (e.g., enlarging the preexisting pores or creating larger pores). These effects can be achieved by hydrating the nail and/or treating the nail with ungual penetration enhancers.

Despite the low content of lipid in the nail plate (<1%), a lipophilic pathway may exist in trans-ungual transport, which allows the penetration of highly hydrophobic permeants (Murdan 2002; Walters et al. 1983). This was evidenced by the increased permeation of a homologous series of alcohols (C8–C12) through human nail plates with increasing lipophilicity of the alcohols (Walters et al. 1983). The nail permeability to these alcohols decreased to a minimum when the alkyl chain length

TABLE 2.2
Permeants Studied in Trans-Ungual Transport *In Vitro*[a]

References	Permeants	Membranes
Walters et al. (1981)	Water; methanol; ethanol	Human nail
Walters et al. (1983)	Water; methanol; ethanol; *n*-propanol; *n*-butanol; *n*-pentanol; *n*-hexanol; *n*-heptanol; *n*-octanol; *n*-decanol; *n*-dodecanol	Human nail
Walters et al. (1985a)	Methanol; ethanol; *n*-propanol; *n*-butanol; *n*-pentanol; *n*-hexanol; *n*-heptanol; *n*-octanol; *n*-decanol; *n*-dodecanol	Human nail
Walters et al. (1985b)	Miconazole	Human nail
Mertin and Lippold (1997b)	Paracetamol; phenacetin; diprophylline; chloramphenicol; iopamidol; amorolfine; bifonazole; ciclopirox; clotrimazole; econazole; griseofulvin; ketoconazole; naftifine; nystatin; tolnaftate	Human nail and bovine hoof
Mertin and Lippold (1997a)	Methyl nicotinate; ethyl nicotinate; butyl nicotinate; hexyl nicotinate; octyl nicotinate; benzyl alcohol; benzoic acid; pyridine	Human nail and bovine hoof
Mertin and Lippold (1997c)	Chloramphenicol	Human nail and bovine hoof
Kobayashi et al. (1998)	5-Fluorouracil; tolnaftate	Human nail
Kobayashi et al. (1999)	5-Fluorouracil; flurbiprofen	Human nail
Kobayashi et al. (2004)	*p*-Hydroxybenzoic acid esters; deuterium oxide; ethanol; pyridine; benzoic acid; antipyrine; 5-fluorouracil; aminopyrine; lidocaine; isosorbide dinitrate; mexiletine; isoproterenol; procaine; croconazole	Human nail
Hui et al. (2002)	Urea; ketoconazole; salicylic acid	Human nail
Hui et al. (2003)	Econazole	Human nail
Hui et al. (2004)	Ciclopirox	Human nail
Hui et al. (2007a)	Oxaboroles; AN2690	Human nail
Hui et al. (2007b)	Panthenol	Human nail
Donnelly et al. (2005)	5-Aminolevulinic acid	Human nail and porcine hoof
Repka et al. (2004)	Ketoconazole	Human nail
Gunt and Kasting (2007b)	Ketoconazole	Human nail
Quintanar-Guerrero et al. (1998)	Miconazole; ketoconazole; itraconazole	Human nail
Khengar et al. (2007)	Mannitol; caffeine	Human nail
Brown et al. (2009)	Caffeine; methylparaben; terbinafine	Human nail
Malhotra and Zatz (2000, 2002)	Water	Human nail
Murthy et al. (2007a)	Glucose; griseofulvin	Human nail
Murthy et al. (2007b)	Salicylic acid	Human nail
Murthy et al. (2009); Nair et al. (2009a, c, d, e)	Terbinafine	Human nail

TABLE 2.2 (*Continued*)
Permeants Studied in Trans-Ungual Transport *In Vitro*[a]

References	Permeants	Membranes
Hao and Li (2008a, b), Hao et al. (2008), Smith et al. (2009, 2010)	Mannitol; urea; tetraethylammonium; water	Human nail
Hao et al. (2009)	Ciclopirox	Human nail
Monti et al. (2005)	Ciclopirox	Bovine hoof
Kim et al. (2001), Myoung and Choi (2003)	Ciclopirox	Porcine hoof

[a] This list is not inclusive of all the nail transport studies.

increased from C1 to C8 and then drastically increased with a further increase in the alkyl chain length to C12, indicating that the transport pathway changed from a hydrophilic pathway to a lipophilic pathway. However, the presence of the lipid pathway in the nail was not supported by later studies (Kobayashi et al. 2004; Mertin and Lippold 1997a). Chloroform/methanol delipidization of the nail led to increased penetration of water, methanol, ethanol, and butanol but decreased the penetration of more lipophilic decanol (Walters et al. 1985a). Essentially the same terbinafine permeation through the normal human nail plate and the "defatted" nail was reported (Nair et al. 2009c).

The dorsal layer of the nail is the main barrier to permeant transport through the nail plate into the nail bed. In an earlier study of penetration of both hydrophilic 5-fluorouracil and hydrophobic flurbiprofen through the dorsal, intermediate, and ventral human nail plate using dorsal-and-ventral-filed nail plate, dorsal-filed nail plate, ventral-filed nail plate, and full-thickness nail plate, the penetration of both drugs decreased in the order of dorsal-and-ventral-filed nail plate, dorsal-filed nail plate, ventral-filed nail plate, and full-thickness nail plate (Kobayashi et al. 1999). The dorsal layer as the main barrier was also supported by other studies. It was shown that 5-aminolevulinic acid penetration through human nail from a bioadhesive patch was improved by filing the dorsal surface of the nail (Donnelly et al. 2005). A recent study on terbinafine transport *in vitro* also showed that terbinafine penetration through human nail after the abrasion of the dorsal layer was approximately four times higher than that through the nail without the abrasion (Nair et al. 2009c). Conversely, the nail permeability to terbinafine was not affected when the ventral layer of the nail was abraded.

2.2.3 Nail Permselectivity

The nail is negatively charged at the physiological pH of 7.4 (Gupchup and Zatz 1999; Murdan 2002). The nail thus favors the permeation of positively charged permeants at pH above the pI and that of negatively charged permeants at pH below the pI. Under physiological conditions, the nail plate has permselectivity to cations over anions. This has been tested in the passive transport experiments of small ions, such as Na and Cl ions, through fully hydrated human nail under different pH conditions (Smith et al. 2010). At pH 7, the passive permeability coefficient for Na ion was higher than that for Cl ion and the permeability coefficient ratio of Cl to Na ions was 0.4. After pH 1 solution treatment of the nail, a prolonged reversible alteration of the nail permselectivity was observed and the permeability coefficient for Cl ion was 25 times more than that for Na ion. The nail permselectivity has also been investigated using electrokinetic theory in iontophoretic transport experiments of neutral and charged permeants of moderate sizes (Dutet and Delgado-Charro 2010; Hao and Li 2008a, b; Murthy et al. 2007a; Smith et al. 2010).

The distribution of charges (and ions) in a pore can be described by an electrical double-layer model (Hiemenz and Rajagopalan 1997). Due to the requirement of electroneutrality, an electrical double layer with an excess of counterions in the adjacent solution of the pore wall in the pore is formed. The counterions distributed in the electrical double layer extend a distance from the charged surface and this distribution is related to the Debye–Huckel thickness $1/\kappa$. The degree of the nail permselectivity therefore depends on the thickness of the electrical double layer within the pore pathways and pore radius. It is related to the ionic strength of the solution in the nail. Under high ionic strength conditions, the electrical double layer extends minimally to the pore radial center and the effect of nail permselectivity is minimal. As the ionic strength decreases, the electrical double layer thickens and overlaps. For negatively charged pores, the nail permselectivity toward a cation is increased at lower ionic strength. This is consistent with the observation of increased partitioning of cationic tetraethylammonium (TEA) into the negatively charged nail with decreasing ionic strength of the equilibrating solutions (pH 7.4) in an *in vitro* partitioning study (Smith et al. 2009). Consequently, the penetration of TEA into and across fully hydrated human nail at pH 7.4 was significantly increased when the solution ionic strength in the donor chamber decreased (Smith et al. 2009). The permeability coefficient of TEA at 0.01 M ionic strength was approximately 2.5-fold of that at 0.15 M in the donor.

2.2.4 Influencing Factors

A number of influencing factors on nail permeability have been studied (Table 2.3). This section will provide a general overview of these influencing factors and how this knowledge can assist in the development of trans-ungual drug delivery systems.

2.2.4.1 Permeant Physicochemical Properties

Molecular size of a permeant is the most important parameter determining the nail permeability of the permeant. Trans-ungual transport is significantly decreased when the molecular weight (MW) of the permeant increases. When the partition coefficient between human nail and vehicle was approximately unity, the thickness normalized permeability coefficient (P_T, expressed in cm^2/s) was related to the MW of nonionic permeants (Figure 2.1): $\log P_T = -7.296 - 0.003708\,\mathrm{MW}$ (or $\log P_T = -0.427 - 3.341 \log \mathrm{MW}$) for human nail plate (Mertin and Lippold 1997b). For bovine hoof membranes, similar correlations were observed, but the permeability of the human nail was more sensitive to the change of molecular size than that of the bovine hoof membrane. In a separate study with human nail plates

TABLE 2.3
Factors Influencing Trans-Ungual Transport

Factors	Effects	References
Permeant molecular weight	Permeability decreases with an increase in molecular weight	Kobayashi et al. (2004), Mertin and Lippold (1997b)
Permeant partition coefficient (octanol/water)	No relationship between lipophilicity and permeability	Mertin and Lippold (1997a), Walters et al. (1983)
Permeant water solubility	Maximum flux generally increases with solubility	Mertin and Lippold (1997a)
Vehicle/cosolvent	Mixed effects	Mertin and Lippold (1997c), Nair et al. (2010), Smith et al. (2011), Walters et al. (1985a)
Vehicle pH	Affect acid dissociation of permeant and nail permselectivity at pH 1–10. Nail structural change at pH >10	Kobayashi et al. (2004), Smith et al. (2010), Walters et al. (1985b)

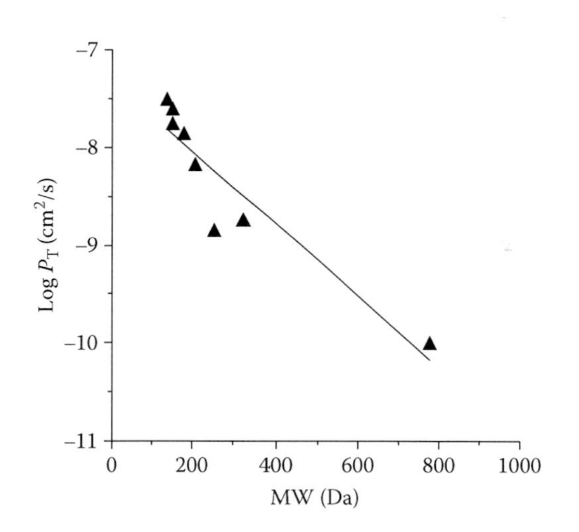

FIGURE 2.1 Relationship between the logarithm of permeability coefficient (P_T) of the human nail and the molecular weight (MW) of the permeant. (Replot using data obtained from Mertin and Lippold, *Journal of Pharmacy and Pharmacology*, 49, 866–72, 1997b.)

(Kobayashi et al. 2004), the relationships between permeability coefficients (expressed in cm/s) of nonionic (P_{non}) and ionic (P_{ion}) forms of permeants with respect to MW were established: $\log P_{non} = -5.260 - 0.00856\,MW$ and $\log P_{ion} = -5.907 - 0.01030\,MW$. For nonionic permeants, the permeability coefficients decreased from 4.6×10^{-6} to 9×10^{-9} cm/s when the MW increased from 20 to 236 Da. For ionic permeants in the MW range of 121–312 Da, the permeability coefficients decreased from 9.1×10^{-8} to 1.7×10^{-9} cm/s when the MW increased.

The effects of permeant lipophilicity on trans-ungual transport have been investigated (Brown et al. 2009; Kobayashi et al. 2004; Mertin and Lippold 1997a; Walters et al. 1983). Human nail permeability to alcohols diluted in normal saline at 37°C decreased with increasing alcohol chain length from methanol (1.5×10^{-6} cm/s) to *n*-octanol (7.5×10^{-8} cm/s) (Walters et al. 1983). This was explained by an increase in lipophilicity and a decrease in partitioning of the permeants into the hydrophilic nails. Later, permeability coefficients of a homologous series of nicotinic acid esters (MW 137–235 Da) of octanol/water partition coefficient (log $K_{o/w}$) between 0.8 and 4.7 through the human nail plate at 32°C were found to be independent of the permeant log $K_{o/w}$ (Mertin and Lippold 1997a). It was suggested that the significant decrease in human nail permeability with increasing lipophilicity of the permeants was attributed to the increased MW, that is, decreased diffusivity of permeants in the nail plate. In another study, a similar relationship between the permeability coefficients of a homologous series of *p*-hydroxybenzoic acid esters (MW 152–222 Da and log $K_{o/w}$ 1.53–4.25) and lipophilicity was reported (Kobayashi et al. 2004). The nail permeability did not increase with an increase in lipophilicity. The multiple regression analysis suggested that permeant MW had a greater contribution to the permeability coefficient than the octanol/water partition coefficient (or the lipophilicity of the permeant).

The relationship between permeant trans-ungual transport and permeant acid dissociation is less conclusive. Contradictory results have been reported. The acid dissociation of miconazole (pK_a 6.65) did not affect its penetration through human nail at a pH range of 3.1–8.2 (Walters et al. 1985b). A few studies found that the acid dissociation of permeants led to a decrease in the total nail permeability for the permeant (Kobayashi et al. 2004; Mertin and Lippold 1997a). The lower permeation of ionized form than nonionized form of a permeant was suggested to be related to the Donnan effect or electrostatic repulsion between the keratins and the charged permeants (Mertin and Lippold 1997a) or the larger apparent molecular size of the permeant in its ionic form due to ion hydration (Kobayashi et al. 2004). In addition to acid dissociation, pH also affects nail permselectivity under moderate pH conditions of $2 < \text{pH} < 10$ and nail structure under extreme conditions such as pH >11 as shown in a recent study (Smith et al. 2010).

The relative solubility of a permeant in a vehicle affects trans-ungual transport. Given that the nail plate can be modeled as a hydrophilic gel membrane, the maximum (steady-state) flux was related to the solubility of a permeant in the vehicle applied onto the nail. For example, the plots of maximum fluxes through human nail plate and bovine hoof membrane versus water solubilities of model permeants of paracetamol and phenacetin (Figure 2.2) gave slopes close to unity in a log-log scale (Mertin and Lippold 1997a). In a separate study, human nail permeability to antifungal drugs was estimated from the permeability of bovine hoof membrane (Mertin and Lippold 1997b). It was found that the difference in drug aqueous solubilities at pH 7.4 can result in 10^5-fold difference in maximum fluxes even though the estimated permeability coefficients (normalized by nail thickness, expressed as cm^2/s) of the antifungal drugs differed only by a factor of 100. This supports the assertion that only drugs having high solubility and small molecular size (e.g., ciclopirox and amorolfine) are good candidates for trans-ungual delivery (Neubert et al. 2006). Drugs of low solubility in water or in the nail (e.g., miconazole nitrate, ketoconazole,

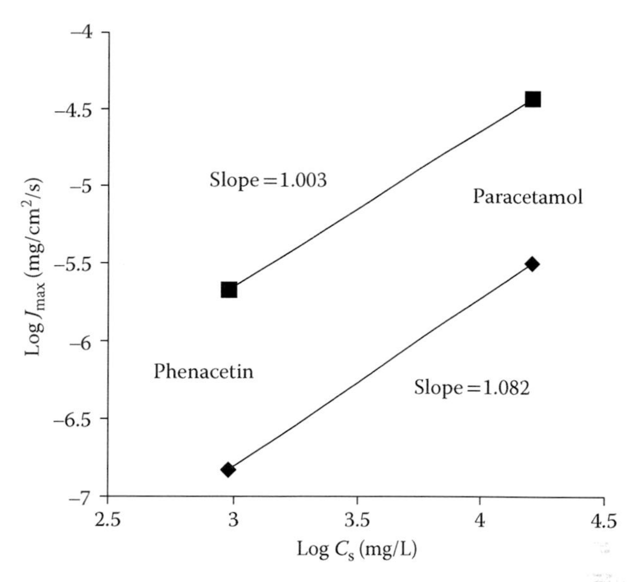

FIGURE 2.2 Relationship between maximum flux (J_{max}) and water solubility of the permeant across human nail plate (diamonds) and bovine hoof membrane (squares). (Replot using data obtained from Mertin and Lippold, *Journal of Pharmacy and Pharmacology*, 49, 30–4, 1997a.)

and itraconazole) were not detected in the receptor in the *in vitro* transport experiments even over a period of 60 days, due to the low permeabilities of the nail for these compounds (Quintanar-Guerrero et al. 1998). Trans-ungual drug penetration was only observed after the nail was pretreated with keratolytic agent papaine (15%, w/v) for 1 day followed by salicylic acid (20%, w/v) for 10 days. Using this method, the steady-state fluxes across the nail were 66 ng/cm^2/s for miconazole nitrate, 11.5 ng/cm^2/s for ketoconazole, and 1.3 ng/cm^2/s for itraconazole.

2.2.4.2 Formulation Factors

Solvents may enhance or retard nail permeability by several mechanisms, such as dehydration of the nail and changes of the solubility or mobility of the permeant in the nail (Walters et al. 1985a). Except for methanol, permeability coefficients of *n*-alcohols (C2–C12) through human nail plate from diluted aqueous solutions (trace amounts of alcohols in normal saline) were generally five times larger than those from neat alcohols. Permeability coefficients from the neat alcohols were a function of alkyl chain length, which showed a similar pattern to that from the diluted aqueous solutions. Increasing the concentration of dimethyl sulfoxide (DMSO) in normal saline decreased the permeability coefficients of both methanol and hexanol. Increasing the concentration of isopropyl alcohol in normal saline up to 50% decreased the permeability coefficients of both methanol and hexanol. Further increases in the percentage of isopropyl alcohol in normal saline increased the permeability of methanol but decreased the permeability of hexanol. In another study (Hui et al. 2002), DMSO-contained formulations increased penetration of salicylic acid, urea, and ketoconazole into the ventral layer of the nail compared to permeants in normal saline. Ethanol in pH 7.4 phosphate buffer (42:58, v/v, final pH 8.1) was found to deswell the bovine hoof membrane and change the pK_a of permeants (Mertin and Lippold 1997b). Smith et al. (2011) studied the effects of organic solvents such as ethanol, propylene glycol, and polyethylene glycol (MW 400 Da) on ungual uptake and transport *in vitro*. It was found that solute partitioning into and transport across the nail decreased with an increase in the organic solvent concentration in the donor media. In another study by Nair et al. (2010), low-MW polyethylene glycols (e.g., MW 200 and 400 Da) moderately enhanced the permeation and loading of terbinafine in nails. These results demonstrated the complicated effects of cosolvents in nail formulations.

2.2.4.3 Nail Conditions

Water content of the nail is important in determining the physical properties and permeability of the nail. The nail permeability increases when the nail hydrates. In a study on the effect of hydration on ketoconazole permeation through human nail *in vitro*, the dorsal side of the nail was sequentially exposed to different RH environments (15–100% RH) in ascending and descending RH orders, and the ventral side of the nail was exposed to 99.5% RH (Gunt and Kasting 2007b). The steady-state flux of ketoconazole increased from 0.175 μg/cm^2/h at 15% RH to 0.527 μg/cm^2/h at 100% RH. Estimated diffusivity of ketoconazole through the nail was a function of water volume fraction in the nail, with the most dramatic increase in the region of 80–100% RH (i.e., water volume fraction of 0.23–0.27) (Gunt and Kasting 2007b). The drastic increase in flux observed in this RH region was consistent with the results of water content and water diffusivity in the nail (Gunt and Kasting 2007a). This suggests the importance of nail hydration in trans-ungual delivery of drugs.

Nail conditions in a disease state can also affect nail permeability. A fungal nail plate, particularly the ventral layer of the nail, generally becomes thicker than a healthy nail plate. It was found that the nail-thickness normalized permeability coefficient of fungal nail plate for 5-fluorouracil was not significantly different from that of healthy nail plate (Kobayashi et al. 2004). A recent study by Nair et al. (2011) also found that terbinafine transport through onychomycotic nails was not significantly different from that of normal healthy nails. This suggests that the fungal nail permeability could be estimated from healthy nail permeability data, taking into account the thickness of the nail. In general, diseased nail permeability is substantially less than that of healthy nail due to the thicker diseased nail as suggested by the limited efficacy of topical antifungals (e.g., ciclopirox) for onychomycosis in the market (Gunt and Kasting 2007b). In addition, the flaky, noncompacted structure of many nails in the later stages of onychomycosis may have fewer continuous diffusion pathways remaining in the nail. The effects of nail disease condition on drug permeability require further investigation.

2.2.5 Nail Permeability to Water

It is known, for example, from nail water uptake studies, that water penetrates easily into the nail plate. Earlier studies on the water permeability of nail plate showed that water flux through partially hydrated nails ranged from 2.0 to 3.0 mg/cm^2/h

(Baden et al. 1973; Spruit 1971). Water flux through fully hydrated human nail plate was 12.6 mg/cm^2/h and the permeability coefficient at 37°C was reported to be 16.5×10^{-3} cm/h (4.6×10^{-6} cm/s) (Walters et al. 1981, 1983). The thickness normalized permeability coefficient (i.e., permeability coefficient × thickness) of water through human nail was approximately 1000 times higher than that through the human stratum corneum. The effective diffusion coefficient of water in the nail calculated from the lag time was 5.4×10^{-7} cm^2/s at 37°C. The water diffusivity in human nail at 32°C was recently determined by the vapor-phase and liquid-phase uptake–desorption studies (Gunt et al. 2007). In the vapor-phase study, the effective diffusion coefficient of water in the nails increased from 8×10^{-10} cm^2/s at 15% RH to 8×10^{-8} cm^2/s at 100% RH, corresponding to the water content of the nail from 0.035 to 0.214 g H_2O/g dry nail (Figure 2.3). The effective diffusivity further increased to 3×10^{-7} cm^2/s when the nail soaked in water and fully hydrated to a water content of 0.536 g H_2O/g dry nail (Gunt et al. 2007). The increase in the water diffusivity was associated with the swelling of the nail; the swollen nail keratin network exerted lower diffusion resistance on the diffusing water molecule.

As water easily penetrates into and across the nail due to its small molecular size, it has been used as a probe to study the factors affecting nail permeation (Malhotra and Zatz 2000) and the effectiveness of ungual penetration enhancers (Malhotra and Zatz 2002). Toenail and fingernail showed different permeabilities to water, suggesting a difference in the toenail and the fingernail structures. The permeability coefficient of water at 37°C significantly increased when the donor had pH >11 and the receptor was water containing 0.5% polyethylene glycol-20-oleyl ether. Similar pH effect on nail permeability to water was also reported (Smith et al. 2010). The permeability of fully hydrated nail to water at 20°C (3.1×10^{-6} cm/s) was not affected by pH in the range of pH 1–10 on the dorsal side of the nail but was drastically increased

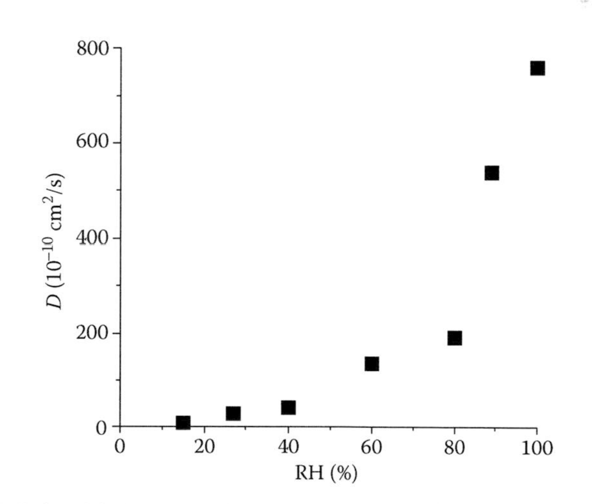

FIGURE 2.3 Relationship between effective water diffusivity (D) and % relative humidity (RH). Water diffusivity was calculated from desorption data of human nail plate. (From Gunt et al., *Journal of Pharmaceutical Sciences*, 96, 3352–62, 2007.)

to 2.4×10^{-5} cm/s at pH 13. Water nail permeability increased with increasing temperature. The activation energy for the permeation of water through the nail was 7.2 kcal/mol determined by the Arrhenius equation, which was closer to the value of delipidized human stratum corneum (6.3 kcal/mol) than of intact stratum corneum (14.3 kcal/mol) (Malhotra and Zatz 2000).

2.2.6 Nail Permeability to Neutral Permeants

Nail permeability to a neutral permeant is controlled by transport hindrance in the permeation pathways across the nail. MW–permeability dependence was observed in nail permeability studies (Hao and Li 2008a). The average passive permeability coefficients of mannitol and urea through fully hydrated nail in PBS were 3×10^{-8} and 5×10^{-7} cm/s, respectively (Hao and Li 2008a, b; Hao et al. 2008). The permeability coefficient ratio of urea to mannitol (~20) was significantly higher than the free diffusion coefficient ratio of the permeants (~2), indicating significant transport hindrance. It is believed that this hindered transport effect is primarily responsible for the low permeability in the fully hydrated nail and long transport lag time. As expected from the water permeability studies discussed in Section 2.2.5, the nail permeability to neutral permeants was not affected by pH at pH 3–9 or ionic strength at 0.04–0.7 M (Hao and Li 2008b). The passive permeability coefficient of glucose through human nail clippings was in the order of 10^{-8} cm/s and was also independent of pH (Murthy et al. 2007a).

Panthenol (MW 205.2, clog P −1.75) is believed to act as a humectant and improve the flexibility and strength of nails. The *in vitro* penetration of panthenol into and through the human nail after daily dosing for 7 days was examined (Hui et al. 2007b). More panthenol was delivered into the deeper layer and nail bed by a film-forming formulation than from an aqueous solution. The film acted as a reservoir of panthenol. It also increased nail hydration and the thermodynamic activity of panthenol. Compared to an aqueous formulation control, the amounts of panthenol delivered from the test formulation into the deeper ventral layer and nail bed were increased by approximately twofold and fourfold, respectively.

2.2.7 Nail Permeability to Charged Permeants

In addition to the hindered transport effect, charge–charge interactions between the permeant and the nail also affect the nail permeability to charged permeants. The passive transport of negatively charged salicylate was affected by pH and ionic strength (Murthy et al. 2007b). Na, Cl, and TEA ions have been selected to study the transport mechanisms of charged permeants through fully hydrated nail (Hao and Li 2008a; Hao et al. 2008; Smith et al. 2009, 2010). In passive transport studies of the charged permeants across fully hydrated human nail, the permeability coefficients for the permeants were consistent with nail permselectivity. The ionic strength of the solution in contact with the nail was found to affect TEA penetration into and across fully hydrated human nail, consistent with electrokinetic theory on the effect of ionic strength upon membrane permselectivity (Smith et al. 2009).

2.2.8 Nail Permeability to Antifungal Agents

Although many antifungal agents (e.g., ciclopirox, amorolfine, terbinafine, griseofulvin, ketoconazole, itraconazole, and off-label fluconazole) have been clinically used in fungal disease treatments, there is limited literature available on nail permeability to these antifungal agents (Tables 2.1 and 2.2). This section provides a brief review of trans-ungual permeation of these antifungal agents.

2.2.8.1 Ciclopirox

Ciclopirox has a MW of 207.3 Da and clog P of 2.0 (Hui et al. 2004, 2007a). It is slightly soluble in water and has a pK_a of 7.2. Penlac®, a nail lacquer of ciclopirox olamine (MW 268.4 Da), and is the only topical product in the U.S. market indicated for the treatment of onychomycosis. In an *in vitro* transport study (Hao et al. 2009), the flux of ciclopirox through human nail plates was about 0.3 μg/h/cm^2. The permeability coefficient was 1.3×10^{-8} cm/s and the apparent lag time was 20 h. In another study (Hui et al. 2004), the *in vitro* human nail penetration of ciclopirox from three different formulations was compared after topical application twice daily for 14 days. The amounts of ciclopirox delivered into the dorsal/intermediate layer of the nail were higher than those in deeper ventral/intermediate layer and the nail bed. It was found that the nature of the formulation played an important role in ciclopirox penetration, but the drug concentration in the formulation had no effect on the drug penetration. In a separate *in vitro* transport study of ciclopirox (Hui et al. 2007a), it was found that only 0.1% of the applied dose penetrated through the nail into the cotton ball support under the nail plate after daily topical application of Penlac for 14 days. The amount of ciclopirox detected in the dorsal section was approximately 7 and 14 times higher than that in the ventral section and the nail bed, respectively.

2.2.8.2 Terbinafine

Terbinafine is an effective antifungal agent and is administered orally in the treatment of onychomycosis. It has a MW of 327.9 Da, pK_a of 7.1, and log $K_{o/w}$ of 3.3. Topical application is preferred due to the side effects associated with oral administration of terbinafine and the long treatment duration. In an *in vitro* transport study, the flux of terbinafine through human nail from saturated terbinafine solution in a mixture of PBS and ethanol (50:50, v/v) was 0.55 μg/cm^2/h and the transport lag time was up to 121 h (Brown et al. 2009). Pretreatment of the nail with penetration enhancers increased the steady-state flux and shortened the lag time. In another transport study of terbinafine using aqueous solution *in vitro* (Nair et al. 2009e), the amount of terbinafine permeated through human nail at 1 h was 0.77 μg/cm^2. When terbinafine was formulated in ethanol/water solution (Nair et al. 2009a), the passive flux of terbinafine was 0.017 μg/cm^2/min (i.e., 1 μg/cm^2/h) similar to the amount delivered in the above study. Nail abrasion to improve terbinafine delivery was also examined. The abrasion of the dorsal layer of the human nail increased terbinafine permeation by approximately fourfold, while the ventral layer abraded nail did not show higher permeation compared to the intact nail without abrasion (Nair et al. 2009c). This suggested that the reduction in total nail thickness alone did not account for the enhancement with the dorsal layer abraded nail. Among the enhancers used in trans-ungual delivery, inorganic salts have been found to act as penetration enhancers for terbinafine using a rapid screening method, TranScreen-N™ (Murthy et al. 2009). Among these penetration enhancers, sodium phosphate was suggested to provide the highest enhancement to terbinafine penetration into and across human nail when it was incorporated into terbinafine formulations. The cumulative amount of terbinafine permeated after 24-h passive transport was 11.4 μg/cm^2, higher than the control without sodium phosphate in the formulation (2.2 μg/cm^2) (Nair et al. 2009d). Salicylic acid pretreatment of the intact nail for 12 h also increased terbinafine permeation by twofold compared to the untreated nail. The synergistic use of the keratolytic agent and iontophoresis for 1 h in the pretreatment stage enhanced terbinafine penetration (Nair et al. 2009c). The enhancement was believed to be a result of iontophoresis-enhanced penetration of the keratolytic agent into the nail. An ex vivo toe model has demonstrated the lack of delivery of a sufficient amount of terbinafine into toenail by passive transport without an effective enhancer (Nair et al. 2009b). A clinical study in patients with onychomycosis showed that the terbinafine concentration was not higher than the minimum inhibitory concentration (MIC) of terbinafine for dermatophytes by passive delivery (Amichai et al. 2010).

2.2.8.3 Amorolfine

Amorolfine has a MW of 317.5 Da and pK_a of 6.6. It is available in its salt form (amorolfine hydrochloride, MW 353.98 Da). Loceryl® nail lacquer of amorolfine is a topical product in Europe to treat nail fungal infections. The level of amorolfine measured in the nail exceeded the MIC of most onychomycosis-causing fungi after 24-h treatment (Polak 1993). Application of 5% amorolfine in either an ethanol or a methylene chloride lacquer resulted in permeation rates through the nail in the range of 20–100 ng/cm^2/h. Absorption of amorolfine was greater from the methylene chloride lacquer than the ethanol lacquer (Franz 1992).

2.2.8.4 Ketoconazole

Ketoconazole has a MW of 531.4 and pK_a values of 2.9 and 6.5. It has poor nail permeability due to its large molecular size and insolubility in water. After 7-day dosing of ketoconazole formulated in saline and a test carrier containing DMSO, ketoconazole was delivered mainly onto the dorsal layer (Hui et al. 2002). DMSO had marginal effect on ketoconazole penetration. In another *in vitro* study, the effective diffusion coefficient of ketoconazole in the nail pretreated with 15% papain for 1 day and subsequent 20% salicylate for 10 days was estimated to be 3.6×10^{-8} cm^2/s when the nail was exposed to ethanol/water (60:40) (Quintanar-Guerrero et al. 1998). Compared to the untreated nail, the nail etched by phosphoric acid gel was more permeable to ketoconazole released from the formulations (Repka et al. 2004). This can be explained by controlled disruption of the resistant dorsal surface and increased diffusion area. It has also been shown that the ketoconazole-loaded film had better bioadhesive properties to the etched nail than untreated nail (Mididoddi and Repka 2007).

2.2.8.5 Econazole

Nail lacquer of econazole was formulated and applied on human nail plate twice daily for 14 days *in vitro* (Hui et al. 2003). The rate of delivery of the drug into the deeper ventral/intermediate layer and nail bed was 0.21 μg/cm^2/h. Incorporation of

2-*n*-nonyl-1,3-dioxolane in the nail lacquer increased the flux and efficacy coefficient (defined as the ratio of flux through the nail plate to the MIC) by sixfold. The drug concentration in the ventral/intermediate layer calculated from the nail density of 1.3 g/cm^3 (Hui et al. 2003; Leider and Buncke 1954) was far above the MICs for most fungi. 2-*n*-Nonyl-1,3-dioxolane had minimal penetration into the nail. It was suggested that the enhancing effect of 2-*n*-nonyl-1,3-dioxolane on drug penetration was related to its effects on the nail lacquer. 2-*n*-Nonyl-1,3-dioxolane functioned as an adhesion promoter and plasticizer to the film-forming polymer of the nail lacquer (Hui et al. 2003).

2.2.8.6 Oxaboroles

Oxaboroles are a new class of antifungal agents containing boron. Among a group of oxaboroles with different MWs and lipophilicities (clog *P* values) screened for nail penetration *in vitro* (Hui et al. 2007a), 5-fluoro-1,3-dihydro-1-hydroxy-2,1-benzoxaborole (AN2690) showed the best penetration into and through the nail, probably due to its small MW (MW 152 Da) and hydrophilicity (clog *P* 1.24). Better penetration of AN2690 from its formulation than Penlac into the ventral layer and the nail bed was reported after daily topical application for 14 days. The total amount of AN2690 penetrated through the nail after 14-day treatment amounted to 16% of the applied dose compared to 0.1% for ciclopirox. The amount of AN2690 in the nail bed was sevenfold higher than that in the dorsal layer of the nail. This further demonstrated the effects of permeant MW and lipophilicity on nail penetration.

2.2.8.7 Oxiconazole

The *in vivo* nail penetration of oxiconazole nitrate from a 1% w/v lotion in fingernails of healthy subjects was evaluated (van Hoogdalem et al. 1997). The fingernails were topically treated with the lotion twice daily for 6 weeks. Nail clippings were collected over 8 weeks after the start of the treatment and assayed for drug amounts in different depths of sectioned nail specimens. The maximum drug level was 790 ng/mg in the upper 0–50-μm layer of the nail. The drug levels decreased drastically in the deeper layers. The maximum drug concentrations were 350 ng/mg in the upper 51–100-μm layer and 104 ng/mg in the upper 101–200-μm layer, respectively. The amounts of drug in each layer of nail samples reached the maximum in about 5 weeks after the start of the treatment. Total amount of the drug penetrated into the nail was less than 0.2% of the topical dose. In the same study, codelivery of a mucolytic agent, *N*-acetylcysteine, significantly increased the penetration of oxiconazole into the upper 0–50-μm layer by disrupting the keratin disulfide bonds in the nail. However, the penetration of oxiconazole into the deeper layers of the nail was not enhanced, probably due to the poor penetration of *N*-acetylcysteine into the deeper layers or difference in the structures between different layers of the nail.

2.3 CONCLUSION

In conclusion, this chapter summarizes the permeability of nail plate to various permeants and the influential factors that affect nail permeability. The molecular size of a permeant is the most important factor determining its nail permeability. Increasing nail hydration through formulations can increase nail permeability. However, the nail plate is a formidable barrier of high intrinsic resistivity to the penetration of permeants. Effective trans-ungual drug delivery requires the use of effective enhancement methods to increase trans-ungual transport, such as chemical and physical enhancement techniques, which will be covered in Chapters 3, 4, and 8 of this book.

REFERENCES

Amichai, B., B. Nitzan, R. Mosckovitz, and A. Shemer. 2010. "Iontophoretic Delivery of Terbinafine in Onychomycosis: A Preliminary Study." *The British Journal of Dermatology* 162: 46–50.

Baden, H. P. 1970. "The Physical Properties of Nail." *The Journal of Investigative Dermatology* 55: 115–22.

Baden, H. P., L. A. Goldsmith, and B. Fleming. 1973. "A Comparative Study of the Physicochemical Properties of Human Keratinized Tissues." *Biochimica et Biophysica Acta* 322: 269–78.

Brown, M. B., R. H. Khengar, R. B. Turner, B. Forbes, M. J. Traynor, C. R. Evans, and S. A. Jones. 2009. "Overcoming the Nail Barrier: A Systematic Investigation of Ungual Chemical Penetration Enhancement." *International Journal of Pharmaceutics* 370: 61–7.

Crank, J. 1975. *The Mathematics of Diffusion*. Oxford: Clarendon Press.

Deen, W. M. 1987. "Hindered Transport of Large Molecules in Liquid-Filled Pores." *AlChE Journal* 33: 1409–25.

Donnelly, R. F., P. A. McCarron, J. M. Lightowler, and A. D. Woolfson. 2005. "Bioadhesive Patch-Based Delivery of 5-Aminolevulinic Acid to the Nail for Photodynamic Therapy of Onychomycosis." *Journal of Controlled Release* 103: 381–92.

Duarte, A. F., O. Correia, and R. Baran. 2009. "Nail Plate Cohesion Seems to be Water Independent." *International Journal of Dermatology* 48: 193–5.

Dutet, J., and M. B. Delgado-Charro. 2009. "In Vivo Transungual Iontophoresis: Effect of DC Current Application on Ionic Transport and on Transonychial Water Loss." *Journal of Controlled Release* 140: 117–25.

Dutet, J., and M. B. Delgado-Charro. 2010. "Transungual Iontophoresis of Lithium and Sodium: Effect of pH and Co-Ion Competition on Cationic Transport Numbers." *Journal of Controlled Release* 144: 168–74.

Edwards, H. G., D. E. Hunt, and M. G. Sibley. 1998. "FT-Raman Spectroscopic Study of Keratotic Materials: Horn, Hoof and Tortoiseshell." *Spectrochimica Acta, Part A: Molecular and Biomolecular Spectroscopy* 54: 745–57.

Egawa, M., T. Fukuhara, M. Takahashi, and Y. Ozaki. 2003. "Determining Water Content in Human Nails with a Portable Near-Infrared Spectrometer." *Applied Spectroscopy* 57: 473–8.

Egawa, M., Y. Ozaki, and M. Takahashi. 2005. "In Vivo Measurement of Water Content of the Fingernail and Its Seasonal Change." *Skin Research and Technology* 12: 126–32.

Finlay, A. Y., P. Frost, A. D. Keith, and W. Snipes. 1980. "An Assessment of Factors Influencing Flexibility of Human Fingernails." *The British Journal of Dermatology* 103: 357–65.

Fleckman, P. 2005. "Structure and Function of the Nail Unit." In *Nails: Diagnosis, Therapy, Surgery*, edited by R. K. Scher, C. R. Daniel III, A. Tosti, B. E. Elewski, P. Fleckman, and P. Rich, 13–25. Philadelphia: Elsevier Saunders.

Franz, T. J. 1992. "Absorption of Amorolfine through Human Nail." *Dermatology* 184 (suppl. 1): 18–20.

Gniadecka, M., O. Faurskov Nielsen, D. H. Christensen, and H. C. Wulf. 1998. "Structure of Water, Proteins, and Lipids in Intact Human Skin, Hair, and Nail." *The Journal of Investigative Dermatology* 110: 393–8.

Gotter, B., W. Faubel, and R. H. Neubert. 2010. "Photothermal Imaging in 3D Surface Analysis of Membrane Drug Delivery." *European Journal of Pharmaceutics and Biopharmaceutics* 74: 26–32.

Gunt, H., and G. B. Kasting. 2006. "Hydration Effect on Human Nail Permeability." *Journal of Cosmetic Science* 57: 183–4.

Gunt, H. B., and G. B. Kasting. 2007a. "Equilibrium Water Sorption Characteristics of the Human Nail." *Journal of Cosmetic Science* 58: 1–9.

Gunt, H. B., and G. B. Kasting. 2007b. "Effect of Hydration on the Permeation of Ketoconazole through Human Nail Plate In Vitro." *European Journal of Pharmaceutical Sciences* 32: 254–60.

Gunt, H. B., M. A. Miller, and G. B. Kasting. 2007. "Water Diffusivity in Human Nail Plate." *Journal of Pharmaceutical Sciences* 96: 3352–62.

Gupchup, G. V., and J. L. Zatz. 1999. "Structural Characteristics and Permeability Properties of the Human Nail: A Review." *Journal of Cosmetic Science* 50: 363–85.

Hao, J., and S. K. Li. 2008a. "Transungual Iontophoretic Transport of Polar Neutral and Positively Charged Model Permeants: Effects of Electrophoresis and Electroosmosis." *Journal of Pharmaceutical Sciences* 97: 893–905.

Hao, J., and S. K. Li. 2008b. "Mechanistic Study of Electroosmotic Transport across Hydrated Nail Plates: Effects of pH and Ionic Strength." *Journal of Pharmaceutical Sciences* 97: 5186–97.

Hao, J., K. A. Smith, and S. K. Li. 2008. "Chemical Method to Enhance Transungual Transport and Iontophoresis Efficiency." *International Journal of Pharmaceutics* 357: 61–9.

Hao, J., K. A. Smith, and S. K. Li. 2009. "Iontophoretically Enhanced Ciclopirox Delivery into and across Human Nail Plate." *Journal of Pharmaceutical Sciences* 98: 3608–16.

Hao, J., K. A. Smith, and S. K. Li. 2010. "Time-Dependent Electrical Properties of Human Nail upon Hydration In Vivo." *Journal of Pharmaceutical Sciences* 99: 107–18.

Hiemenz, P. C., and R. Rajagopalan. 1997. *Principles of Colloid and Surface Chemistry*. New York: Marcel Dekker.

Hui, X., S. J. Baker, R. C. Wester, S. Barbadillo, A. K. Cashmore, V. Sanders, K. M. Hold, et al. 2007a. "In Vitro Penetration of a Novel Oxaborole Antifungal (AN2690) into the Human Nail Plate." *Journal of Pharmaceutical Sciences* 96: 2622–31.

Hui, X., T. C. Chan, S. Barbadillo, C. Lee, H. I. Maibach, and R. C. Wester. 2003. "Enhanced Econazole Penetration into Human Nail by 2-*n*-Nonyl-1,3-Dioxolane." *Journal of Pharmaceutical Sciences* 92: 142–8.

Hui, X., S. B. Hornby, R. C. Wester, S. Barbadillo, Y. Appa, and H. Maibach. 2007b. "In Vitro Human Nail Penetration and Kinetics of Panthenol." *International Journal of Cosmetic Science* 29: 277–82.

Hui, X., Z. Shainhouse, H. Tanojo, A. Anigbogu, G. E. Markus, H. I. Maibach, and R. C. Wester. 2002. "Enhanced Human Nail Drug Delivery: Nail Inner Drug Content Assayed by New Unique Method." *Journal of Pharmaceutical Sciences* 91: 189–95.

Hui, X., R. C. Wester, S. Barbadillo, C. Lee, B. Patel, M. Wortzmman, E. H. Gans, and H. I. Maibach. 2004. "Ciclopirox Delivery into the Human Nail Plate." *Journal of Pharmaceutical Sciences* 93: 2545–8.

Jemec, G. B., T. Agner, and J. Serup. 1989. "Transonychial Water Loss: Relation to Sex, Age and Nail-Plate Thickness." *The British Journal of Dermatology* 121: 443–6.

Jemec, G. B., and J. Serup. 1989. "Ultrasound Structure of the Human Nail Plate." *Archives of Dermatology* 125: 643–6.

Khengar, R. H., S. A. Jones, R. B. Turner, B. Forbes, and M. B. Brown. 2007. "Nail Swelling as a Pre-Formulation Screen for the Selection and Optimisation of Ungual Penetration Enhancers." *Pharmaceutical Research* 24: 2207–12.

Kim, J. H., C. H. Lee, and H. K. Choi. 2001. "A Method to Measure the Amount of Drug Penetrated across the Nail Plate." *Pharmaceutical Research* 18: 1468–71.

Kitamori, K., M. Kobayashi, H. Akamatsu, A. Hirota-Sakashita, M. Kusubata, S. Irie, and Y. Koyama. 2006. "Weakness in Intercellular Association of Keratinocytes in Severely Brittle Nails." *Archives of Histology and Cytology* 69: 323–8.

Kobayashi, Y., T. Komatsu, M. Sumi, S. Numajiri, M. Miyamoto, D. Kobayashi, K. Sugibayashi, and Y. Morimoto. 2004. "In Vitro Permeation of Several Drugs through the Human Nail Plate: Relationship between Physicochemical Properties and Nail Permeability of Drugs." *European Journal of Pharmaceutical Sciences* 21: 471–7.

Kobayashi, Y., M. Miyamoto, K. Sugibayashi, and Y. Morimoto. 1998. "Enhancing Effect of N-Acetyl-l-Cysteine or 2-Mercaptoethanol on the In Vitro Permeation of 5-Fluorouracil or Tolnaftate through the Human Nail Plate." *Chemical & Pharmaceutical Bulletin (Tokyo)* 46: 1797–802.

Kobayashi, Y., M. Miyamoto, K. Sugibayashi, and Y. Morimoto. 1999. "Drug Permeation through the Three Layers of the Human Nail Plate." *Journal of Pharmacy and Pharmacology* 51: 271–8.

Kronauer, C., M. Gfesser, J. Ring, and D. Abeck. 2001. "Transonychial Water Loss in Healthy and Diseased Nails." *Acta Dermato-Venereologica* 81: 175–7.

Leider, M., and C. M. Buncke. 1954. "Physical Dimensions of the Skin; Determination of the Specific Gravity of Skin, Hair, and Nail." *A. M. A. Archives of Dermatology and Syphilology* 69: 563–9.

Malhotra, G. G., and J. L. Zatz. 2000. "Characterization of the Physical Factors Affecting Nail Permeation Using Water as a Probe." *Journal of Cosmetic Science* 51: 367–77.

Malhotra, G. G., and J. L. Zatz. 2002. "Investigation of Nail Permeation Enhancement by Chemical Modification Using Water as a Probe." *Journal of Pharmaceutical Sciences* 91: 312–23.

Marshall, R. C. 1983. "Characterization of the Proteins of Human Hair and Nail by Electrophoresis." *The Journal of Investigative Dermatology* 80: 519–24.

Mertin, D., and B. C. Lippold. 1997a. "In-Vitro Permeability of the Human Nail and of a Keratin Membrane from Bovine Hooves: Influence of the Partition Coefficient Octanol/Water and the Water Solubility of Drugs on Their Permeability and Maximum Flux." *Journal of Pharmacy and Pharmacology* 49: 30–4.

Mertin, D., and B. C. Lippold. 1997b. "In-Vitro Permeability of the Human Nail and of a Keratin Membrane from Bovine Hooves: Prediction of the Penetration Rate of Antimycotics through the Nail Plate and Their Efficacy." *Journal of Pharmacy and Pharmacology* 49: 866–72.

Mertin, D., and B. C. Lippold. 1997c. "In-Vitro Permeability of the Human Nail and of a Keratin Membrane from Bovine Hooves: Penetration of Chloramphenicol from Lipophilic Vehicles and a Nail Lacquer." *Journal of Pharmacy and Pharmacology* 49: 241–5.

Mididoddi, P. K., and M. A. Repka. 2007. "Characterization of Hot-Melt Extruded Drug Delivery Systems for Onychomycosis." *European Journal of Pharmaceutics and Biopharmaceutics* 66: 95–105.

Mohorcic, M., A. Torkar, J. Friedrich, J, Kristl, and S. Murdan. 2007. "An Investigation into Keratinolytic Enzymes to Enhance Ungual Drug Delivery." *International Journal of Pharmaceutics* 332: 196–201.

Monti, D., L. Saccomani, P. Chetoni, S. Burgalassi, M. F. Saettone, and F. Mailland. 2005. "In Vitro Transungual Permeation of Ciclopirox from a Hydroxypropyl Chitosan-Based, Water-Soluble Nail Lacquer." *Drug Development and Industrial Pharmacy* 31: 11–7.

Murdan, S. 2002. "Drug Delivery to the Nail Following Topical Application." *International Journal of Pharmaceutics* 236: 1–26.

Murdan, S., D. Hinsu, and M. Guimier. 2008. "A Few Aspects of Transonychial Water Loss (TOWL): Inter-Individual, and Intra-Individual Inter-Finger, Inter-Hand and Inter-Day Variabilities, and the Influence of Nail Plate Hydration, Filing and Varnish." *European Journal of Pharmaceutics and Biopharmaceutics* 70: 684–9.

Murthy, S. N., S. R. Vaka, S. M. Sammeta, and A. B. Nair. 2009. "TranScreen-N: Method for Rapid Screening of Trans-Ungual Drug Delivery Enhancers." *Journal of Pharmaceutical Sciences* 98: 4264–71.

Murthy, S. N., D. C. Waddell, H. N. Shivakumar, A. Balaji, and C. P. Bowers. 2007a. "Iontophoretic Permselective Property of Human Nail." *Journal of Dermatological Science* 46: 150–2.

Murthy, S. N., D. E. Wiskirchen, and C. P. Bowers. 2007b. "Iontophoretic Drug Delivery across Human Nail." *Journal of Pharmaceutical Sciences* 96: 305–11.

Myoung, Y., and H.-K. Choi. 2003. "Permeation of Ciclopirox across Porcine Hoof Membrane: Effect of Pressure Sensitive Adhesives and Vehicles." *European Journal of Pharmaceutical Sciences* 20: 319–25.

Nair, A. B., B. Chakraborty, and S. N. Murthy. 2010. "Effect of Polyethylene Glycols on the Trans-Ungual Delivery of Terbinafine." *Current Drug Delivery* 7: 407–14.

Nair, A. B., H. D. Kim, B. Chakraborty, J. Singh, M. Zaman, A. Gupta, P. M. Friden, and S. N. Murthy. 2009a. "Ungual and Trans-Ungual Iontophoretic Delivery of Terbinafine for the Treatment of Onychomycosis." *Journal of Pharmaceutical Sciences* 98: 4130–40.

Nair, A. B., H. D. Kim, S. P. Davis, R. Etheredge, M. Barsness, P. M. Friden, and S. N. Murthy. 2009b. "An Ex Vivo Toe Model Used to Assess Applicators for the Iontophoretic Ungual Delivery of Terbinafine." *Pharmaceutical Research* 26: 2194–201.

Nair, A. B., S. M. Sammeta, H. D. Kim, B. Chakraborty, P. M. Friden, and S. N. Murthy. 2009c. "Alteration of the Diffusional Barrier Property of the Nail Leads to Greater Terbinafine Drug Loading and Permeation." *International Journal of Pharmaceutics* 375: 22–7.

Nair, A. B., S. M. Sammeta, S. R. Vaka, and S. N. Murthy. 2009d. "A Study on the Effect of Inorganic Salts in Transungual Drug Delivery of Terbinafine." *Journal of Pharmacy and Pharmacology* 61: 431–7.

Nair, A. B., S. R. Vaka, and S. N. Murthy. 2011. "Transungual Delivery of Terbinafine by Iontophoresis in Onychomycotic Nails." *Drug Development and Industrial Pharmacy* 37: 1253–8.

Nair, A. B., S. R. Vaka, S. M. Sammeta, H. D. Kim, P. M. Friden, B. Chakraborty, and S. N. Murthy. 2009e. "Trans-Ungual Iontophoretic Delivery of Terbinafine." *Journal of Pharmaceutical Sciences* 98: 1788–96.

Neubert, R. H., C. Gensbugel, A. Jackel, and S. Wartewig. 2006. "Different Physicochemical Properties of Antimycotic Agents are Relevant for Penetration into and through Human Nails." *Die Pharmazie* 61: 604–7.

Nuutinen, J., I. Harvima, M. R. Lahtinen, and T. Lahtinen. 2003. "Water Loss through the Lip, Nail, Eyelid Skin, Scalp Skin and Axillary Skin Measured with a Closed-Chamber Evaporation Principle." *The British Journal of Dermatology* 148: 839–41.

Polak, A. 1993. "Kinetics of Amorolfine in Human Nails." *Mycoses* 36: 101–3.

Quintanar-Guerrero, D., A. Ganem-Quintanar, P. Tapia-Olguin, Y. N. Kalia, and P. Buri. 1998. "The Effect of Keratolytic Agents on the Permeability of Three Imidazole Antimycotic Drugs through the Human Nail." *Drug Development and Industrial Pharmacy* 24: 685–90.

Repka, M. A., P. K. Mididoddi, and S. P. Stodghill. 2004. "Influence of Human Nail Etching for the Assessment of Topical Onychomycosis Therapies." *International Journal of Pharmaceutics* 282: 95–106.

Smith, K. A., J. Hao, and S. K. Li. 2009. "Effects of Ionic Strength on Passive and Iontophoretic Transport of Cationic Permeant across Human Nail." *Pharmaceutical Research* 26: 1446–55.

Smith, K. A., J. Hao, and S. K. Li. 2010. "Influence of pH on Transungual Passive and Iontophoretic Transport." *Journal of Pharmaceutical Sciences* 99: 1955–67.

Smith, K. A., J. Hao, and S. K. Li. 2011. "Effects of Organic Solvents on the Barrier Properties of Human Nail." *Journal of Pharmaceutical Sciences* 100: 4244–57.

Spruit, D. 1971. "Measurement of Water Vapor Loss through Human Nail In Vivo." *Journal of Investigative Dermatology* 56: 359–61.

Stern, D. K., S. Diamantis, E. Smith, H. Wei, M. Gordon, W. Muigai, E. Moshier, M. Lebwohl, and P. Spuls. 2007. "Water Content and Other Aspects of Brittle versus Normal Fingernails." *Journal of the American Academy of Dermatology* 57: 31–6.

Susilo, R., H. C. Korting, W. Greb, and U. P. Strauss. 2006. "Nail Penetration of Sertaconazole with a Sertaconazole-Containing Nail Patch Formulation." *American Journal of Clinical Dermatology* 7: 259–62.

Tudela, E., A. Lamberbourg, M. Cordoba Diaz, H. Zhai, and H. I. Maibach. 2008. "Tape Stripping on a Human Nail: Quantification of Removal." *Skin Research and Technology* 14: 472–7.

van Hoogdalem, E. J., W. E. van den Hoven, I. J. Terpstra, J. van Zijtveld, J. S. C. Verschoor, and J. N. Visser. 1997. "Nail Penetration of the Antifungal Agent Oxiconazole after Repeated Topical Application in Healthy Volunteers, and the Effect of Acetylcysteine." *European Journal of Pharmaceutical Sciences* 5: 119–27.

Vejnovic, I., L. Simmler, and G. Betz. 2010. "Investigation of Different Formulations for Drug Delivery through the Nail Plate." *International Journal of Pharmaceutics* 386: 185–94.

Walters, K. A., G. L. Flynn, and J. R. Marvel. 1981. "Physicochemical Characterization of the Human Nail: I. Pressure Sealed Apparatus for Measuring Nail Plate Permeabilities." *Journal of Investigative Dermatology* 76: 76–9.

Walters, K. A., G. L. Flynn, and J. R. Marvel. 1983. "Physicochemical Characterization of the Human Nail: Permeation Pattern for Water and the Homologous Alcohols and Differences With Respect To the Stratum Corneum." *Journal of Pharmacy and Pharmacology* 35: 28–33.

Walters, K. A., G. L. Flynn, and J. R. Marvel. 1985a. "Physicochemical Characterization of the Human Nail: Solvent Effects on the Permeation of Homologous Alcohols." *Journal of Pharmacy and Pharmacology* 37: 771–5.

Walters, K. A., G. L. Flynn, and J. R. Marvel. 1985b. "Penetration of the Human Nail Plate: The Effects of Vehicle pH on the Permeation of Miconazole." *Journal of Pharmacy and Pharmacology* 37: 498–9.

Wessel, S., M. Gniadecka, G. B. Jemec, and H. C. Wulf. 1999. "Hydration of Human Nails Investigated by NIR-FT-Raman Spectroscopy." *Biochimica et Biophysica Acta* 1433: 210–6.

3 Topical Nail Formulations

H. N. Shivakumar, Michael A. Repka, Sudaxshina Murdan, and S. Narasimha Murthy

CONTENTS

3.1 TOPICAL UNGUAL FORMULATIONS

Currently, most nail diseases are treated by oral or topical administration of therapeutic agents. Following oral administration, the actives are absorbed into the systemic circulation, and a fraction of them are found to be distributed to the nail apparatus. However, oral administration of drugs has been associated with a number of adverse effects and drug interactions. It is also observed that the treatment durations have been long, and the recurrence rate of the disease has been high with oral therapy. In this context, topical therapy is preferred as it is known to minimize most of the adverse effects associated with oral administration. However, topical therapy has been partly successful owing to poor penetration of the drug to the nail bed and nail matrix. Therefore, in most cases, topical administration has mostly been recommended as adjunct treatment to the oral therapy.

Topical monotherapy is usually indicated in cases of moderate to mild infections ($\leq$50% of the distal end affected) where few nails are affected, but the nail matrix is not involved (Lecha et al. 2005). The topical treatment has been the first choice of treatment owing to its higher efficacy in children under 2 years where the nails are too thin. Topical therapy has been the only treatment option in cases where systemic treatment is contraindicated or declined. The advantage of this treatment modality is that fewer adverse effects are expected when compared to oral therapy. Therefore,

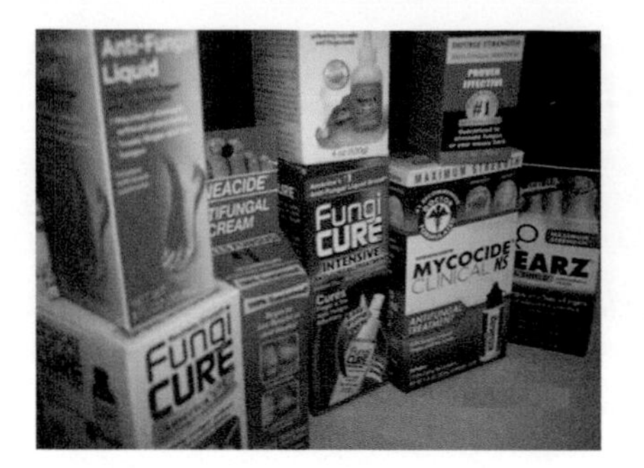

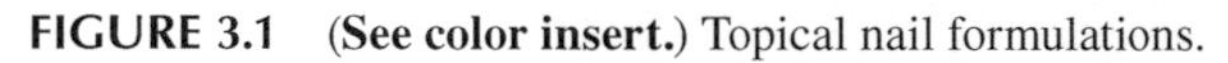
FIGURE 3.1 (**See color insert.**) Topical nail formulations.

there is extensive research going on in the area of topical delivery of drugs to provide more efficient therapy. Some of the topical nail formulations available in the market are portrayed in Figure 3.1.

The various topical nail formulations used for the treatment of different nail disorders include the following:

1. Nail solutions
2. Nail lacquers
3. Semisolids
   a. Gels
   b. Ointments
   c. Creams
4. Films and patches
5. Nail powders

3.1.1 Nail Solutions

Nail solutions are highly concentrated solutions of one or more drugs in a suitable solvent. The solutions are applied to the surface of the nail with the help of a brush and allowed to dry. The solution so applied in many instances, may not form a film on the surface of the nail, unlike conventional nail lacquers. Some of the solutions indicated for the topical treatment of onychomycosis include tioconazole solution, solution of undecylenic acid, and solution of salicylate. When the nail solution is a concentrated aqueous or alcoholic antimicrobial solution formulated in a volatile vehicle, it is called a paint.

Tioconazole is an imidazole derivative indicated in topical monotherapy for onychomycosis that does not involve the nail matrix. A concentrated solution of tioconazole (28%) was used for topical application to the nail plate for 6–12 months in the treatment of onychomycosis (de Berker 2009a). Clinical and mycological cure rates of 22% were reported from a clinical trial involving 27 patients on application of tioconazole solution for 12 months. It was also noted that subjects with fingernail involvement responded well to the topical treatment with the solution (Hay et al. 1985). However, the lack of a control group proved to be a serious limitation of the study reported. Topical tioconazole solution displayed a significantly improved efficacy in combination with low-dose griseofulvin in treatment of onychomycosis as reported by a double-blind randomized study (Hay et al. 1987).

Trosyl® nail solution is a well-known brand of tioconazole available in the United Kingdom, Ireland, and Canada. The clear yellow-colored solution containing tioconazole (283 mg/mL) is manufactured by Pfizer, Sandwich, United Kingdom. The nail solution is indicated for topical treatment of nail infections caused by dermatophytes, yeasts, and molds. The solution needs to be applied to the affected nails twice a day using an applicator brush supplied by the manufacturer. On application, the nail solution takes 10–15 min to dry and forms a slightly greasy film on the nail that allows the drug to penetrate the nail bed and reach the site of infection. A treatment duration of 6–12 months has been suggested to clear up the nail infections. Cure rates varying from 20% to 70% have been reported from several published clinical studies with Trosyl® nail solution.

The depth of penetration of topically applied therapeutic agents is found to depend on the solvents constituting the nail solutions. The *in vitro* penetration of three different drugs into the human fingernail plate was investigated from three different test formulations containing dimethyl sulfoxide (DMSO) (Hui et al. 2002). The concentrations of the three radiolabeled actives, namely, [^{14}C]-urea, [^{14}C]-salicylic acid, and [^{3}H]-ketoconazole in the three test formulations were 0.002%, 0.068%, and 0.154%, respectively. Similar concentrations of [^{14}C]-urea, [^{14}C]-salicylic acid, and [^{3}H]-ketoconazole were maintained in the three different control solutions composed of normal saline that was used as a control. The test or the control solutions were applied twice daily on the human nail plate mounted on a one-chamber diffusion cell for 7 days. After the dosing period, the outer and inner sections of the nail plate were separated using a sampling instrument composed of a microdrill and assayed for the drug content. The results showed that the radioactivity contents of the three actives in the inner sections of the nail plate were about twofold greater for the test formulations compared to the control solutions. In the saline control, salicylic acid in particular displayed a greater binding affinity to the outer surface of the nail plate, making it unavailable for absorption into the deeper layers of the nail. In contrast, more amounts of salicylic acid were delivered to deeper areas of the nail plate with the carrier formulation composed of DMSO. The study demonstrated that penetration of topically applied therapeutic agents can be improved by using solvents such as DMSO in the topical formulation.

The *in vitro* penetration of [^{14}C]-labeled panthenol through the human nail from a nonlacquer film-forming formulation and an aqueous solution was evaluated (Hui et al. 2007a). The test formulation was made of 2% [^{14}C]-panthenol in 98% nail lacquer base containing acrylate copolymer and phytantriol in ethanol (Neutrogena Corporation, Los Angeles, CA). The studies were performed by mounting human nail plate on a one-chamber vertical diffusion cell. The dorsal surface of the nail plate was open to the air, while the ventral surface was in contact with a small cotton ball that acted as a supporting bed. For the penetration studies, aliquots of 15 μL of the test formulation or the aqueous solution containing 0.07–0.011 μCi radioactivity were applied to the dorsal surface of the nail plate every day for a period of 7 days. After a week, the inner ventral section of the nail plate was separated from the outer dorsal surface using a microdrill, and [^{14}C]-panthenol content in each layer and in the supporting bed was determined. The panthenol concentrations in the dorsal layer, the ventral layer, and the supporting bed were found to be 10.05 ± 3.52, 4.97 ± 0.90, and 0.80 ± 0.33 mg eq./g, respectively for the test nail formulation. On the contrary, the concentrations in the dorsal layer, the ventral layer, and the supporting bed were found to be 13.51 ± 6.16, 2.09 ± 1.99, and 0.28 ± 0.15 mg eq./g, respectively for the aqueous solution. The penetration studies indicated the potential of the test formulation to deliver higher amounts of the active into the deeper layers of the nail plate compared to the aqueous solution.

The daily flux rate of panthenol penetrating into/through the nail plate from the formulations was further determined by kinetic studies. For these studies, 15 μL of the test solutions was applied to seven groups of nail plates ($n = 3$) for 7 days. Samples of each group were collected and the dorsal layer, the ventral layer, and the supporting bed of each nail plate were assayed to determine the drug penetration rates and flux. The average flux in the interior nail and the supporting cotton bed was found to be 10.25 ± 2.75 and 1.47 ± 0.79 μg eq./cm^2/h, respectively. A significantly higher amount of [^{14}C]-panthenol was delivered into the interior nail ($p < 0.001$) and the supporting bed ($p < 0.003$) from the nail formulation when compared to the aqueous solution. The better penetration of panthenol from the test formulation compared to aqueous solution was most likely due to the formation of a drug-concentrated film on the dorsal nail surface following evaporation of the formulation. The film not only acted as a drug reservoir but also increased the nail hydration and the thermodynamic activity of panthenol, thereby enhancing the drug transport and penetration.

A nail solution was found to exhibit a better *in vitro* penetration through the nail plate than three other nail lacquer formulations. The penetration of a lead oxaborole antifungal agent 5-fluoro-1,3-dihydro-1-hydroxy-2,1-benzoxaborole (AN2690) from four different formulations was evaluated in a pilot study for clinical development (Hui et al. 2007b). Among the four formulations that contained the antifungal agent (10% w/w), three were lacquer-based preparations made of different film-forming agents, while the fourth formulation was a nail solution. The lacquer-based formulations were typically composed of a film-forming agent and a plasticizer in a suitable solvent. Formulation A was composed of poly(vinyl methyl ether-alt-maleic acid monobutyl ester) (20%) as a film former in ethanol (70%). The formulation was found to leave a water-insoluble durable film that was resistant to damage. Since it was thought that the insoluble film would hamper the penetration of the subsequent doses, the film had to be removed before reapplication of further doses. In order to overcome this limitation, formulations B and C, which formed easily removable films, were developed. Formulation B, composed of poly(2-hydroxyethyl methacrylate) (15%) as a film former and dibutyl sebacate (5%) as a plasticizer in a solvent system of ethanol (56%) and water (14%), formed a water washable film. Instead, formulation C was made of polyvinyl acetate (15%) as a water-insoluble film former and dibutyl sebacate (5%) as a plasticizer in a solvent system of ethyl acetate (15%) and ethanol (55%). However, the water-insoluble film formed in this case was not as durable as formulation A, allowing easy removal of the film by peeling or scratching. Formulation D was a simple solution containing the active in a solvent system of propylene glycol (20% w/w) and ethanol (70%).

For the permeation studies, cadaver fingernail plates were mounted on a one-cell diffusion chamber with the dorsal surface open to the air and the ventral surface in contact with a small cotton ball that was used as a supporting bed. About 10 μL of the lacquer formulations were applied on the dorsal surface of the nail plate, while the supporting cotton bed was wetted with normal saline to moisten the nail plate. After the dosing period of 14 days, the ventral layer of the nail plate (0.3–0.4 mm) was cut using a cutter drill and assayed for the drug content. The drug content in the ventral layer of the nail plate was found to be 3.62 ± 0.95, 4.02 ± 2.05, 5.56 ± 2.46, and 8.96 ± 4.06 μg/mg for formulations A, B, C, and D, respectively. Likewise, the amount of the antifungal agent in the supporting bed was found to be 294 ± 92, 257 ± 119, 418 ± 207, and 752 ± 695 μg/sample for formulations A, B, C, and D, respectively. Though formulations A, B, and C most likely formed supersaturated lacquer films on the nail surface, the films might have interfered with the penetration of the active into the deeper layers of the nail. The studies demonstrated the greater efficacy of formulation D to enhance the *in vitro* penetration of the antifungal agent to the deeper layers of the nail plate compared to the other three formulations.

In the final pivotal study, the *in vitro* nail penetration of active from formulation D was compared with the ciclopirox 8% topical solution (Penlac®, Dermik Laboratories, Berwyn, PA). The concentrations of antifungal agent in the dorsal and ventral nail layers were found to be significantly higher than that of ciclopirox ($p < 0.002$). The amount of active in the supporting cotton bed was found to be 250-fold higher than that of the ciclopirox, suggesting a superior *in vitro* penetration efficacy of formulation D compared to the commercial lacquer. However, several other parameters that would have been responsible for the superior *in vitro* penetration of the model antifungal agent compared to ciclopirox were not discussed in the paper.

A solution of undecylenic acid and salicylate in DMSO was indicated for onychomycotic infections caused by dermatophytes, nondermatophytes, and yeasts (Heimisch 1970). Solution of undecylenic acid (12.5% and 25%) was available over the counter (OTC) for topical treatment of onychomycosis (Moossavi and Scher 2001). Monphytol® paint was one of the well-known brands manufactured by Laboratories for Applied Biological Limited, London, United Kingdom. The topical paint composed of chlorobutanol, 3%; methyl undecenoate, 5%; propyl undecenoate, 0.7%; salicylic acid, 3%; methyl salicylate, 25%; and propyl salicylate, 5%, in solvent system of methyl alcohol and propyl alcohol. The colorless solution was a rapidly drying nongreasy paint that could be easily applied to the affected area. The product was indicated for infections caused by tinea pedis, tinea circinata, and tinea unguium. But the product was discontinued in the United Kingdom, citing the problems associated with the actives. However, solutions of undecylenic acid (12.5% and 25%) continue to be available as OTC products in the United States in the brand names of Equate, FUNGICURE, Fungi-Nail, and so on. These solutions can be used for topical application in different fungal infections as a thin layer using a brush applicator twice a day.

Phytex® paint is one of the brands available in the United Kingdom that is manufactured by Pharmax, Bexley, United Kingdom. The solution contains salicylate (1.45% w/v) along with tannic acid (4.89% w/v), boric acid (3.12% w/v), and methyl salicylate (0.53% v/v) in a solvent mixture composed of acetic acid and ethyl acetate. Salicylic acid functions as an antifungal that also aids in penetration of the actives deep into the nail. Tannic acid and boric acid prevent fungal growth, while methyl salicylate eases any irritation in the skin. Following application, the solvents evaporate and a clear film is formed over the applied area. The paint is recommended for twice-a-day application for several weeks until the infection subsides. However, due to lack of published clinical reports, Monphytol® and Phytex® have been indicated only in certain cases, but not considered for the first line of treatment of onychomycosis (Murdan 2002).

The other topical solutions reported for the treatment of onychomycosis are solutions of clotrimazole (Weuta 1972) and benzalkonium chloride (Wadham et al. 1999). A 10% solution of glutaraldehyde was indicated for the topical treatment of superficial onychomycosis (Suringa 1970). Topical therapy with 1% 5-fluorouracil solution in propylene glycol has been used in the treatment of onychomycosis (Bagatell 1977).

Generally, the efficacy of the topical solutions was thought to be hampered by the poor penetration of the actives into the nail unit. As a result, most of these agents are suggested for use following chemicomechanical nail avulsion procedures to improve efficacy. Clinical experience has indicated that vigorous debridement involving filing the dorsal surface of the nail plate before application of a therapeutic agent would increase the success rate of the topical therapy. Debridement of the nail surface is accomplished with a nail clipper, a nail file, or a nail burr. Decreasing the amount of the infected nail and reducing the nail barrier thickness are the likely reasons for the improved efficacy of topical formulations following debridement.

3.1.2 Nail Lacquers

The conventional nail lacquers or enamels have been used for years in cosmetics to protect and beautify the nail. Of all the nail formulations, nail lacquers are the most preferred topical formulations due to their advantages like longer residence time on the nail, aesthetically pleasing appearance, and patient familiarity with the product (Murdan et al. 2002).

The conventional nail lacquers typically consist of solvent-extenders, film-forming polymers (15%), thermoplastic resin (7%), a plasticizer (7%), solvents, a suspending agent (1%), a colorant (1%), and pearlescent materials, if needed. The commonly used film former is nitrocellulose that forms a durable waterproofing component, while the traditional thermoplastic resin would be toluene sulfonamide-formaldehyde resin, which helps in the adhesion of the lacquer to the nail. The plasticizers like dibutyl phthalate or camphor are added to impart flexibility and durability to the film and thereby prevent shrinkage. The solvent-extenders included toluene, butyl or ethyl acetate, or isopropyl alcohol, which evaporate quickly leaving behind an elastic film that adheres well to the nail plate. The colorants that enhance the product's aesthetic appeal included insoluble color lakes like red barium lake and yellow aluminum lake, iron oxide, mica, and titanium oxide. The suspending agents such as bentonite maintain the viscosity, spreadability, and flowability of the lacquer.

The medicated nail lacquers are the most convenient topical preparations used in the treatment and management of nail disorders. These nail lacquers usually contain a therapeutic agent in a suitable lacquer base. The medicinal lacquers, when applied to the nail plate with a brush, dry in few minutes to deposit a water-insoluble polymeric film on the surface of the nail plate. The concentration of the drug in the dried polymeric film would be much higher than its concentration in the original lacquer due to the evaporation of the solvent. The high drug concentration in the supersaturated polymeric film promotes the trans-ungual delivery owing to the higher diffusion gradient created across the nail plate (Marty 1995). It is also noted that the occlusive polymeric film formed on evaporation of the solvent reduces the water loss from the surface of the nail plate to the atmosphere (Murdan et al. 2008; Spruit 1972). This is likely to cause the hydration of the upper layers of the nail plate and thereby promote the drug diffusion into the nail bed (Gunt and Kasting 2007). The increased hydration of the nail plate by occlusive nail lacquers is also likely to lead to the germination of drug-resistant fungal spores into drug-susceptible hyphae at the target infection sites, thereby facilitating fungal eradication (Flagothier et al. 2005).

The medicated nail lacquer formulation has to be chemically and physically stable, the ingredients must be compatible, and the viscosity should allow the lacquer to flow freely into the grooves of the nail from the container for easy application (Murdan 2002). Following topical application, the lacquer must dry in as few as 3–5 min and form a film that adheres well to the nail plate. Ideally, the film formed must not get disturbed or dissipated during daily routine activities, yet should be easily removed on application of a suitable lacquer remover. Further, the lacquers must be colorless and nonglossy in order to be acceptable to the male patients as well.

The polymeric film containing the drug can be considered as a matrix-controlled release system in which the drug is uniformly dissolved or dispersed in the polymeric film (Murdan 2002). The drug release from the film in such a case will be governed by Fick's law of diffusion. The drug release across a planar surface of unit area can be represented by

$$\mathrm{J} = -D\frac{\mathrm{d}c}{\mathrm{d}x} \tag{3.1}$$

where D is diffusion coefficient of the drug in the polymeric film and $\mathrm{d}c$ is the difference in the concentration across the diffusional path length of $\mathrm{d}x$. The thickness ($\mathrm{d}x$) or the diffusional path length increases with time as the film surface adjacent to the nail becomes depleted of the drug. Generally, the drug diffusion into the nail plate following topical application of the nail lacquer is said to depend on solubility of the drug in the polymeric film, solubility of the drug in the nail, diffusion coefficient of the drug in the polymeric film, diffusion coefficient of the drug in the nail plate, and the drug content in the polymeric film. Therefore, these previously mentioned factors need to be considered during the formulation development of a nail lacquer in order to optimize the ungual and trans-ungual delivery of drugs.

The amount of drug permeated through the nail plate and the drug uptake by the nail was found to depend on the solvent constituting the lacquer. A higher flux of amorolfine through human nail was reported with methylene chloride lacquer when compared to a lacquer formulated using ethanol as a solvent (Franz 1992). On soaking, the drug uptake by the nail was greater for the lacquer formulated with methylene chloride (2.9 ± 0.6 μg/mg) when compared to the lacquer formulated with ethanol (1.2 ± 0.4 μg/mg). On the contrary, a contradicting report is obtained from a study where amorolfine concentrations in the human nail layers were higher following 24-h contact with ethanol lacquer compared to methylene chloride lacquer (Polak 1993).

The drug concentration in the deposited lacquer film following application is found to influence the drug permeation and uptake into the nail plate. Increasing the concentration of amorolfine from 1% to 5% was found to increase the drug uptake into the nail plate following a treatment period of either 6 h or 6 days (Pittrof et al. 1992). Varying the concentrations of chloramphenicol in the lacquer solutions influenced the *in vitro* drug permeation across bovine hoof membrane used as a barrier (Mertin and Lippold 1997). Bovine hoof membrane is a widely accepted model used to predict the permeation across human nail plate. The lacquer formulation was composed of the chloramphenicol, Eudragit® RLPO 20%, dibutyl sebacate 2%, and methanol to 100%. Increasing the drug concentration from 2.2% to 31.3% increased the amount of chloramphenicol diffused across the hoof membrane, while the amount permeated relative to the drug content of the lacquer remained constant. It was noted that the lacquer containing 31.3% of the drug was found to form thermodynamically unstable supersaturated solid solution. A further increase in the drug concentration to 47.6% formed a suspension matrix that failed to further increase the amount of chloramphenicol permeated through the hoof membrane. The studies showed that the relative permeation rates of the drug depended on the solid state of the drug in the lacquer film. The relative permeation rates from the solution matrices were found to be independent of the drug concentration in the lacquers, while the relative permeation rates decreased when the lacquers changed to a suspension matrix.

The presence of penetration enhancer in the nail lacquer solution is found to increase the penetration into the nail plate. The ability of 2-*n*-nonyl- 1,3-dioxolane to enhance the *in vitro* penetration of econazole to the deeper layers of the nail plate from EcoNail™ lacquer was assessed (Hui et al. 2003). The test and the control lacquer formulations were reformulated from EcoNail™ lacquer. The test formulation contained Econazole (5% w/w), Eudragit® RLPO in ethanol with 2-*n*-nonyl-1,3-dioxolane (18% w/w) as a penetration enhancer. The composition of the control formulation remained the same but did not contain the penetration enhancer. The test or the control formulations were applied to human fingernails twice a day for 2 weeks. After the dosing period, the inner ventral section of the nail plate was isolated using a micrometer-controlled nail sampling/powder removal instrument and assayed for the absorbed econazole. The drug content in the ventral part of the nail plate was found to be 11.1 $\pm$ 2.6 μg eq./mg for the test formulation and 1.78 $\pm$ 0.32 μg eq./mg for the control formulation. The study demonstrated the role of 2-n-nonyl-1-3-dioxolane as a nail penetration enhancer. The enhancer was thought to exert a plasticizing effect on the drug-loaded lacquer film, which would have promoted the film adhesion to the surface of the nail plate, which in turn might have facilitated the drug penetration. It was also observed that the dorsal surface of the nail was found to contain higher amounts of econazole when the lacquer was free from the enhancer. The amount of econazole that percolated to the supporting bed under the nail was 47.55 $\pm$ 21.99 μg eq./mg of the sample for the test lacquer and 0.24 $\pm$ 0.09 μg eq./mg of the sample for the control lacquer. The studies demonstrated that the presence of 2-n-nonyl-1-3-dioxolane in the lacquer film was found to deliver sixfold more econazole through the nail plate compared to similar formulation devoid of the enhancers. Since the subsequent radiolabeling studies ruled out the penetration of 2-*n*-nonyl- 1,3-dioxolane deeper into the nail plate, the mechanism of drug penetration was thought to prevail at the formulation–nail interface. The econazole concentration in the deeper layers of the nail plate for the test formulation was found to be 14,000 times higher than the minimum inhibitory concentration (MIC) values for dermatophytes. However, the concentrations observed in the nail plate following the *in vitro* permeation studies may not represent the concentrations in the nail matrix *in vivo*.

Most of the medicated nail lacquers produce a hard water-resistant film on the nail surface that may need to be removed either by mechanical means or using a suitable organic solvent. These procedures may damage the newly growing nail and the adjacent skin areas, rendering them less resistant to the infecting fungi. Some of the water-insoluble films may not adhere well to the surface of the nail and therefore fail to assist the drug transport into the nail plate. However, some reports have demonstrated the better efficacy of these formulations compared to the water-insoluble film-forming lacquers in enhancing the drug penetration into the nail plate. The *in vitro* efficacy of a novel experimental nail lacquer that formed a water-soluble film (P-3051, Polichem SA, Lugano, Switzerland) and a commercial nail lacquer (Penlac™, Aventis Pharma, Deutschland, Germany) was assessed (Monti et al. 2005). The experimental lacquer was composed of ciclopirox (8%), hydroxypropyl chitosan (HPCH) (1%), cetostearyl alcohol (1%), ethanol 95° (73%), ethyl acetate (4%), and purified water (13%). The experimental formulation formed a water-soluble lacquer, whereas the commercial formulation formed a water-insoluble lacquer. The *in vitro* permeation studies were performed in vertical permeation cells using bovine hoof membrane with an average thickness of 150 μm as a barrier. The ciclopirox flux at the steady state was found to be 3.05 ± 0.63 and 4.70 ± 0.60 μg/cm^2/h for the commercial lacquer and the experimental lacquer, respectively. Similarly, the lag time required to achieve the steady state was found to be 12.48 ± 1.31 and 3.36 ± 0.46 h for Penlac™ and the experimental lacquer, respectively. The ability of the experimental lacquer to enhance the steady-state permeation and decrease the lag time was tentatively attributed to a possible higher affinity of the positively charged chitosan derivative for the negatively charged surface of the nail. The better drug transport across the hoof membrane could be due to the interaction of the hydroxypropyl groups of the HPCH with the keratin of the nail by formation of hydrogen bonds that could have resulted in strong adhesion of the water-soluble lacquer film to the nail surface.

Though the hydrophilic nail lacquers tend to adhere well to the nail plate and demonstrate a superior drug penetration compared to their hydrophobic counterparts, they are prone to getting easily washed off from the nail surface during routine daily activities. In this context, a bilayered nail lacquer of terbinafine hydrochloride (TH) was proposed to overcome the limitations of the water-soluble hydrophilic lacquers for the treatment of onychomycosis (Shivakumar et al. 2010). The composite bilayer-forming lacquer on application formed an underlying drug-loaded hydrophilic layer and overlying hydrophobic vinyl layer (Figure 3.2).

The hydrophilic lacquer is expected to adhere well to the dorsal surface of the nail plate, while the vinyl layer is intended to protect the drug-loaded lacquer film from being washed off during routine day-to-day activities. To formulate the hydrophilic lacquer, the drug was dissolved in a mixture of ethanol (60% v/v) in water (pH 3.0), to

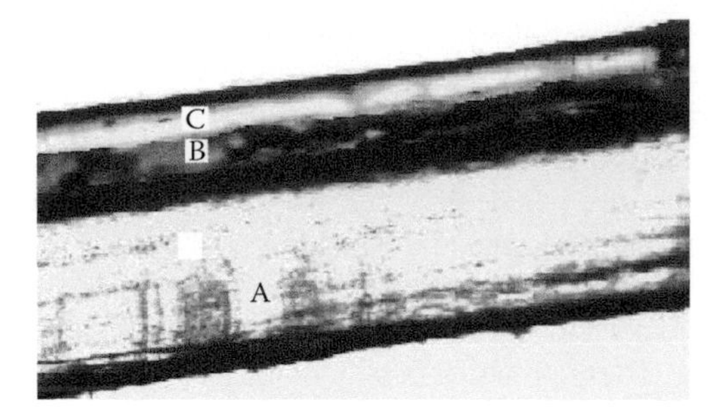

FIGURE 3.2 **(See color insert.)** Cross-section of the nail (A) applied with the bilayered nail lacquer showing the drug-loaded hydrophilic lacquer (B) overlaid with the hydrophobic vinyl lacquer (C). (Reproduced from Shivakumar et al., *Journal of Pharmaceutical Sciences* 99 (10): 4267–76, 2010. With kind permission from Wiley Interscience®.)

which Hydroxypropyl methylcellulose E-15 (HPMC E-15) (6% w/v) soaked overnight in the same solvent system was added and sonicated. The hydroxypropyl groups of the HPMC are expected to interact with the keratin of the nail by hydrogen bonding that could eventually result in strong adhesion of the water-soluble lacquer film to the nail surface. Polyethylene glycol 400 (PEG 400) was incorporated as a nail permeability enhancer in the hydrophilic lacquer. The pH, viscosity, and drying time of the hydrophilic nail lacquer were found to be ~4.0, ~500 cps, and 300 ± 75 sec, respectively. The hydrophobic nail lacquer was produced by dissolving 4-vinyl phenol (10% w/v) in ethyl acetate to which dibutyl phthalate (4% v/v) was added as a plasticizer. Drug-loaded hydrophilic lacquer that was devoid of PEG 400 was referred to as "control lacquer," while the drug-loaded hydrophilic lacquer containing PEG 400 was termed as "monolayer lacquer." *In vitro* permeation studies through human cadaver nails were performed in vertical Franz diffusion cells (Logan Instruments Ltd., Somerset, NJ). The nail plates having an average thickness of 0.5 mm were mounted on a nail adapter having an active diffusional area of 0.2 cm^2 (PermeGear, Inc., Hellertown, PA). The whole unit was sandwiched between the two chambers of Franz diffusion cell. About 20 μL of the drug-loaded hydrophilic nail lacquer was initially applied on the surface of the nail plate and allowed to completely dry. The same amount of the vinyl lacquer was applied on top of the drug-loaded lacquer to form the "bilayered lacquer." *In vitro* permeation studies through the human cadaver nails were also performed with the monolayer and the control lacquers. The studies indicated that the amount of TH permeated across the human cadaver nail in 6 days was 0.32 ± 0.14, 1.12 ± 0.42, and 1.42 ± 0.53 μg/cm^2 from control, monolayer, and bilayered nail lacquers, respectively. A higher nail drug load was seen following the *in vitro* permeation studies with the bilayered lacquer (0.59 ± 0.13 μg/mg) as compared to monolayer (0.36 ± 0.09 μg/mg) and control (0.28 ± 0.07 μg/mg) lacquers.

The ability of the bilayered lacquer to withstand multiple washings was assessed by employing a washing protocol developed and validated in-house in an effort to simulate the routine day-to-day activities. The bilayered nail lacquer was applied on the nail plates that were mounted on Franz diffusion cells with the help of nail adapters. The donor compartment of the diffusion cell was charged with 500 μL of distilled water (pH 3.0). The water was allowed to stand for 1 min and collected to determine the drug content. Each washing step with distilled water was repeated 50 times, and the drug content of each washing was determined to finally compute the cumulative amount of TH lost from the lacquers. The drug loss despite multiple washings was significantly lower ($p < 0.001$) for the bilayered lacquer when compared to the monolayer or the control lacquer owing to the protective vinyl coating (Figure 3.3). A small sample clinical study ($n = 6$) successfully demonstrated the superior efficacy of bilayered lacquer to achieve better drug load in the nail plate (1.27 ± 0.184 μg/mg) compared to monolayer (0.67 ± 0.18 μg/mg) and control (0.21 ± 0.04 μg/mg) lacquers.

A number of clinical trials have demonstrated the efficacy of topically applied nail lacquer formulations in treatment of different nail disorders. An *in vitro* study was initially undertaken to assess the penetration of [^{14}C]-labeled ciclopirox into the nails avulsed as a result of onychomycosis (Bohn and Kraemer 2000). The drug concentration in different layers of the avulsed nail was determined 24 h after topical application of ciclopirox nail lacquer (8%). The mean concentrations were found to range from 7.8 μg/mg in the uppermost layer to 0.034 μg/mg in the lowermost layer. The *in vivo* drug penetration following topical application of the nail lacquer was also studied in a small number of healthy human volunteers. Ciclopirox nail lacquer was applied to five subjects every day for a treatment period of 45 days with intermittent removal of the nail lacquer once a week. The drug concentration in the nail was found to increase throughout the treatment period and reached a maximum of 6.8 μg/mg by the end of the treatment period. Though the residual concentrations were detected in the nail 14 days posttreatment, the drug concentrations in the nail declined significantly. These studies indicated that ciclopirox nail lacquer was successfully able to achieve and maintain effective drug concentrations in the nail during the treatment period.

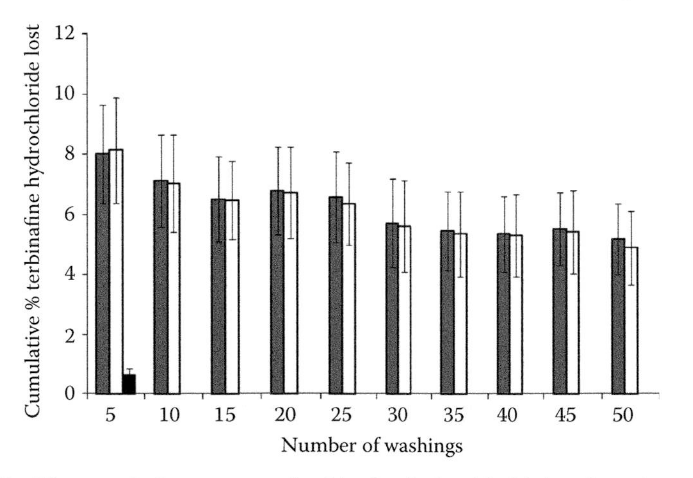

FIGURE 3.3 The cumulative amounts of terbinafine hydrochloride lost from the control (gray filled column), monolayer (unfilled columns), and bilayered (black filled column) lacquers during multiple washings. (Reproduced from Shivakumar et al., *Journal of Pharmaceutical Sciences* 99 (10): 4267–76, 2010. With kind permission from Wiley Interscience®.)

A film-generating solution similar to nail lacquer (Loceryl™) was developed to deliver effective concentrations of amorolfine to the nail plate and reduce the number of applications (Marty 1995). The solution typically composed of amorolfine (5% w/w), a water-insoluble polymer (methacrylic acid copolymers/vinyl polymers), a plasticizer (triacetin/dibutyl phthalate), and a volatile solvent (ethanol, ethyl/butyl/methyl acetate, methylene chloride, methyl ethyl ketone, or isopropanol). The lacquer is intended for application using an applicator 1–2 times weekly to filed nail plates, for up to 6 months for fingernails and for 9–12 months for toenails. On application, the solution formed a film on the surface of the nail that had a drug concentration of 25% w/w. The resulting supersaturated lacquer film was known to enhance the trans-ungual permeation of amorolfine. After single application *in vitro*, the concentration in the nail was found to be 56 μg/g/cm^2 by the end of 6 h after application and 188 μg/g/cm^2 after 7 days. The efficacy of the nail lacquer in treatment of onychomycosis was further demonstrated by *in vivo* studies in patients following a treatment period of 1 month. Two weeks postapplication of the lacquer, the subungual tissue concentration remained well above the MIC for the test organisms.

Penlac® nail lacquer solution contains ciclopirox (8%) in a solution base consisting of butyl monoester of poly[methylvinyl ether/maleic acid] as a water-insoluble film former in solvent system of ethyl acetate NF and isopropyl alcohol USP. Poly [methylvinyl ether/maleic acid] is a water-insoluble film former, while ethyl acetate and isopropyl alcohol are the solvents that vaporize when topically applied. The clear, colorless to pale yellow-colored solution is indicated in patients with mild to moderate onychomycosis of fingernail and toenail that does not involve the lunula. It is recommended to be used once daily at bedtime for a period of up to 48 weeks during which the lacquer film needs to be removed once every week with alcohol before reapplication of fresh lacquer.

3.1.3 Semisolids

Semisolid formulations in the form of gels, ointments, creams, and lotions have been investigated for topical treatment of nail disorders. Though nail lacquer formulation is a popular choice for nail topical antifungal treatment, other conventional nail formulations like a gel, cream, or lotion can provide the necessary chemical gradient to deliver the drug to the target-infected sites. Application under occlusion and periodic debridement of the dorsal surface of the nail plate prior to application has been shown to improve the efficacy of these topical preparations.

3.1.3.1 Gels

Gels are the most preferred semisolid formulation in the treatment of different nail disorders. Gels are known to hold a considerable amount of water and therefore have the ability to hydrate the nail plate. The hydration level attained by a hydrogel is likely to be much higher than that achieved by a nail lacquer. The swelling and hydration of the nail plate is said to enhance the drug permeation as a consequence of formation of a less dense structural keratin matrix with large pores. Gels are the most suitable formulations for iontophoresis, a minimally invasive technique used to deliver charged molecules across the nail plate using an electric field. However, the limitation associated with these formulations is the tendency to get quickly dissipated from the surface of the nail plate during routine day-to-day activities (Repka et al. 2004). The other disadvantage with these formulations is that they tend to undergo cross-linking that may increase their hydrophobicity and decrease the drug permeation through the nail.

In vitro studies have demonstrated better efficacy of gel formulations compared to lacquer in enhancing the permeation of antifungals across the nail plate. The penetration of ciclopirox from a marketed gel, experimental gel, and a marketed lacquer through human nail was studied (Hui et al. 2004). The marketed gel used in the study was Loprox® Gel (Medicis, The Dermatology Company, Scottsdale, AZ), while the marketed lacquer used in the study was Penlac™. The marketed gel contained ciclopirox (0.77%), octadecanol, dimethicone copolyol 190, Carbomer®980, isopropyl alcohol, sodium hydroxide, docusate sodium, and purified water, while the experimental gel was made of ciclopirox (2%), propylene glycol, ethylenediaminetetraacetic acid (EDTA), urea, citric acid, butylated hydroxy toluene (BHT), hydroxypropyl cellulose, ethyl alcohol, and purified water. Healthy human fingernail plates were mounted on a one-cell diffusion chamber with the dorsal surface exposed to the air and the ventral surface in contact with a small cotton ball that was used as a supporting bed. About 10 μL of the formulations (containing 0.34 μCi of radioactivity) were applied on the dorsal surface of the nail plate every 8 h for 14 days. After 14 days of application, the amounts of residual drug remaining on the surface, drug within the infection-prone area, and that penetrated through the nail were determined. The ciclopirox content equivalent in the ventral/intermediate surface of the nail plate and the supporting bed for the marketed gel was found to be 0.55 ± 0.28 μg eq./mg and 46.26 ± 4.08 μg eq./mg sample, respectively, which was significantly higher compared to other formulations ($p < 0.05$). The corresponding ciclopirox concentration in the deeper ventral layers of the nail plate for the marketed gel was found to be 727 ± 372 μg/cm^3. The dorsal surface of the nail contained more unabsorbed drug for the marketed lacquer compared to the gel formulations. This study demonstrated the ability of the gel formulations to increase the drug penetration into the keratinized nail plate. It was therefore concluded that the type of formulation played a crucial role in enhancement of ciclopirox delivery, whereas the drug concentration in the formulation may not have been the determining factor. Ciclopirox gel (0.77%) has been recommended in infections with tinea pedis/tinea cruris or tinea corporis. The gel has to be applied twice a day for 4 weeks (Gupta and Cooper 2008). The cure rates with these infections are high, with infections resolving within 2–4 weeks with topical treatment.

Ketoconazole (0.125%) gel was prepared and used to evaluate the effect of “nail etching” on the permeability of human nail plate (Repka et al. 2004). The model antifungal drug and methyl paraben were dissolved in ethanol, while Carbopol 974P NF was blended in a mortar and pestle into the hydroalcoholic dispersion. The drug solution was added to the dispersion of Carbopol followed by the addition of required amounts of triethanolamine to maintain the desired pH. The permeation of ketoconazole from the gel was determined across etched and unetched human nails in vertical Franz diffusion cells. Full-thickness nail plates were sandwiched between polypropylene adapters and mounted on the diffusion cells with the dorsal surface of the nail plate (diffusional surface area of 0.7 cm^2) facing the donor compartment.

The nail plates with the dorsal surface pretreated with phosphoric acid gel (10%) for 60 sec were considered to be "etched," whereas the untreated nail plate acted as control. The receiver fluid constituting 0.5% of Brij 58 in phosphate buffered saline was maintained at a temperature of 37 ± 0.5°C under continuous stirring. The amount of drug permeated across etched nail was found to be 60% higher compared to that across unetched nail. The increase in the permeation of ketoconazole across etched nail surface was related to disruption of the dorsal surface of the nail plate that decreased the effective barrier thickness for drug permeation. The increase in the nail surface area due to etching was ascribed to development of microporosities that were found to facilitate interpenetration and bonding of a gel, thereby promoting drug diffusion. The studies also suggested the fact that gels could be the most ideal formulations for etched nails as they could easily flow into the microporosities and establish good contact with the contours of nail surface at microscopic levels.

The penetration of tritiated water that was used as a marker through the human nail plate from gel formulations containing various penetration enhancers was studied (Malhotra and Zatz 2002). The water permeation parameters of the gels containing the potential penetration enhancers were compared with those without enhancers. Gels containing the enhancers were prepared in either aqueous or hydroalcoholic vehicles or DMSO. Hydroxypropyl cellulose was used as a gelling agent in hydroalcoholic vehicles or DMSO, while hydroxyethyl cellulose was used as a gelling agent in water. The penetration enhancers, such as mercaptan compounds, sulfites, keratolytic agents, and surfactants, were used either alone or in combination with each other. The gels were prepared by dissolving the enhancers in appropriate solvents with vortexing. The solutions were spiked with tritium to which the respective polymers were added. The final solutions were agitated on a mechanical shaker and maintained at 37°C overnight to facilitate the gel formation.

The permeation studies were performed in Franz diffusion cells specially designed to hold the human nails. The donor compartment of the cells was loaded with 200 μL of the gel formulation and covered with parafilm. The receiver fluid containing 0.5% PEG-20-oleyl ether in water was maintained at 37°C under constant stirring. The best enhancement was obtained with *N*-(2-mercaptopropionyl) glycine (MPG), a mercaptan derivative of amino acid, in combination with urea. Gels containing 10% MPG resulted in approximately 2.5-fold increase in the normalized water flux when compared to the control gels that did not contain any enhancers. Though 20% urea alone resulted in a mere 1.16-fold increase in the flux enhancement, a combination of 20% urea and 10% MPG produced an approximately 3.5-fold increase in the normalized water flux. The mercaptan levels were found to be more critical than the urea levels in enhancing the penetration of water across the nail plate. Barrier integrity of the nails was found to be compromised after treatment with effective chemical modifiers. It was also observed that the changes induced in the nail keratin matrix by treatment with effective chemical modifiers were irreversible. However, the other chemicals failed to increase the penetration of the tritiated water through the nail plate.

A poloxamer gel containing sodium phosphate as a penetration enhancer has been proposed for delivery of terbinafine hydrochloride (TH) across the nail plate (Nair et al. 2009). The topical gel formulation was prepared by the cold process.

The drug solution (1 mg/mL, pH 3.0) was added slowly to the aqueous dispersion of poloxamer (21% w/v) at a temperature of 4–5°C with constant agitation. Sodium phosphate (0.5 M) was added to the dispersion by mixing, and the pH was readjusted to 3.0 with 0.1 M HCl. The dispersion was mixed and kept in a refrigerator overnight. Five hundred milligrams of the gel (with or without sodium phosphate) was charged into the donor compartment of Franz diffusion cell, and permeation studies were performed through the human nail plate for 24 h. The cumulative amount of drug permeated through the nail plate in the stipulated time from the gel (9.70 ± 0.93 μg/cm²) was comparable to the amount permeated from the aqueous solution (11.44 ± 1.62 μg/cm²). The permeation profile of TH from the poloxamer gels across the human nail plate is shown in Figure 3.4. The study demonstrated that gels would be the ideal formulation for delivery of TH to the nail in which water-soluble penetration enhancers such as inorganic salts could be easily incorporated. Moreover, the gel formulation containing the sodium phosphate was successful in delivering the high molecular weight, poorly water-soluble active in amounts comparable to that from aqueous solution.

A double-blind, randomized, vehicle-controlled study demonstrates the use of tazarotene 0.1% gel in the treatment of fingernail psoriasis (Scher et al. 2001). The gel was found to provide some benefit when applied under occlusion to overcome the proximal nail fold barrier. The study suggests that the gel may also be used when applied to the free edge for 24 weeks, though the treatment was associated with moderate irritation of the nail folds.

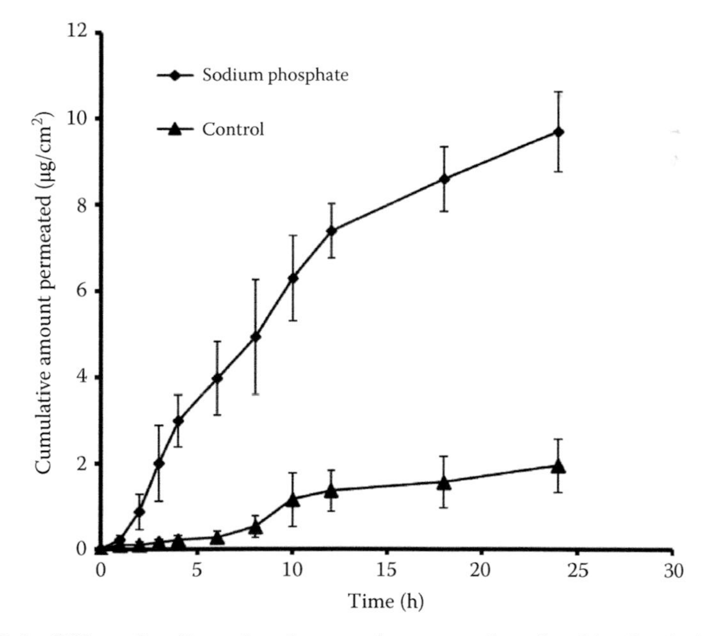

FIGURE 3.4 Effect of sodium phosphate on the permeation of terbinafine hydrochloride from poloxamer gel across the nail plate. ◆ - sodium phosphate; ▶ - control. (Reproduced from Nair et al., *Journal of Pharmacy and Pharmacology*, 61: 431–7, 2009. With kind permission from Wiley Interscience®.)

3.1.3.2 Ointments

Ointments have been used in topical treatment of various nail disorders like onychomycosis and psoriasis. These topical formulations, owing to their hydrophobic nature, have found limited applications in treatment of nail disorders when compared to gels. However, several clinical studies have reported the efficacy of ointments in treatment of different nail disorders. The materials used for preparing ointments could include hydrocarbons like paraffins, oils, fatty acids, higher alcohols, and waxes.

Ointments were used in the treatment of subungual hyperkeratosis, which does not involve clipping of the nail. Calcipotriol ointment was as effective as combined betamethasone dipropionate and salicylic acid ointment in the treatment of subungual hyperkeratosis (Obadiah and Scher 2002). The efficacy was assessed after twice-a-day application in 58 patients with nail bed psoriasis for 3–9 months. Both groups resulted in similar reduction in the nail thickness although changes in other markers of the disease were not defined. To obtain the best results, the nail plates need to be mechanically trimmed to the point of separation of nail bed before application of these topical preparations.

Bifonazole (1%), in combination with urea (40%), was used in ointment as well as cream for the topical treatment of onychomycosis. Fresh ointment was applied and dressings renewed every day until the diseased nail adequately softened for atraumatic removal. After removal of the diseased nail, the cream was applied for 4 weeks or longer. Cure rates ranging from 22% to 68% have been reported following a treatment period of 4 months (Fleckmann 2002). An improved result of bifonazole–urea ointment in combination with short-duration griseofulvin was reported in a randomized double-blind comparison study. A combination of tolnaftate (2%) in urea 20% ointment under occlusive dressing was used successfully for the treatment of onychomycosis fingernails and toenails (Ishii et al. 1983).

Anthralin ointment was used in topical treatment in refractory nail psoriasis (Yamamoto et al. 1998). Anthralin ointment (0.4–2%) was applied as short contact topical therapy to the nail bed for 30 min before washing. Moderate improvement was documented in 60% of patients after the therapy that spanned 5 months.

Cyclosporine has been difficult to incorporate into topical preparations for the nail. Some success has been reported with a 10% oil preparation used for several months (Tosti et al. 1990). However, subsequent studies demonstrated the product was deemed to be unstable, which in most cases was the reason for the poor results.

3.1.3.3 Creams

Creams are emulsion-based topical formulations that may be used in the treatment of different nail disorders. The creams would typically be water in oil (w/o) type or oil in water (o/w) type emulsions depending on the solubility of the actives that need to be incorporated. The base materials used for preparing creams could include hydrocarbons such as paraffins, oils, fatty acids, higher alcohols, and waxes along with a suitable emulsifier. Clinical experts do not find any good rationale for treating onychomycosis with antifungal creams applied to the intact nail, except possibly in cases of white onychomycosis where the fungal infection would be superficial (de Berker 2009a).

Short courses of the creams have been found to be successful in the treatment of psoriatic nails (Schissel and Elston 1998). One percent 5-fluorouracil in propylene glycol (Frederiksson 1974) or in urea 20% cream (de Berker 2009b) was used when pitting and nail thickening were the major problems. However, the studies have suggested that the preparations might be inappropriate when onycholysis is a major feature.

Topical steroids are known to elicit a clear response in the psoriatic changes of the nail fold, which then arrests the secondary matrix inflammation and the subsequent ridging. Treatment usually requires a highly potent steroid like clobetasol propionate. However, the usage of these steroids for more than 3–4 months can cause local skin atrophy. Therefore, the duration of treatment has to be decided based on the risk-to-benefit ratio. Calcipotriol cream was used once a day for 5 days every week in combination with clobetasol propionate cream that was applied once daily for the other 2 days of the week in the treatment of subungual hyperkeratosis (Rigopoulos et al. 2002).

A comparative open-label single-blind study involving 90 patients revealed that the clearance of the infecting fungi from toenail onychomycosis by oral griseofulvin increased from 27% to 46% when it was combined with isoconazole cream, and to 43% when administered with a keratolytic cream (Arenas et al. 1991).

Surgical avulsion of the affected toenails (three patients) or the fingernails (five patients) was performed in a group of eight patients with dermatophyte onychomycosis (Baden 1994). This was followed by application of ketoconazole 2% cream or ciclopirox 8% cream under occlusion with a polyethylene wrap at night until the nail regrew. Clearance from the fungal infection was observed in an 18-month follow-up period.

A study was undertaken to determine the efficacy of clotrimazole 1% cream in the treatment of onychomycosis in association with mechanical reduction of nail plate (Davis 2006). Twenty-six subjects in the age range of 60–78 years, comprising 16 males and 10 females, were included in the study. A total of 92 infected nails were diagnosed by microscopy and positive culture. The nails were thinned using a burr, thereby reducing the nail barrier to improve the drug absorption. The subjects were instructed to apply the cream to the infected nail two times a day. Subjects were advised to return every 2 weeks for further thinning of the nails and mapping of the infected area. After 12 weeks of application, an average improvement of 96.2% was seen, with the infection completely resolved in 80% of the nails. The study demonstrated the fact that mechanical reduction of the nail plate would improve the efficacy of topical preparations in treatment of onychomycosis.

Loceryl® cream, available in tubes of 20 g with a screw cap (Loceryl®, Galderma, Amersham, United Kingdom), contains 0.25% amorolfine hydrochloride with other excipients like 2-phenoxyethanol, polyoxyl 40 stearate, stearyl alcohol, liquid paraffin, white soft paraffin, carbomer, sodium hydroxide, disodium EDTA, and purified water. The cream has been indicated for treatment of dermatomycoses and tinea infections and has to be continued until complete cure is achieved. Loprox Gel, referred to earlier, is also available as a cream. Loprox cream, available in tubes of 45 g, contains ciclopirox olamine (1%) with benzyl alcohol, cetyl alcohol, cocoamide diethanolamine, lactic acid, mineral oil, myristyl alcohol, octyldodecanol, polysorbate 60, purified water, sorbitan monostearate, and stearyl alcohol. These creams have been indicated for infections of the nail caused by dermatophytes, yeasts, and fungi.

3.1.3.4 Lotions

Nail lotions are preparations meant for topical applications to the dorsal surface of the nail plate for treatment of various nail disorders. The lotions are concentrated solutions or dispersions of one or more actives in aqueous or hydroalcoholic vehicles. Nail lotions may or may not deposit a film on the surface of the nail following topical applications. The term "nail lotions" has been sometimes synonymously used in the literature to refer to nail solutions.

There are few clinical investigations undertaken to demonstrate the efficacy of topically applied lotions in the treatment of nail disorders. The *in vivo* penetration of oxiconazole through the nail plate from 1% w/v lotion was determined in six healthy human subjects (van Hoogdalem et al. 1997). However, the composition of the nail lotion used for the study has not been disclosed in the report. *N*-acetylcysteine (acetylcysteine) was used as an ungual penetration enhancer in the lotion to improve the penetration of the drug through the nail plate. The topical treatment with the lotion without acetylcysteine resulted in a drug level of 120–1420 ng/mg in the upper 0–50 μm, while the concentrations decreased in the deeper layers. The application of the lotion containing acetylcysteine was found to increase the mean drug levels from 790 ± 420 to 1570 ± 820 ng/mg in the upper 0–50 μm of the nail clippings, suggesting the enhancing effect of acetylcysteine on the penetration of oxiconazole into the upper nail layers. The application of the lotion containing acetylcysteine also prolonged the mean residence time of the drug in the upper 51–100 μm nail layers from 3.7–4.7 to 4.1–6.4 weeks, indicating the increased retention of the drug within the nail. These effects were attributed to the increased binding of oxiconazole with the keratin of the dorsal surface of the nail in presence of acetylcysteine that would have increased the drug uptake or reduced the drug elimination. However, the same effect was not evident in the lower 101–200 μm layers indicating the failure of acetylcysteine to penetrate into the deeper layers due to a difference in the nail matrix structure. The tendency of acetylcysteine to increase the oxiconazole content was noted only in the upper layer of the nail, though this effect was not consistent in all the subjects. Nevertheless, the enhancing effect of acetylcyteine on the upper layer of the nail plate was found to be insufficient to increase the total drug uptake of the nail.

Loprox® lotion is one of the topical preparations available for the treatment of different nail disorders. The lotion containing ciclopirox olamine (1%) is a product of Medicis. Loprox® lotion has the same composition as the Loprox® cream referred to in Section 3.1.3.3. The topical application was one of the first antifungals for superficial dermatophytes, yeasts, and fungal infections.

3.1.4 Films and Patches

The conventional topical dosage forms used in the treatment of nail disorders have inherent limitations in that these formulations need to be applied frequently, and constant shedding of these dosage forms often fails to maintain therapeutic drug levels at the desired site. A film or patch is a novel drug delivery system that is more likely to maintain therapeutic drug concentrations at the target-infected sites for prolonged time periods.

Techniques such as hot-melt extrusion and film-casting have been proposed to produce drug loaded films. The physicochemical properties of hot-melt extruded films were assessed for topical treatment of nail disorders such as onychomycosis (Repka et al. 2004). The Killion extruder (Model KLB 100) was used to produce the hot-melt extruded films. The mixture of ketoconazole (20%) and polymer was blended thoroughly and dried at 50°C for 24 h before extrusion. The dry blend was extruded at 115–120°C to form homogenous films having a width of 5 ± 0.25 inch. Films with thickness in the ranges of 9–13 mm or 0.23–0.33 mm were produced by the process of hot-melt extrusion. The films produced were found to demonstrate good content uniformity, drug content, bioadhesion, and permeation through the nail plate. The findings of the investigations paved the way for the development of films and patches for the topical treatment of different nail disorders.

The physicochemical properties of the hot-melt extruded films containing ketoconazole were further investigated in order to establish their trans-ungual applications (Mididoddi and Repka 2007). The influence of tartaric acid (4%) on the bioadhesive and mechanical properties of hot-melt extruded films of hydroxylpropyl cellulose containing ketoconazole was reported later (Mididoddi et al. 2006). The drug and the polymer with other excipients were blended and dried at 40–50°C before being extruded. The extruder temperature was maintained between 150°C and 160°C during the process of extrusion. Homogenous films having a thickness of 0.23–0.33 mm were produced by the process. Tartaric acid was found to function as an effective plasticizer by increasing the percent elongation and decreasing the tensile strength of the hot-melt extruded films. The report further confirmed the transnail application of the films produced by hot-melt extrusion.

A bioadhesive patch was designed for ungual delivery of 5-aminolevulinic acid (ALA), which could be useful in the treatment of onychomycosis (Donnelly et al. 2005). ALA is a naturally occurring precursor of photosentasizer protoporphyrin IX (PpIX). Topical application of excess ALA is known to bypass the negative feedback inhibition and builds up photosensitizing concentrations of PpIX at the site of application. ALA-based photodynamic therapy (PDT) has been extensively investigated for topical treatment of skin neoplasias. In this context, feasibility studies were undertaken to assess if sufficient amount of ALA could be delivered to the nail matrix and nail bed following topical application of a patch.

Bioadhesive patches containing 50 mg/cm^2 of ALA and spiked with radiolabeled ALA were prepared by a film-casting technique. The patches were produced from aqueous solution of ALA, methy vinyl ether and maleic anhydride copolymer (20% w/w), tripropylene glycol methyl ether (10% w/w), and ethanol (30% w/w). An appropriate amount of ALA was dissolved in the aqueous solutions to obtain the required strength. The radiolabeled ALA was added by stirring to the so-formed gel before casting into the mold placed on a leveled surface. The gel is dried under a warm air flow to produce a patch having a dimension of 2.5 × 3 cm. The backing layer that was attached to the surface of the formed patches consisted of medical grade polyvinyl chloride plasticized with dibutyl phthalate (Plastisol®).

For the *in vitro* permeation studies, circular sections of human nail plates (average thickness 0.08 ± 0.01 cm) were mounted on a vertical Franz diffusion cell with the help of a Perspex support using a silicone elastomer mix cured with stannous oxide (0.5% w/w). The ventral surface of the nail was aligned to the support allowing intimate contact with the receiver fluid. Drug-loaded bioadhesive patches were applied to the dorsal surface of the nail plate using 10 μL deionized water. The receptor fluid comprising 0.1 M borate buffer, pH 5.0, was maintained at 37°C and a stirring speed of 600 rpm. The receptor fluid was assayed to determine the amount permeated.

The penetration of the drug into the neonate porcine hoof was determined at 37°C using the vertical Franz diffusion cells. Excised hoofs were cut into cylinders (5 mm diameter and 4 mm thick) and inserted into a custom-made stainless steel washer having a circular hole of identical dimension to the cylinders. The cylinders were supported on Cuprophane® membrane that rested on a stainless steel grid. The receiver was filled with phosphate buffered saline, pH 7.4. The circular patch was applied to the uppermost layer of the hoof cylinders and the adhesion was triggered using 5 μL deionized water. The hoof cylinders were separated from the patch after the studies and flash-frozen in atmosphere of nitrogen before sectioning. The hoof cylinders were maintained at −25°C using a cryostatic microtome. Five sections measuring 50 μm were taken parallel to the upper surface of the hoof cylinder to which the patch was applied. Each section was assayed for the radiolabeled ALA content in a scintillation counter.

ALA concentration of 2.8 mM was achieved on the ventral side of the nail plate on application of the patch for 24 h, while almost 90% of ALA permeated the nail plate in 72 h. ALA concentration of 0.1 mM was achieved at a mean depth of 2.375 mm of the neonate porcine hoof where application time had no significant influence on ALA concentrations. It was proposed that the drug penetration from the patch could be further improved using penetration enhancers and by filing the dorsal surface of the nail plate.

A patch containing sertaconazole in an adhesive layer was developed to treat onychodystrophy and onychomycosis. The patch developed by Trommsdorff Arzneimittel GmbH & Co KG, Germany, had an area of 2.2 cm^2 and contained 3.63 mg of sertaconazole as the active. The inactive ingredients used to fabricate the patch were lactate, aerosol, aluminum acetylacetonate, and Durotak®. The patch was designed to provide a constant drug supply to the desired site by the mechanism of occlusion in order to rapidly improve and cure onychomycosis. Following application for six weeks (the patch being replaced at weekly intervals), drug concentrations above the minimum inhibitory concentrations were found in the nail plate, and as expected due to the low nail plate permeability, no systemic absorption was observed (Susilo et al. 2006). The occlusive patch system was likely to hydrate the nail plate, which would eventually help to reconstitute the structural changes of the nail plate seen in onychodystrophy. The patch was intended to be applied once a week, thereby preventing the soiling of the clothes as the drug is not dissipated. The above-mentioned benefits of the transungual patch were thought to greatly enhance patient compliance. The onychomycotic drug, ciclopirox, has also been formulated into patches; the nature of the pressure-sensitive adhesive and of the vehicles was found to influence drug permeation across hoof membranes (used as model for nail plate) (Myoung and Choi 2003).

3.1.5 Powders

Some reports of the use of topical azoles or allylamines as powders for prophylaxis in onychomycosis are seen (Gupta and Cooper 2008). However, miconazole powder failed to show any success as a prophylactic to fungal nail infections. The use of miconazole powder (2%) twice a week for 2 years in shoes and on feet did not show any significant effect in the prevention of reinfection in a small double-blind study (Warshaw and St Clair 2005).

3.2 NAIL INJECTIONS

Nail injections are included in this chapter as they are intended for regional delivery into the nail apparatus in the treatment and management of nail psoriasis, though not exactly for topical applications. Psoriasis is a disease of the nail matrix that is characterized by pitting and surface ridging of the nail. Pitting represents psoriatic pathology with parakeratosis and loss of loose scale from the surface of the nail plate. Surface ridging is a secondary phenomenon related to the transmission of inflammation from psoriasis into the adjacent tissues. Potent steroids have been effective in managing both of the symptoms, although they are found to respond at different thresholds. However, topical steroids fail to penetrate the deeper nail matrix to treat pitting. In this context, steroidal injections into the proximal nail fold may prove to be beneficial to treat nail psoriasis (de Berker 2009b). Intralesional injections with corticosteroids have been considered a standard treatment for nail psoriasis by many practicing physicians. The steroid can be injected directly into the psoriatic digit. One can use a needleless injector (Dermo-jet, Robbins Instruments, Chatham, NJ) or a syringe with a needle. An insulin syringe with a fine-gauge needle has been considered a suitable alternative. The injection site is determined by the apparent origin of dystrophy. The most common site of injection would be the nail matrix, where small doses can be administered into the middle and the lateral nail folds. These injections may not require anesthesia, though patients may vary in their temperament and pain threshold. Thickened nails may need deeper injections, which may be very painful and require preliminary anesthesia.

The intralesional therapy requires repeated steroid injections since the benefit does not seem to last beyond 2–9 months depending on the dose. Important considerations in intralesional therapy are the frequency of injection and the risk of atrophy. It has been generally considered safe to treat a single digit every 3 months over a prolonged period where any sign of atrophy is considered a warning to the physicians. Triamcinolone has been the most commonly used steroid injection in the dose ranging from 2.5 to 10 mg. Nail ridging and subungual hyperkeratosis are known to respond better to high concentrations of intralesional injections than pitting and onycholysis. Triamcinolone (10 mg/mL) was administered as four injections (0.1 mL), two into the proximal nail fold to target the nail matrix, and two into the lateral nail fold to reach the nail bed. The treatment cleared 19 patients from subungual hyperkeratosis in 46 affected nails (de Berker and Lawrence 1998). Improvement in ridging, thickening, onycholysis, and pitting was seen in 93%, 83%, 50%, and 45% of the affected nails, respectively. The regimen involving administration of high doses of steroidal injections less frequently is considered to be as effective as low doses of injections given often.

3.3 HERBAL NAIL PRODUCTS

Cuticle creams are emollients that condition the cuticle, proximal nail fold, and the nail plate. They are used between manicures to maintain a healthy appearance of the nail unit. Cuticle creams may not be cream but are formulated into lotions, oils ointments, or waxy substances (Moossavi and Scher 2001). Botanical oils such as wheat germ oil, sunflower oil, apricot kernel oil, primrose oil, tea tree oil, grapeseed oil, castor oil, and avocado oil are some of the popular emollients used in the formulation of nail care products. Ingredients added to these products can also include aloe vera, retinol, and beta-carotene. Decoration of the nails includes application of complex designs and patterns using an airbrush or stencil, or hand painting with tiny bristle brushes at a nail salon. Self-sticking henna tattoos are available for decoration of the nail.

Some of the herbs have been used as antifungals in the topical treatment of nail disorders such as onychomycosis. The efficacy of tea tree oil (*Melaleuca alternifolia*) in topical treatment of onychomycosis has been compared with clotrimazole (Buck et al. 1994). In a double-blind, randomized placebo-controlled study undertaken to treat toenail onychomycosis, 2% butenafine and 5% tea tree oil in a cream was applied three times a day with occlusion for 8 weeks. All the toenails that responded were able to be mechanically removed at the end of the treatment period. An 80% cure rate was seen in those treated with butenafine in *M. alternifolia* oil after 36 weeks of observation (Syed et al. 1999).

The antifungal activity of extracts from leaves of Piper regnellii against dermatophytes was investigated (Koroishi et al. 2008). The hydroalcoholic extract of the leaves was found to display a strong activity against *Trichophyton mentagrophytes, Trichophyton rubrum, Microsporum canis*, and *Microsporum gypseum* with a MIC of 15.62, 15.62, 15.62, and 62.5 µg/mL, respectively. Light microscopy and scanning electron microscopy revealed well-formed, extensive mycelial growth in nail fragments not exposed to the extract. On the contrary, growth was not visible in nail fragments exposed to the extracts having concentrations greater than 1.2 mg/mL. The active chloroform fraction was lyophilized and chromatographed, and the structures were established as eupometanoid-3 and eupometanoid-5. The MIC for eupometanoid-3 and eupometanoid-5 against *T. rubrum* was found to be 50 and 6.2 µg/mL, respectively, indicating the potential usefulness of the plant in treating dermatophyte infection of the nail.

A clinical trial was undertaken to assess the efficacy of the extract of *Ageratina pichinchensis* for the topical treatment of onychomycosis (Ofelia et al. 2009). The therapeutic effectiveness and tolerability of two different concentrations (12.6% and 16.8%) of the extract on patients with mild to moderate onychomycosis was evaluated. The dried aerial part of the plant was extracted by maceration with a mixture of hexane and ethyl acetate (7:3), concentrated and analyzed by high-performance liquid chromatography (HPLC) to standardize the potency. The MIC was determined against *T. rubrum* (125 µg/mL) and *T. mentagrophytes* (250 µg/mL) employing agar dilution method. A cosmetic nail lacquer (Perlamex 59, Mexico) was used as a base to prepare the medicated lacquers. Two lacquers of two different strengths (one with 12.6% and the other with 16.8%) were prepared by adding the appropriate quantity of the extract to the lacquer base. About 122 patients diagnosed with mild to moderate onychomycosis were divided into two groups. Group 1 received the lacquer with 12.6% of the extract, while the other group received application of the lacquer with 16.8% of the extract. A mycological diagnosis was performed to identify the fungi by microscopy before and after treatment. After the treatment period of 6 months, the therapeutic efficacy of the 12.6% extract was found to be 67.2%, while the efficacy of

16.8% extract was found to be 79.1%. Both the treatments exhibited 100% tolerability as no side effects were noticeable during the treatment period.

A herbal fungitoxic cream for preventing and curing fungal infections in human nail has been patented (Bindra et al. 2001). The cream is composed of extract of *Juglans regia* (walnut hulls) and pulverized roots of *Nardostachys jatamansi* or *Vetiveria zizanioides* or *Catharanthus roseus*. The dry extract of walnut hulls in acetone, alcohol, or butanol when applied to human nails having mild to severe fungal infections displayed poor efficacy in prolonged treatment. However, when the extract was synergized with pulverized roots of *V. zinanioides* or *C. roseus* or *N. jatamansi*, it showed improved efficacy in curing the fungal infection and prevention of recurrence. The base material nitrocellulose/amyl nitrate/food-grade gelatin with the nonionic emulsifiers allowed the formulation to adhere to the surface of the nail, while linseed oil helped in drying of the formulation on application. When the formulation was applied on 10 human subjects with fungal infections, the fungitoxic effects started after 4 days, and the complete recovery was observed after 10 days.

3.4 CONCLUSIONS

The two most common infectious diseases that affect the nails are onychomycosis and nail psoriasis. These nail disorders usually affect diabetics, the elderly, and the immunosuppressed population. Oral treatment has been associated with a number of adverse effects and drug interactions. Moreover, the treatment durations have been long, and relapse of the disease has been common with oral treatment. The topical therapy is known to circumvent most of the problems associated with oral administration. However, topical therapy so far has been partially successful owing to the poor penetration of the drug to the nail bed, which is the residing site of most of the pathogens in the diseased conditions. In this context, the topical therapy has been recommended as a monotherapy in the initial stages or as a support to the oral therapy in the advanced stages of most of the nail diseases. However, topical therapy has been the only resort in cases where systemic treatment is contraindicated or declined.

Though innumerable papers have demonstrated the clinical efficacy of commercially available nail formulations, the literature on development of topical nail formulations has been scarce. Discovery of new nail penetration enhancers and development of novel formulations have shown to offer better prospects for successful topical therapy of nail disorders. However, designing an optimal formulation is a key to delivering therapeutic concentrations of drugs across the nail plate into the nail bed.

REFERENCES

Arenas, R., G. Fernandes, and L. Dominguez. 1991. "Onychomycosis Treated with Itraconazole or Griseofulvin Alone with and without Topical Antimycotic or Keratolytics Agent." *International Journal of Dermatology* 30: 586–9.

Baden, H. P. 1994. "Treatment of Distal Onychomycosis with Avulsion and Topical Antifungal Agents under Occlusion." *Archives of Dermatology* 130: 558–9.

Bagatell, F. K. 1977. "Topical Therapy for Onychomycosis (Letter)." *Archives of Dermatology* 113: 378.

Bindra, R. L., A. K. Singh, A. S. Shawl, and S. Kumar. 2001. Antifungal Herbal Formulation for Treatment of Human Nail Fungus and Process There Off. US Patent US 6296838.

Bohn, M., and K. T. Kraemer. 2000. "Dermatopharmacology of Ciclopirox Nail Lacquer Topical Solution 8% in the Treatment of Onychomycosis." *Journal of the American Academy of Dermatology* 43: S57–69.

Buck, D., D. Nidorf, and J. Addino. 1994. "Comparison of Two Topical Preparations for the Treatment of Onychomycosis: *Melaleuca alternifolia* Oil and Clotrimazole." *The Journal of Family Practice* 38: 601–5.

Davis, K. J. 2006. "Study to Determine the Efficacy of Clotrimazole 1% Cream for the Treatment of Onychomycosis in Association with the Mechanical Reduction of the Nail Plate." *The Foot* 16: 19–22.

de Berker, D. 2009a. "Fungal Nail Diseases." *The New England Journal of Medicine* 360 (20): 2108–6.

de Berker, D. 2009b. "Management of Nail Psoriatic Nail Disease." *Seminars in Cutaneous Medicine and Surgery* 28 (1): 39–43.

de Berker, D., and C. M. Lawrence. 1998. "A Simplified Protocol of Steroid Injection for Psoriatic Nail Dystrophy." *British Journal of Dermatology* 138: 90–5.

Donnelly, R. F., P. A. McCarron, J. M. Lightowler, and A. D. Woolfson. 2005. "Bioadhesive Patch-Based Delivery of 5-Aminolevulinic Acid to the Nail for Photodynamic Therapy of Onychomycosis." *Journal of Control Release* 103: 381–92.

Flagothier, C., C. Piérard-Franchimont, and G. E. Pierard. 2005. "New Insights into the Effect of Amorolfine Nail Lacquer." *Mycoses* 48: 91–5.

Fleckmann, P. 2002. "Onychomycosis: Diagnosis and Topical Therapy." *Dermatologic Therapy* 15: 71–7.

Franz, T. J. 1992. "Absorption of Amorolfine through Human Nail." *Dermatology* 184 (suppl. 1): 18–20.

Frederiksson, T. 1974. "Topically Applied Fluorouracil in the Treatment of Psoriatic Nails." *Archives of Dermatology* 110: 735.

Gupta, A. K., and E. A. Cooper. 2008. "Update in Antifungal Therapy of Dermatophytosis." *Mycopathologia* 166: 353–67.

Gunt, H. B., and G. B. Kasting. 2007. "Effect of Hydration on the Permeation of Ketoconazole through Human Nail Plate In Vitro." *European Journal of Pharmaceutical Sciences* 32: 254–60.

Hay, R. J., Y. M. Clayton, and M. K. Moore. 1987. "A Comparison of Tioconazole 28% Nail Solution Versus Base As an Adjunct to Oral Griseofulvin in the Treatment of Onychomycosis." *Clinical and Experimental Dermatology* 12: 175–7.

Hay, R. J., R. M. Mackie, and Y. M. Clayton. 1985. "Tioconazole Nail Solution—An Open Study of its Efficacy in Onychomycosis." *Clinical and Experimental Dermatology* 10: 111–5.

Heimisch, I. 1970. "Erfahrungen mit der lokalen anwendung eines in dimethyl- sulfoxyd (DMSO) incorporierten antimycoticums bei onycho mykosen." *Mykosen* 13: 175–7.

Hui, X., S. J. Baker, R. C. Wester, S. Barbadillo, A. K. Cashmore, V. Sanders, K. M. Hold, et al. 2007b. "In Vitro Penetration of Novel Oxabarole Antifungal (AN2690) into the Human Nail Plate." *Journal of Pharmaceutical Sciences* 96 (10): 2622–31.

Hui, X., T. C. K. Chan, S. Barbadillo, C. Lee, H. I. Maibach, and R. C. Wester. 2003. "Enhanced Econazole Penetration into Human Nail by 2-n-Nonyl-1-3-Dioxolane." *Journal of Pharmaceutical Sciences* 92 (1): 142–8.

Hui, X., S. B. Hornby, R. C. Wester, S. Barbadillo, Y. Appa, and H. Maibach. 2007a. "In Vitro Human Nail Penetration and Kinetics of Panthenol." *International Journal of Cosmetic Science* 29: 277–2.

Hui, X., Z. Shainhouse, H. Tanojo, A. Anigbogu, G. E. Markus, H. I. Maibach, and R. C. Wester. 2002. "Enhanced Human Nail Drug Delivery: Nail Inner Drug Content Assayed by a Unique Method." *Journal of Pharmaceutical Sciences* 91 (1): 189–95.

Hui, X., R. C. Wester, S. Barbadillo, C. Lee, B. Patel, M. Wortzmann, E. H. Gans, and H. I. Maibach. 2004. "Cilcopirox Delivery into Human Nail Plate." *Journal of Pharmaceutical Sciences* 93 (10): 2545–8.

Ishii, M., T. Hamanda, and Y. Asai. 1983. "Treatment of Onychomycosis by ODT Therapy with 20% Urea Ointment and 2% Tolnaftate Ointment." *Dermatologica* 167: 273–9.

Koroishi, A. M., S. R. Foss, D. A. G. Cortez, T. Ueda-Nakamura, C. V. Nakamura, and B. P. D. Filho. 2008. "In Vitro Antifungal Activity of Extracts and Neolignans from Piper Regnellii against Dermatophytes." *Journal of Ethnopharmacology* 117 (2): 270–7.

Lecha, M., I. Effendy, M. Feuilhade de Chauvin, N. Di Chiacchio, and R. Baran. 2005. "Treatment Options Development of Consensus Guidelines." *Journal of the European Academy of Dermatology and Venereology* 19 (suppl. 1): 25–33.

Malhotra, G. G., and J. L. Zatz. 2002. "Investigation of Nail Penetration Enhancement by Chemical Modification Using Water As a Probe." *Journal of Pharmaceutical Sciences* 91 (2): 312–3.

Marty, J.-L. 1995. "Amorolfine Nail Lacquer: A Novel Formulation." *Journal of the European Academy of Dermatology and Venereology* 4 (suppl. 1): S17–21.

Mertin, D., and B. C. Lippold. 1997. "In Vitro Permeability of the Human Nail and of a Keratin Membrane for Bovine Hooves: Penetration of Chloramphenicol from Lipophilic Vehicles and a Nail Lacquer." *Journal of Pharmacy and Pharmacology*, 49: 241–5.

Mididoddi, P. K., S. Prodduturi, and M. A. Repka. 2006. "Hot Melt Extruded Hydroxyl Propyl Cellulose Films for Human Nail." *Drug Development and Industrial Pharmacy* 32: 1059–66.

Mididoddi, P. K., and M. A. Repka. 2007. "Characterization of Hot Melt Extruded Drug Delivery Systems for Onychomycosis." *European Journal of Pharmaceutics and Biopharmaceutics* 66: 95–105.

Monti, D., P. Saccomani, S. Chetoni, S. Burgalassi, and M. F. Saettone. 2005. "In Vitro Transungual Permeation of Ciclopirox from a Hydroxypropyl Chitosan-Based Water-Soluble Nail Lacquer." *Drug Development and Industrial Pharmacy* 31: 11–7.

Moossavi, M., and R. K. Scher. 2001. "Nail Care Products." *Clinics in Dermatology* 19: 445–8.

Murdan, S. 2002. "Drug Delivery to the Nail Following Topical Application." *International Journal of Pharmaceutics* 236: 1–26.

Murdan, S., D. Hinsu, and M. Guimier. 2008. "A Few Aspects of Transungual Water Loss (TOWL): Inter-Individual and Intra-Individual Inter-Finger, Inter-Hand and Inter-Day Variabilities and the Influence of Nail Plate Hydration, Filing and Varnish." *European Journal of Pharmaceutics and Biopharmaceutics* 70: 684–9.

Myoung, Y., and H. K. Choi. 2003. Permeation of ciclopirox across porcine hoof membrane: Effect of pressure sensitive adhesives and vehicles. *European Journal of Pharmaceutical Sciences* 20 (3): 319–325.

Nair, A. B., S. M. Sammeta, S. K. Vaka, and S. N. Murthy. 2009. "A Study on the Effect of Inorganic Salts in Transungual Delivery of Terbinafine." *Journal of Pharmacy and Pharmacology* 61: 431–7.

Obadiah, J., and R. Scher. 2002. "Nail Disorders: Unapproved Treatments." *Clinics in Dermatology* 20: 643–8.

Ofelia, R.-C., R.-R. Ruben, Z. Alejandro, E. J.-F. Jesus, R.-B. Gabriela, and T. Jaime. 2009. "Clinical Trial to Compare the Effectiveness of Two Concentrations of the Argentina Pichinchensis Extract in Topical Treatment of Onychomycosis." *Journal of Ethnopharmacology* 126: 74–8.

Pittrof, F., J. Gerhards, W. Erni, and G. Klecak. 1992. "Loceryl® Nail Lacquer-Realization of a New Galenical Approach to Onychomycosis Therapy." *Clinical and Experimental Dermatology* 17 (suppl. 1): 26–8.

Polak A. 1993. "Amorolfine: A review of its mode of action and in vitro and in vivo antifungal activity." *Journal of American Medical Association/Southeast Asia* 9 (suppl. 4): 11–8.

Repka, M. A., P. K. Mididoddi, and S. P. Stodghill. 2004. "Influence of Human Nail Etching for the Assessment of Topical Onychomycosis Therapies." *International Journal of Pharmaceutics* 282: 95–106.

Rigopoulos, D., D. Ioannides, N. Prastitis, and A. Katsambas. 2002. "Nail Psoriasis: A Combined Treatment Using Calcipotriol Cream and Clobetasol Propionate Cream." *Acta Dermato-Venereologica* 82: 140.

Scher, R. K., M. Stiller, and Y. I. Zhu. 2001. "Tazarotene 0.1% Gel in the Treatment of Fingernail Psoriasis: A Double-Blind, Randomized, Vehicle-Controlled Study." *Cutis* 68: 355–8.

Schissel, D. J., and D. M. Elston. 1998. "Topical 5 Fluorouracil Treatment for Psoriatic Trachyonychia." *Cutis* 62: 27–8.

Shivakumar, H. N., S. R. K. Vaka, N. V. S. Madhav, H. Chandra, and S. N. Murthy. 2010. "Bilayered Nail Lacquer of Terbinafine for Treatment of Onychomycosis." *Journal of Pharmaceutical Sciences* 99 (10): 4267–76.

Spruit, D. 1972. "Effect of Nail Polish on the Hydration of Finger Nails." *American Cosmetics and Perfumery* 87: 57–58.

Suringa, D. W. 1970. "Treatment of Superficial Onychomycosis with Topically Applied Glutaraldehyde. A Preliminary Study." *Archives of Dermatology* 102: 163–7.

Susilo, R., H. C. Korting, W. Greb, and U. P. Strauss. 2006. "Nail Penetration of Sertaconazole with Sertaconazole Containing Nail Patch." *American Journal Dermatology* 7 (4): 259–62.

Syed, T. A., Z. A. Qureshi, S. M. Ali, S. Ahmad, and S. A. Ahmad. 1999. "Treatment of Toe Nail Onychomycosis with 2% Butenafine and 5% *Melaleuca alternifolia* (Tea Tree) Oil in Cream." *Tropical Medicine & International Health* 4: 284–7.

Tosti, A., L. Guerra, F. Bardazzi, and M. Lanzarini. 1990. "Topical Ciclosporin in Nail Psoriasis (Letter)." *Dermatologica* 180: 110.

van Hoogdalem, E. J., W. E. van den Hoven, I. J. Trepstra, J. van Zijtveld, J. S. C. Verschoor, and J. N. Visser. 1997. "Nail Penetration of Antifungal Agent Oxiconazole after Repeated Topical Application in Healthy Volunteers, and the Effect of Acetylcysteine." *European Journal of Pharmaceutical Sciences* 5: 119–27.

Wadham, P. S., J. Griffith, P. Nikravesh, and D. Chososh. 1999. "Efficacy of a Surfactant, Allantoin, and Benzalkonium Chloride Solution for Onychomycosis. Preliminary Results of Treatment with Periodic Debridement." *Journal of the American Podiatric Medical Association* 89: 124–30.

Warshaw, E. M., and K. R. St Clair. 2005. "Prevention of Onychomycosis Reinfection for Patients with Complete Cure of all 10 Toenails: Results of a Double-Blind, Placebo-Controlled, Pilot Study of Prophylactic Miconazole Powder 2%." *Journal of the American Academy of Dermatology* 53: 717–20.

Weuta, V. H. 1972. "Clotrimazole cream und losung klinische prufung im offenen versuch." *Drug Research* 22: 1295–99.

Yamamoto, T., I. Katayama, and K. Nishioka. 1998. "Topical Anthralin Therapy for Refractory Nail Psoriasis." *Journal of Dermatology* 25: 231–33.

4 Approaches to Enhance Ungual and Trans-Ungual Drug Delivery

H. N. Shivakumar, Abhishek Juluri, Michael A. Repka, and S. Narasimha Murthy

CONTENTS

4.1 UNGUAL DRUG DELIVERY ENHANCERS

The nail plate is composed of many strands of keratin that are held together by disulfide linkages. The nail is known to behave like a hydrogel rather than a lipophilic membrane, unlike most of the other biological barriers. Owing to its unique properties, particularly the thickness and relatively compact nature, the nail acts as a formidable barrier to the penetration of topically applied therapeutic agents.

Moreover, the binding of the drug to the keratin in the nail plate reduces the free (active) drug, thereby diminishing the concentration gradient and limiting the penetration into deeper tissues (Murthy et al. 2007). Therefore, on topical application, the drug concentration fails to reach a therapeutically effective concentration in the ventral layer of the nail plate in many cases.

Most of the approaches that promote ungual and trans-ungual delivery of drugs rely heavily on the use of nail permeation enhancers. The drug penetration into the nail plate can be promoted using agents that break the physical and chemical bonds that maintain the integrity of the nail keratin. The disulfide, peptide, hydrogen and polar bonds in the keratin have been identified as potential soft targets, which could be acted upon by chemical enhancers (Wang and Sun 1999). The disulfide bonds of globular interfilamentous proteins, which are known to act as a glue to hold the keratin fibers together, would be targeted in most of the cases.

The ungual penetration enhancers identified thus far fall under the following categories:

1. Solvents
   a. Water
   b. Dimethyl sulfoxide (DMSO)
   c. Methanol
   d. Other solvents
2. Keratolytic agents
   a. Urea, salicylic acid, papain, and so on
3. Compounds that cleave the disulfide bonds
   a. Thiols: *N*-acetylcysteine (NAC), thioglycolic acid (TGA), 2-mercaptoethanol (MPE), and so on
   b. Sulfites: Sodium sulfite, sodium metabisulfite, and so on
   c. Hydrogen peroxide
4. Enzymes
   a. Keratinase
5. Miscellaneous class of penetration enhancers
   a. Inorganic salts: Sodium phosphate
   b. Hydrophobins
   c. Dioxolane
6. Etchants
   a. Phosphoric acid (PA), tartaric acid (TTA), and so on

The penetration enhancers for nail applications could be coapplied with the penetrant or can be used prior to application of the penetrant. The coapplication of nail penetration enhancers would be advantageous in avoiding the tedious pretreatment protocols and would be practically more feasible. However, the pretreatment regimen would be valuable in overcoming any possible incompatibility issues between the penetration enhancer and the permeant. A number of chemical penetration enhancers investigated to date have the potential to increase the permeation of large, charged molecules that otherwise exhibit extremely poor permeation rates through the nail.

4.1.1 Solvents as Permeation Enhancers

4.1.1.1 Water

Water is known to be the principal plasticizer in the nail plate, imparting a certain degree of flexibility to the nail. The water content, under normal conditions, is found to vary from 18% to 20% depending on the relative humidity (RH) to which it may be exposed (Baden 1970). The water content was found to increase to 25% for an excised nail that was equilibrated at 100% RH. The degree of hydration of the nail is the most important factor governing the physical properties of healthy human nails.

Nail plate has the tendency to hydrate and swell similar to hydrogels on contact with aqueous solutions. The swelling would increase the distance between the keratin fibers, resulting in the formation of larger pores through which the permeating molecules can easily diffuse (Murdan 2002). Water is known to play a facilitating role in increasing the permeation of water-soluble permeants through the human nail plate. This was clearly illustrated in the studies where water increased the diffusion of alcohols through the nails from hydroalcoholic solutions. The *in vitro* permeation of a series of homologous alcohols in pure and diluted forms through fingernail plates was determined. The neat alcohols were found to demonstrate a similar permeation trend as the alcohols diluted with saline. However, the permeability coefficients of the alcohol diluted with saline was found to be fivefold higher than the pure form, as shown in Table 4.1 (Walters and Flynn 1983). The increase in the permeability coefficient of the alcohols on dilution with saline clearly showed the role of water in increasing the transport of water-soluble permeants across the nail plate.

TABLE 4.1
Nail Permeability Coefficients of Neat *N*-Alkanols and Their Diluted Counterparts

		Permeability Coefficient (cm/h × 10^{-3})	
Sl. No	Permeant	Alcohols Diluted with Saline	Neat Alcohols
1	Methanol	5.6 (1.2)	9.73 (5.06)
2	Ethanol	5.8 (3.1)	0.91 (0.34)
3	Propanol	0.83 (0.15)	—
4	Butanol	0.61 (0.27)	0.13 (0.15)
5	Pentanol	0.35 (0.07)	—
6	Hexanol	0.36 (0.23)	0.07 (0.03)
7	Heptanol	0.42 (0.12)	—
8	Octanol	0.27 (0.03)	0.05 (0.02)
9	Decanol	2.5 (1.7)	0.59 (0.13)
10	Dodecanol	4.1 (2.7)	—

Source: Reproduced from Walters and Flynn, *International Journal of Cosmetic Science*, 5, 231–46, 1983.

Note: The values in parentheses represent standard deviations.

The transport of solutes through the nail plate was directly influenced by the proportion of water in the binary solution used as a donor. The permeation of methanol, a hydrophilic alcohol, and *n*-hexanol, a hydrophobic alcohol, was reduced when the proportion of water in the donor solution was depleted. This was clearly evident in a study where the proportions of DMSO or isopropanol were gradually increased in the donor solution composed of binary mixtures with water (Walters et al. 1985). The permeability coefficient of *n*-hexanol across the nail was found to decrease by around fivefold as the concentration of DMSO in the binary mixture increased from traces to 86%. A similar magnitude of decrease in partition coefficient of *n*-hexanol was noted when the proportion of isopropanol in the binary mixture was increased. The decrease in the permeation of the two solutes on depletion of the water in the donor clearly indicated the facilitating role of water in increasing the permeation of compounds of varied polarity through the nail plate.

The hydration of the nail was thought to play a key role in topical delivery of actives through the nail plate. The effect of hydration on the *in vitro* permeation of ketoconazole, a poorly water-soluble drug, through excised human nails was assessed (Gunt and Kasting 2007). The dorsal surface of the nail plate mounted on a Franz diffusion cell was dosed three times with 10 μL of 10% [^{3}H]-labeled ketoconazole in ethanol following which the ethanol was evaporated. The ventral surface of the nail plate was in contact with pH 7.4 phosphate buffer containing polyoxyethylene-20 oleyl ether (Oleth-20) as a surfactant, while the dorsal surface of the nail plate was sequentially exposed to RH of 15%, 40%, 80%, and 100% for a period of 40 days. The average water content of the nails increased from 0.22 to 0.27 as the RH increased from 15% to 100%. At the end of the study, about 300 μg/cm^2 of ketoconazole was found to penetrate the nail plate with an average flux of 0.30 μg/cm^2/h. The steady-state flux of the radiolabeled ketoconazole increased nearly by threefold from 0.175 μg/cm^2/h at 15% RH to 0.527 μg/cm^2/h at 100% RH. The effect of hydration on the penetration of ketoconazole was most pronounced at RH values of more than 80%. A drastic increase in the flux by twofold was noted when the RH increased from 80% to 100%. Considering the poor aqueous solubility of ketoconazole, it was most unlikely that the increased permeation of ketoconazole at higher RH was due to the transport through a solution. The increased flux could be most likely due to the increased flexibility and structural expansion of the keratin matrix, thus allowing the high molecular weight compound (molecular weight: 531.44 Da) to diffuse across the nail plate with ease. Based on the results, it was concluded that formulations or treatment modalities that improve hydration of the nail have the potential to improve the penetration of topically applied therapeutic agents. The study showed that hydration of the nail plate would facilitate the transport of a poorly water-soluble high molecular weight compound.

4.1.1.2 Dimethyl Sulfoxide

In dermal delivery, DMSO is known to interact with the lipid domains of the stratum corneum by increasing their fluidity or facilitating the drug partitioning into the skin. However, DMSO is not anticipated to demonstrate the same efficacy as an ungual penetration enhancer, considering the trace amount of lipids (<1%) in the nail plate. Despite this, there are few reports on increase of drug penetration into nail when DMSO was used as an enhancer. DMSO increased the penetration of topically applied antimycotics (Stuttgen and Bauer 1982). Pretreatment of the nail with DMSO was found to increase the penetration of the antifungal agent amorolfine (Franz 1992). Human subject studies demonstrated a maximum penetration when DMSO was used as a penetration enhancer compared to other lipophilic solvents (Kligman 1965). The depth of the penetration of a fluorescent marker was assessed by scraping the nail plate with a knife. A maximum penetration depth of one-fourth the depth of the nail plate was seen with DMSO.

The penetration of radiolabeled urea, salicylic acid, and ketoconazole across the human nail plate from test carrier formulation containing DMSO was assessed (Hui et al. 2002). The test carrier formulation and the saline formulation were applied twice daily for a week on the surface of human nail plates. The inner section of the nail plate was assayed using a unique drilling process. The radioactivity content of the three compounds in the ventral surface of the nail plate was approximately twofold higher in cases of formulations containing DMSO compared to the saline formulations. With salicylic acid in particular, greater binding of the drug to the outer dorsal surface of the nail plate was evident, with the saline formulation making it unavailable to the deeper ventral nail areas. On the contrary, more salicylic acid was delivered to deeper areas of the nail plate with the formulation containing DMSO. However, no attempt was made in the study to elucidate the mechanism of action of DMSO as a trans-ungual penetration enhancer.

Recently, Vejnovic et al. (2010a) have demonstrated an increase in permeation of caffeine when DMSO was used as an enhancer in formulations. The test formulations contained 2% w/v caffeine as a model drug with DMSO (5%) in either water or a binary mixture of 20% v/v ethanol. The reference formulations had same compositions but without DMSO. The *in vitro* permeation studies were performed in vertical Franz diffusion cells using human nail plates as barrier. Following the permeation studies, the nail plates were milled to detect the amount of caffeine retained. The permeability coefficient of caffeine through the human cadaver nail plate was found to increase from 1.56×10^{-8} to 5.12×10^{-8} cm/s (~3.3-fold compared to reference) when DMSO was used in hydroalcoholic system. A twofold increase in permeability coefficient compared to reference was also noted when DMSO was used in water alone. The changes in the permeability coefficient of caffeine when DMSO was used as a enhancer are shown in Figure 4.1. DMSO was also successful in increasing the drug retention in the nail plate, especially when used in hydroalcoholic systems. The percentage of caffeine remaining in the nail plate following the permeation studies was 1.55-fold and 1.18-fold higher than the reference formulations in hydroalcoholic and aqueous systems, respectively when DMSO was used. Visioscan VC 98 was used to characterize the surface of the nail before and after permeation studies. A clear difference in the nail surface treated with test formulations was evident when compared with the nail surface treated with reference formulations. Though DMSO has proved to be an effective transdermal permeation enhancer, contradicting reports are seen in the literature, pertaining to its role as an ungual permeability enhancer. The mechanism of action of DMSO in improvement of penetration through the nail plate continues to be uncertain, though some researchers argue that alteration of the lipid concentration of the nail plate is a possible reason for the improved trans-ungual penetration enhancement. However, this argument does not appear to be totally convincing considering the trace amounts of lipid domains present in the nail plate.

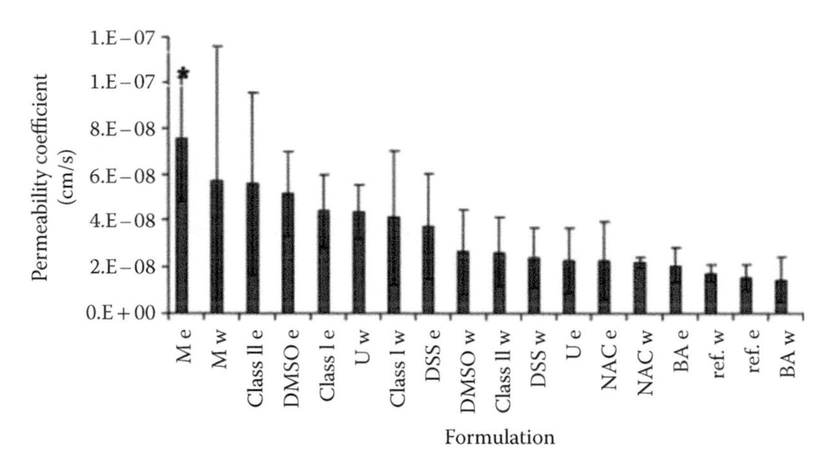

FIGURE 4.1 Permeability coefficients of caffeine through the nail plate delivered using topical formulations containing different enhancers. BA, boric acid; DMSO, dimethyl sulfoxide; DSS, docusate sodium salt; M, methanol; NAC, *N*-acetylcysteine; U, urea; (w), water as solvent; (e), 20% ethanol in water, as solvent; class I and class II are hydrophobins; ref., reference. (Reproduced from Vejnovic et al., *International Journal of Pharmaceutics*, 386, 185–94, 2010a. With kind permission from Elsevier Inc.)

4.1.1.3 Methanol

Methanol was found to be an effective trans-ungual penetration enhancer in formulations containing caffeine as a model permeant (Vejnovic et al. 2010a). The test formulations contained 2% w/v caffeine with methanol (5%) in either water or a binary mixture of 20% v/v ethanol. Formulations with similar compositions, but devoid of methanol, served as reference. The *in vitro* permeation studies were carried out in vertical Franz diffusion cells across human nail plates. Following the permeation studies, the nail plates were milled to detect the amount of caffeine retained. Methanol was found to significantly increase the permeation of caffeine across the nail plate in hydroalcoholic solvent ($P < 0.05$). In presence of methanol, the permeability coefficient of caffeine through the human cadaver nail plate increased from 1.56×10^{-8} cm/s for the reference to 7.5×10^{-8} cm/s with an enhancement factor of 4.8. An enhancement factor of 3.2 was also noted when methanol was used in aqueous system. The permeability coefficients of the various enhancers tested in either water or 20% ethanol are represented in Figure 4.1. The caffeine content in the nail plate after the permeation studies increased from $0.36 \pm 0.09\%$ to $0.61 \pm 0.14\%$ when methanol was used in hydroalcoholic system. The amount of caffeine retained in the nail plate increased by 1.61-fold when methanol was used in aqueous system. The topographical changes of the nail surface before and after permeation studies were evaluated using Visioscan VC 98. Formulations with methanol induced structural changes on the dorsal nail surface, increasing the roughness when compared to the reference formulations. The changes in the nail surface structure after the experiments with methanol are shown by the 3D graph represented in Figure 4.2. These structural changes of the nail plate were assumed to increase the penetration of caffeine across the nail plate. Though the authors attributed the structural changes

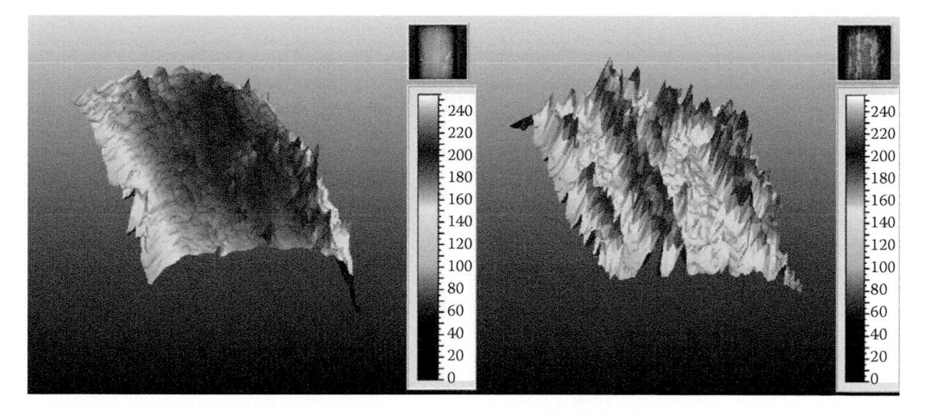

FIGURE 4.2 Changes in the nail surface structure illustrated on a 3D graph. On the left image is dry nail before and on the right, after the experiment with methanol in 20% (v/v) ethanol/water formulation. The images were captured using Visioscan VC 98 machine. (Reproduced from Vejnovic et al., *International Journal of Pharmaceutics*, 386, 185–94, 2010a. With kind permission from Elsevier Inc.)

of the nail surface to the methanol-induced lipid depletion, no direct evidence supporting the depletion was reported in the investigation. Though methanol could be a good solvent for lipophilic drugs, its toxicity profile would limit its further use.

4.1.2 Keratolytic Agents

Keratolytic agents are thought to disrupt the tertiary structure and the hydrogen bonds in the keratin, thereby promoting penetration through the nail plate. The dismantling of the disulfide bonds promotes the disruption of the nail plate barrier and helps in the drug permeation. The other mode of action proposed for keratolytic agents is that they act by softening and hydrating the nail plate, especially in aqueous vehicles (Kobayashi et al. 1998). The swelling and hydration of the nail plate would enhance the drug permeation as a consequence of formation of a less dense structure with large pores.

The effect of keratolytic agents like papain, urea, and salicylic acid on the *in vitro* permeation of imidazole antimycotics such as miconazole nitrate, ketoconazole, and itraconazole through human nail was studied (Quintanar-Guerrero et al. 1998). Papain is an endopeptidase enzyme of plant origin that contains highly reactive sulfhydryl groups. The permeation studies across nail plates were performed in side-by-side diffusion cells containing 60% ethanol in water as donor and receiver fluid. Since no permeation of the three antimycotics was observed over a period of 60 days in the absence of keratolytic agents, permeation studies were performed following pretreatment. The two protocols employed were "single-step pretreatment" with salicylic acid alone (20%) for 10 days and "two-step pretreatment" with papain (15%) for 1 day followed by salicylic acid (20%) for 10 days. The "single-step pretreatment" or even addition of urea (40%) in the donor solution during "single-step pretreatment" failed to induce the transport of any of the antimycotics. However, the "two-step pretreatment" was successful in allowing the permeation of the three compounds. The "two-step" procedure resulted in a flux of 6.66×10^5, 1.15×10^5, and 0.13×10^5 mg/cm^2/s for miconazole nitrate, ketoconazole, and itraconazole, respectively and effective diffusion constants of 6.29×10^8, 3.60×10^8, and 3.00 cm^2/s, respectively. The effective diffusion constants (D_{eff}) were calculated from

$$D_{eff} = \frac{h^2}{6T_L} \tag{4.1}$$

where h represents the thickness of the nail plate and T_L stands for the lag time.

The lag times for miconazole nitrate, ketoconazole, and itraconazole were found to be 32.15 ± 3.47, 56.22 ± 7.08, and 67.5 ± 10.53 min, respectively. Scanning electron microscopic observations revealed that the "two-step pretreatment" procedure was found to damage and fracture the nail surface. The fractured surface was thought to create pathways for drug penetration.

Concentrated solutions of urea and salicylic acid are used for hydrating and softening the nail, thereby assisting the topical treatment of diseased nail. Urea has been employed at high concentrations (40%) to chemically avulse the diseased nails. Urea in combination with salicylic acid was found to effectively increase the penetration of bifonazole into the nail plate in clinical trials (Torres-Rodriguez et al. 1991). The aqueous solutions of urea and sodium salicylate would result in a pH of 7.2 and 6.2, respectively. It is likely that these alkaline conditions would retain the basic drug, such as bifonazole, in an undissociated state, thereby favoring its passive permeation through the nail plate.

Urea and salicylic acid are known to increase the permeation of tritiated water across human nail in conjunction with *N*-(2-mercaptopropionyl) glycine (MPG) from aqueous gel formulations (Malhothra and Zatz 2002). Urea in combination with other cysteine derivatives has proved to enhance the penetration of permeants from aqueous formulations through human nail (Wang and Sun 1999). Cysteines are thiols that act on disulfide bonds in the nail keratin, while urea was likely to act on the hydrogen bonds and facilitate the structural breakdown. Urea in combination with NAC was found to enhance the *in vitro* permeation of miconazole nitrate through the nail plate (Sun et al. 1997). Miconazole is an antifungal agent that can be used in topical treatment of fungal infections of the nail. The formulation containing 2% miconazole nitrate and 20% urea with varying amounts of NAC (5% or 10%) was used as a donor in a study that spanned 3 weeks. It was noted that increasing the NAC concentration from 5% to 10% resulted in a 2- to 2.5-fold increase in the drug permeated into and through the nail. The concentration of miconazole nitrate in the nail exceeded the minimum inhibitory concentrations. Generally, the keratolytic agents are found to be more effective in aqueous vehicles that usually promote hydration and swelling of the nail plate. The keratolytic structural breakdown, combined with the swelling of the nail plate, is thought to promote the transport of poorly water-soluble, high molecular weight antimycotics in most cases.

4.1.3 Compounds That Cleave the Disulfide Bonds

4.1.3.1 Thiols

Thiols are a group of compounds containing sulfhydryl groups (–SH) that have shown some promise as trans-ungual penetration enhancers (Figure 4.3a). The mechanism involved in the enhancement of the permeation across the nail is the reduction of

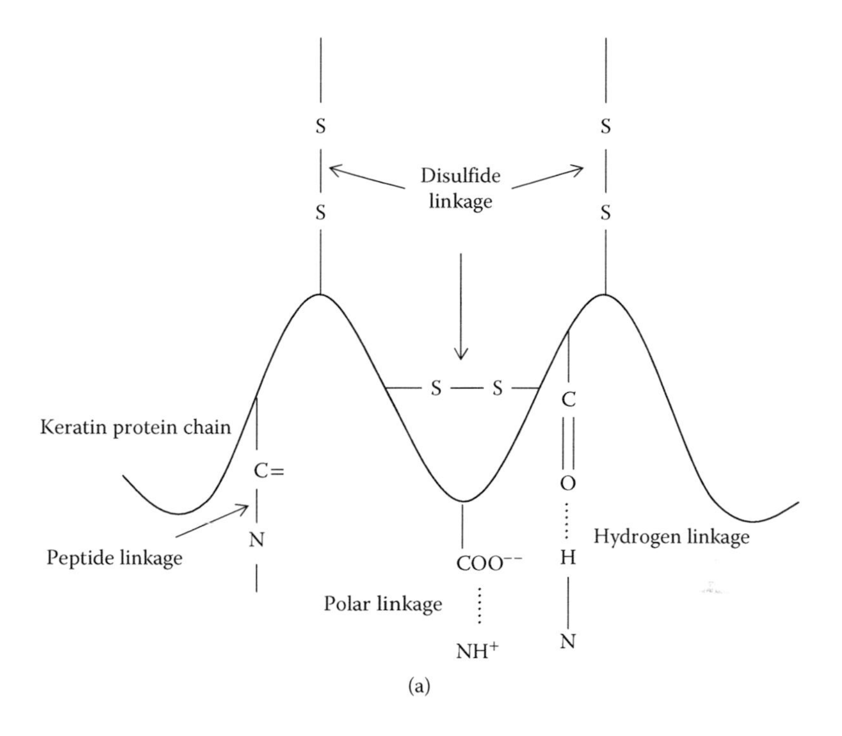

Nail–S–S–Nail + R–SH ↔ 2 Nail–SH + R–S–S–R

(b)

FIGURE 4.3 (a) Disulfide linkages and other bonds in nail keratin. (b) Mechanism of cleavage of disulfide links in the keratin by thiol compounds. "Nail-S-S-Nail" represents the disulfide linkage of the nail plate and "R-SH" stands for a thiol. (Redrawn from Sun et al., *Percutaneous Absorption: Drugs-Cosmetics-Mechanisms-Methodology*, Marcel Dekker, Inc, New York, 1999.)

the disulfide linkage in the keratin matrix of the nail as shown in Figure 4.3b (Sun et al. 1999). These compounds tend to undergo oxidation while reducing the disulfide linkage of the nail keratin.

The reduction of the keratin would destabilize the keratin, compromise the barrier integrity of the nail plate, and promote the trans-ungual permeation (Murdan 2002). Once the disulfide bonds have been broken, they may not be reformed in the dead nail plate. Therefore, one can say that the action of the thiols as trans-ungual penetration enhancers would be irreversible. Most of the thiol compounds are used for pretreating the surface of the nail plate prior to application of the therapeutic agents. Some of the thiol compounds that have been tested as trans-ungual penetration enhancers include pyrithione (PTO), MPG, NAC, cysteine, MPE, and TGA. Thiols are found to be more effective in aqueous or hydroalcoholic vehicles that tend to promote the swelling and hydration of the nail plate. Thiols have proved to be particularly useful for increasing the permeability of large, charged molecules through the nail plate.

Permeation of tritiated water across human cadaver nails was promoted by thiol compounds like MPG and PTO (Malhothra and Zatz 2002). MPG is a mercaptan derivative of the amino acid glycine. The compound is used as a hepatoprotectant, mucolytic, and antidote in heavy metal poisoning. MPG proved to be an effective enhancer in aqueous gels, whereas PTO was found to be effective in lipophilic gels. The two enhancers were used by coapplication along with the permeant during the study. Aqueous gels containing MPG were prepared using hydroxypropyl cellulose as a gelling agent in a solvent system of propylene glycol and water. The test formulation contained 10% MPG as a penetration enhancer, while the control formulation was devoid of any enhancer. The permeation of tritiated water from the gel formulations into and across the nail plate was studied in specialized Franz diffusion cells. Gel formulations containing MPG enhanced the permeation of tritiated water by 2.49-fold. By virtue of its small size, the glycine derivative was thought to get well-incorporated in the nail matrix, thereby increasing the flux of tritiated water. When urea (20%) was added to the gel containing MPG, the transport of tritiated water was further enhanced. MPG was found to act synergistically with urea to increase the permeation of tritiated water by 3.54-fold compared to the control. It was noted that the mercaptan levels were found to be more critical than the urea levels in increasing the permeation of tritiated water through the nail plate. Decreasing the mercaptan levels was found to decrease the enhancement effect, whereas decreasing the urea concentrations hardly affected the penetration enhancement to a large extent.

PTO is 2-mercaptopyridine-1-oxide, which finds its application as a fungicidal and bactericidal agent. The sodium and the zinc derivatives of PTO are known to possess fungicidal properties. The zinc derivative (ZPTO) is commonly used in antidandruff topical formulations. PTO, on reducing the disulfide bonds of the nail keratin, would be oxidized to a dipyrithione, a dimer that possesses antifungal activity. The gels of PTO were composed of hydroxypropyl cellulose as a gelling agent in DMSO. The test gel formulation contained 10% PTO as a penetration enhancer, whereas the control gel formulation did not contain any enhancer. *In vitro* permeation studies performed in Franz diffusion cells indicated that the gel containing PTO increased the permeation of the tritiated water by 2.59-fold compared to the control. Normalized water flux determined before and after treatment with different chemical enhancers was used as an index to determine the barrier integrity of the nail plates. Barrier integrity studies indicated that the structure of keratin matrix would have been irreversibly altered after treatment with effective enhancers. Disruption of the keratin matrix is assumed to result in better permeation of the tritiated water through the human nail. Despite the encouraging results obtained with MPG and PTO in promoting the permeation of a marker such as tritium, the efficacy of these thiol compounds in enhancing the trans-ungual permeation of other chemical entities is yet to be established.

Sulfhydryl compounds such as NAC and MPE are found to increase the permeation of drug with varied polarity across the nail plate. This was illustrated in the case of *in vitro* permeation studies of 5-fluorouracil and tolnaftate across the human nail plate (Kobayashi et al. 1998). 5-Fluorouracil is a hydrophilic drug having a molecular weight of 130 Da and aqueous solubility of 17.1 mg/mL. On the other hand, tolnaftate was a lipophilic drug with a molecular weight of 307 Da and an aqueous solubility of 0.39 μg/mL. The aqueous systems had the drug with either NAC (3%) or menthol (3%) or MPE (3%) in 40% ethanol–water, whereas the lipophilic systems contained the drug with the same enhancers in 10% ethanol in isopropyl myristate. In addition to these, the other aqueous systems were composed of the drug in urea (8 M) or sodium salicylate (40%). The drug permeation through the nail plate was determined in side-by-side diffusion cells maintained at 37°C. The dorsal surface of the nail plate was exposed to the drug suspension, while the ventral surface was in contact with the receiver solution made of 40% polyethylene glycol. Both the enhancers were found to be more efficacious in aqueous systems compared to the lipophilic counterparts, The steady-state flux through the nail pieces and the permeability coefficient were computed. NAC and MPE were found to increase the flux of 5-fluorouracil by 13-fold and 16-fold, respectively from aqueous vehicles compared to water that served as a control. Likewise, NAC and MPE were successful in increasing the flux of 5-fluorouracil by 6.7-fold and 8.4-fold, respectively compared to the control in lipophilic solvents. The extent of permeation enhancement of 5-fluorouracil was more pronounced in aqueous vehicles compared to the lipophilic counterparts, as water most likely facilitated the action of the enhancers by promoting the swelling and softening of the nail plate. Though it appeared that the two enhancers affected the barrier integrity of the nail plates in either of the vehicles, the enhancers seem to penetrate well and get incorporated better into the nail plate in aqueous vehicles. The better penetration of these enhancers into the nail plate from aqueous systems would eventually result in increased diffusion of the soluble drug through the nail plate. On the contrary, the two enhancers were less effective in lipophilic systems owing to the inability of these vehicles to penetrate the hydrophilic nail plate. The effect of different concentrations of NAC (0%, 0.1%, 0.5%, 1%, 3%, 5%, 10%) on the *in vitro* permeation of 5-fluorouracil across the nail plate from aqueous vehicles was also determined. The permeation flux was found to be directly proportional to the concentrations of NAC.

The amounts of tolnaftate permeating through the nail plate were not detectable from any solvent systems devoid of the penetration enhancers. Detectable amounts were only seen when NAC and MPE were used as enhancers. The flux of the tolnaftate from aqueous solvents containing NAC and MPE was found to be 0.137 ± 0.08 and 0.058 ± 0.017 μg/cm^2/h, respectively, whereas the drug flux from lipophilic systems containing NAC and MPE was found to be 0.053 ± 0.009 and 0.223 ± 0.032 μg/cm^2/h, respectively. The poor permeation of tolnaftate in the nail plate was attributed to its high molecular weight and poor aqueous solubility. Both sulfhydryl enhancers used in the study were likely to disrupt the nail plate barrier. The disruption of the nail plate might have improved the trans-ungual transport of the tolnaftate. However, the barrier property of the nail plate was not restored following the permeation studies.

The relationships between octanol/vehicle partition coefficients and the permeability coefficients in different solvents systems were elucidated as a part of the study. Tolnaftate displayed high permeability coefficients in aqueous solvent systems, while 5-fluorouracil showed high values from lipophilic systems. The higher thermodynamic activities of the two drugs in systems with contrasting polarities was most likely responsible for the high permeability coefficients. Solvent systems containing NAC and MPE were found to increase the permeability of both the drugs, but the permeability coefficients were independent of the octanol/water partition coefficient.

TGA is a reducing agent containing a thiol moiety that would facilitate its reaction with the disulfide linkage of keratin. TGA or the salts have found their applications in depilatory products at a concentration of 5%. TGA is also used in hair waving and hair strengthening products at a concentration of 8–11%. TGA has shown some promise as a trans-ungual penetration enhancer, especially in aqueous or hydroalcoholic solutions.

Pretreatment of the nail surface with aqueous solution of TGA increased the permeation of topically applied drugs (Khenger et al. 2007). The dorsal surfaces of nail sections mounted on vertical Franz diffusion cells were exposed to 5% solution of TGA in 20% ethanol. After 20 h, the solution in the donor was removed and replaced with the solution of permeant. Pretreatment of human nails with TGA was found to increase the permeation of radiolabeled mannitol by 2.8-fold, compared to control nails that were pretreated with TGA-free solvent, in a study that extended for 120 h. The same study demonstrated that the steady-state flux of the other permeant, caffeine, through pretreated nail plates was found to be 3.8 times higher compared to control nails. The steady-state flux values from nails pretreated with TGA were found to be 14.10 ± 3.42 μg/cm^2 compared to control nail plates that displayed a steady-state flux value of 3.64 ± 1.54 μg/cm^2. The barrier property of the nails, which was modified by pretreatment with TGA, was not reversible after the study that spanned 165 h. The preliminary studies undertaken indicated that pretreatment with TGA could help in the transport of small molecular weight compounds such as mannitol and caffeine across the nail plate.

Pretreatment with TGA was found to increase the water content of the nail plate and enhance the passive permeation of model permeants such as mannitol and urea through the nail plate (Hao et al. 2008). Mannitol and urea are polar neutral molecules with a molecular weight of 182 and 60 Da, respectively. Pretreatment with TGA for 24–48 h was found to increase the water content of the nail plate. The average water content of the nail clippings in phosphate buffered saline was found to be 39%, which, on pretreatment with 0.5, 1.8, and 3.7 M TGA, increased to 47%, 59%, and 60%, respectively. However, higher concentrations of TGA failed to further increase the water content of the nail plate. The results indicated that TGA had the tendency to increase the hydration capability of the nail clippings, which in fact depended on the concentrations of the TGA in the saline. The permeability coefficient of mannitol and urea across nail plates pretreated with 0.5 M TGA was found to increase fourfold and twofold, respectively, when compared to the control. Pretreatment with 1.8 M TGA was found to further increase the flux of mannitol and urea to around eightfold and fourfold, respectively, when compared to the control. A higher TGA concentration was thought to increase the number of pores of similar size, resulting in higher permeability. Based on the observations, two different mechanisms were proposed by which TGA could increase the permeation across the nail plate. The enhancer could increase the size of the preexisting pores, thereby decreasing the transport hindrance and enhancing the trans-ungual permeability. The second assumption was that TGA could act without disturbing the transport hindrance by creating higher numbers of smaller pores. The changes induced by TGA in the nail plate were found to be irreversible after the pretreatment.

TGA (5%) was found to increase the trans-ungual flux of drugs of varied polarity on pretreatment (Brown et al. 2009). The permeation of caffeine (log $P = -0.07$, molecular weight = 194 Da), methylparaben (log P = 1.96, molecular weight = 152 Da), and terbinafine hydrochloride (log P = 3.3, molecular weight = 328 Da) was studied in Franz diffusion cells using human nail clippings as barrier. Pretreatment with TGA for 20 h increased the flux of caffeine and methylparaben across the nail plate by 3.87-fold and 2.07-fold, respectively, compared to untreated control. The pretreatment was also found to decrease the lag time needed to achieve the steady state by 6.6-fold and 1.7-fold for caffeine and methylparaben, respectively, compared to the control. Though it was hypothesized that TGA reduced the disulfide bonds of the nail keratin leading to structural expansion of the nail, the hypothesis was not supported with any microscopic evidence. The structural changes were assumed to facilitate the uptake of large quantities of water that most likely increased the drug mobility leading to an improvement in caffeine and methylparaben flux rates. Though pretreatment with TGA reduced the lag time of terbinafine hydrochloride by 3.61-fold, it failed to increase the trans-ungual flux of the drug. The high lipophilicity of terbinafine hydrochloride can account for its poor penetration through the hydrophilic nail plate. In addition, the binding tendency of terbinafine hydrochloride to the nail keratin could be the other possible factor that can contribute to the poor flux of the drug. Though pretreatment with aqueous or hydroalcoholic solutions of TGA has proved to enhance the transport of small molecular weight compounds, its efficacy in increasing the permeation of high molecular weight, charged compounds across the nail plate is yet to be established.

4.1.3.2 Sulfites

Sulfites are thought to react with the disulfide bonds in proteins or peptides to produce thiols and thiosulphates. The reaction between sodium sulfite and the keratin of the nail is as per the following reaction (Murdan 2008):

$$R-S-S-R+NaSO_3 \rightarrow R-S-H+R-S-SO_3H$$

The thiol formed would reduce the disulfide linkage in the keratin matrix of the nail infringing on the barrier property and promoting the trans-ungual permeation.

Sodium sulfite was found to increase the penetration of 5,6-carboxyfluorescein through nail clippings of healthy human volunteers (Ng et al. 2007). Sodium sulfite was found to enhance the transnail permeation on pretreatment of the nail plate. The enhancement was also observed when sodium sulfite was incorporated in the donor solution with the permeant.

Sodium metabisulfite ($Na_2S_2O_5$) was investigated as a potential trans-ungual penetration enhancer (Malhotra and Zatz 2002). One mole of sodium metabisulfite is known to react with water to produce 2 mol of sodium bisulfite ($NaHSO_3$). The bisulfite formed was in turn anticipated to cleave the disulfide linkages in the nail keratin. However, sodium metabisulfite alone failed to enhance the permeation of water across the nail plate. Bisulfites, being a milder reducing agent than mercaptans, would have failed to significantly increase the permeation of water. The failure was attributed to the inability of the bisulfites to completely reduce the disulfide bonds and hence breach the integrity of the nail plate.

Nevertheless, the permeation of tritium across the nail plates from gels containing sodium metabisulfite (10%) in combination with urea (20%) was found to result in an enhancement factor of 1.20 compared to enhancer-free control. Considering the ineffectiveness of the bisulfite, the marginal improvement in the permeation of tritium by the combination of the two enhancers could most likely be attributed to the keratolytic action of urea.

Recently, Nair et al. (2009) have demonstrated an improvement in the permeation of terbinafine hydrochloride when sodium sulfite and other inorganic salts were used as enhancers. Permeation studies were performed across human cadaver nail plates mounted on nail adapters in vertical Franz diffusion cell. Terbinafine hydrochloride solution (1 mg/mL, 500 μL) was filled in the donor compartment with or without 0.5 M sodium sulfite, while the receptor contained 5 mL of acidified water (pH 3.0). The cumulative amount of drug permeated after 24 h was found to be 6.93 ± 0.72 and 2.24 ± 0.72 $\mu g/cm^2$ in the presence and the absence of sodium sulfite, respectively. Sodium sulfite was also found to increase the nail drug load in the active diffusional and the peripheral area by about 4-fold and 5.4-fold, respectively. Though it was hypothesized that the inorganic salts acted as penetration enhancers by increasing the hydration of the nail plate and the thermodynamic activity of the drug, the interaction of sodium sulfite with the disulfide linkages of the nail keratin cannot be ruled out, as no structural changes were documented during the studies.

4.1.3.3 Hydrogen Peroxide

Hydrogen peroxide is the only oxidizing agent that has shown some promise as a trans-ungual penetration enhancer. The nails tend to soften on exposure to aqueous solution of hydrogen peroxide. Hydrogen peroxide is thought to act by oxidizing and cleaving the disulfide bonds, thereby enhancing the permeability across the nail plate. Currently, the compound finds its application in the hair care industry and forms a part of tooth bleaching products.

Preliminary screening studies have indicated that hydrogen peroxide was able to increase the weight of the human nail and the hoof membrane by an extent of $69 \pm 7\%$ and $93 \pm 13\%$, respectively (Khenger et al. 2007). Pretreatment of nails with aqueous solution of hydrogen peroxide (35%) for 20 h was found to increase the permeation of mannitol by 3.2-fold during the study that extended for 120 h. Though the enhancement with hydrogen peroxide seemed to be encouraging, its efficacy in trans-ungual penetration had to be further demonstrated with other chemical entities.

Hydrogen peroxide was used in combination with other trans-ungual penetration enhancers like urea to improve the permeation across the nail plate. Urea hydrogen peroxide is a hydrogen-bonded convenient adduct that is found to have better stability than hydrogen peroxide alone. The compound is a strong oxidizing agent that releases hydrogen peroxide in a controlled fashion, causing the nail to swell. However, pretreatment of nail plates with urea hydrogen peroxide failed to increase the permeation of caffeine, methylparaben, and terbinafine hydrochloride (Brown et al. 2009). The failure of the compound to enhance the permeability was attributed to the marked stability of the disulfide linkage against oxidation as opposed to their reactivity with the thiol group. In this context, a "sequential pretreatment" protocol was followed, which involved pretreatment with TGA (5%) followed by urea hydrogen peroxide (17.5%). The "sequential pretreatment" increased the permeation of caffeine and terbinafine hydrochloride by ~4-fold and ~19-fold, respectively, compared to untreated nail plate that acted as the control. The "sequential pretreatment" procedure was found to almost double the terbinafine flux across the nail plate compared to the single pretreatment with TGA alone. The permeation profiles of terbinafine hydrochloride after pretreatment with single and sequential procedures are shown in Figure 4.4. It was thought that the high concentration of urea hydrogen peroxide used for the sequential pretreatment (15–17.5%) would have promoted the oxidation of mixed sulfides or would have even reformed the disulfide bonds, leading to the production of thiosulfinates, thiosulfonates, and eventually, sulfonic acid. The "sequential pretreatment" would most likely expand the keratin network increasing the "swellability" of the nail plate, thereby rendering it more porous. The increase in the size of the pores was the likely reason for the increased trans-ungual permeation of high molecular weight terbinafine hydrochloride in particular. Though the results obtained for "sequential pretreatment" were encouraging, the protocol employed was lengthy and tedious and therefore needed further optimization to suit routine applications in clinical settings.

The sequential pretreatment procedure was put into clinical use when urea hydrogen peroxide was an essential component of MedNail® (MedPharm Ltd., Guildford, UK) technology (Murdan 2008). The technique involved pretreatment of the nail with a reducing agent such as TGA followed by an oxidizing agent like urea hydrogen peroxide. The pretreatment was found to increase the flux of terbinafine by about 18-fold and enhanced the fungal clearance by topical applications like Penlac and Loceryl.

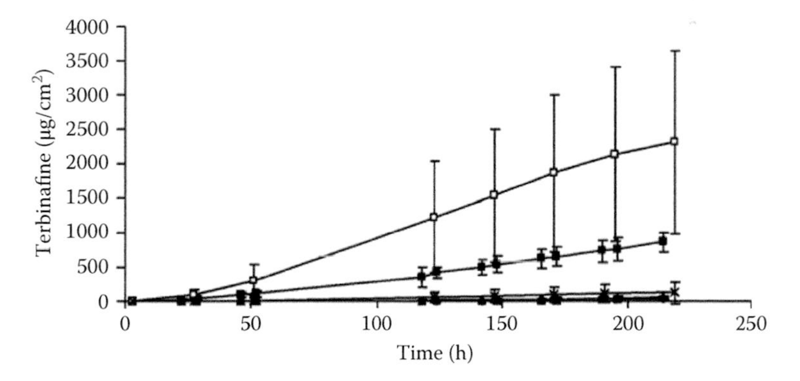

FIGURE 4.4 Permeation profiles of terbinafine hydrochloride following pretreatment with single and sequential enhancing systems. Single treatment consisted of 20-h pretreatment with either thioglycolic acid (TGA) (■) or urea hydrogen peroxide (urea H_2O_2) (▶). Sequential treatment systems were either 20-h TGA pretreatment followed by 20-h urea H_2O_2 (□) or urea H_2O_2 20-h pretreatment followed by 20-h TGA pretreatment (×) (the data for this is not easily visible on the graph as the results overlap with control data). Control (♦) represents no penetration enhancer. (Reproduced from Brown et al., *International Journal of Pharmaceutics*, 370, 61–7, 2009. With kind permission from Elsevier Inc.)

4.1.4 Enzymes

Specific enzymes called proteases are known to hydrolyze the keratin present in the nail plate, affecting its barrier integrity. Papain is a protease of plant origin that was found to degrade the surface of the nail and increase the permeability of topically applied drugs. The effect of papain pretreatment on the *in vitro* permeation of imidazole antimycotics through human nail is illustrated in Section 4.1.2. Keratinase is a keratolytic enzyme that has the ability to hydrolyze the keratin matrix of the nail plate, weakening its barrier properties and enhancing the trans-ungual permeation (Mohorcic et al. 2007). To assess the effect of the keratinase, fresh nail clippings from healthy human volunteers were placed in glass vials filled with 1 mL of Tris–HCl buffer (pH 7.5) containing the enzyme at a concentration of 1–10 mg/mL. Control experiments were performed by incubating the nail clippings in enzyme-free buffer for the same period of time. The buffers with the nail clippings were incubated at 35°C for 48 h following which the clippings were rinsed with distilled water, cut into three pieces, and observed under a scanning electron microscope (SEM). The SEM revealed that keratinase acted on the intercellular matrix separating the corneocytes of the nail plate from one another. The surface view and the cross-sectional view of the nail plate after incubation with keratinase (10 mg/mL) are captured in Figure 4.5. The enzyme was found to corrode the surface of the corneocytes resulting in a striated surface. The cross-section of the nail plate revealed that a small portion of the nail plate was severely disrupted, while the deeper nail layers were unaffected or moderately affected. The inability of the enzyme to affect the deeper nail layers was attributed to the poor diffusion of the keratinase molecules into the nail plate owing to the high molecular weight of the keratin (~30,000 Da). However, disruption of the dorsal surface of the nail plate could be sufficient to enhance the ungual drug delivery since the dorsal surface is known to act as the main barrier to drug diffusion. It was also noted that the action of the keratinase was irreversible and found to be concentration-dependent, causing more damage at higher concentrations.

To further demonstrate the efficacy of keratinase on the barrier property of keratin, the permeation of the model drug metformin hydrochloride across enzyme-pretreated bovine hoof membrane was assessed in Franz diffusion cells. Bovine hoof membrane having a thickness of 150–200 μm was used as a barrier considering its structural similarity with the human nail plate. The flux through the untreated and the enzyme-pretreated hoof membrane was found to be $2.5 \pm 0.6 \times 10^{-3}$ and $5.7 \pm 2.9 \times 10^{-3}$ μg/mm²/min, respectively, while the permeability coefficient for the untreated and the enzyme-pretreated hoof membrane was found to be $1.2 \pm 0.3 \times 10^{-3}$ and $2.9 \pm 1.5 \times 10^{-3}$ mm/min, respectively. The increase in the membrane permeability by more than twofold following enzyme pretreatment suggested that keratinase could be a potential trans-ungual enhancer. Though the flux and the permeability coefficient through the enzyme-pretreated membrane were significantly higher than the untreated membrane, no significant increase in the diffusion coefficient was noted (*t*-test, $P < 0.05$). The diffusion coefficient failed to increase as the enzymes were found to attack only the outer surfaces of the exposed hoof membrane but not the deeper layers. The enhanced permeability was therefore

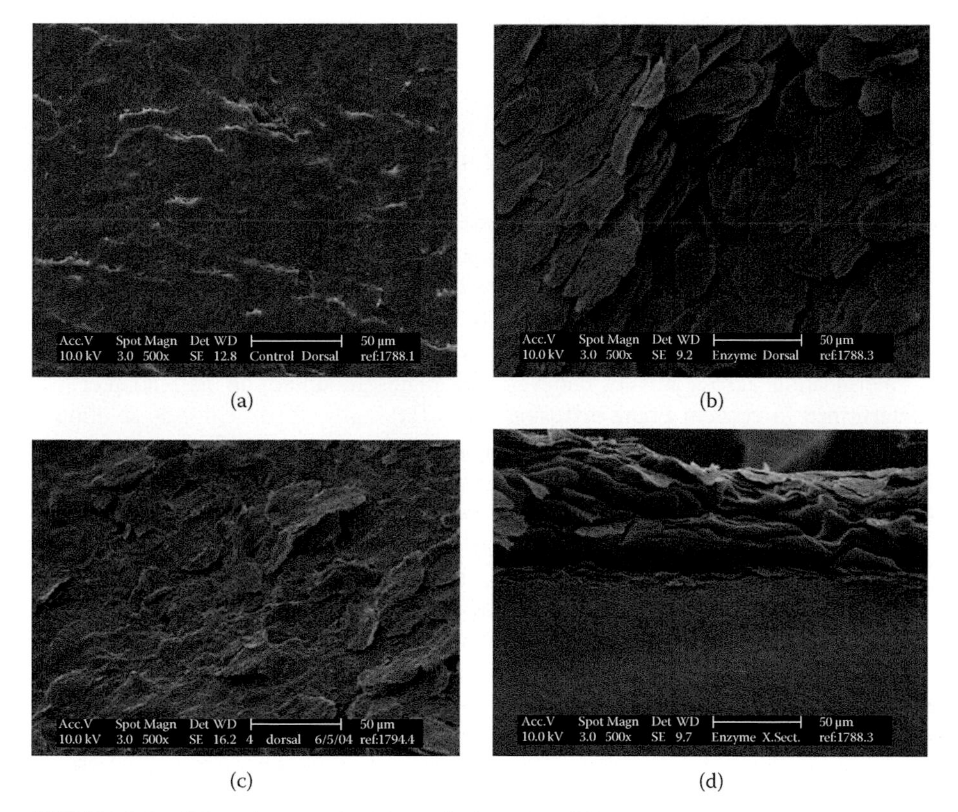

FIGURE 4.5 SEM micrograph of the nail plate after different treatments. (a) The nail plate incubated in buffer (control). (b) Upon incubation with keratinase (10 mg/mL), the corneocytes on the dorsal nail surface separate from one another, as keratinase acts on the intercellular matrix. (c) Upon incubation with a lower concentration of keratinase (1 mg/mL), fewer corneocytes on the dorsal nail surface separate from one another, compared to (b). (d) A cross-section view confirms nail cell separation upon incubation with keratinase. (Reproduced from Mohorcic et al., *International Journal of Pharmaceutics*, 332, 196–201, 2007. With kind permission from Elsevier Inc.)

attributed to the increased partitioning of the drug into the hoof membrane. The studies demonstrated that keratinase could irreversibly disrupt the nail barrier and thereby act as ungual permeation enhancer. Though bovine hooves membrane is a widely accepted model to predict the permeation, the membrane is found to be more permeable compared to the human plate. The extent of swelling on incubation with water was 36% for the hoof membrane against 27% for human nail. Moreover, hoof membranes are known to contain significantly lesser amounts of half-cysteine and disulfide linkages compared to the human nail, which makes them more resistant to compounds that tend to destabilize the disulfide linkages. These differences in the chemical constitution and the barrier properties of the hoof membrane warrant further investigations using human nail plate as a barrier to demonstrate the efficacy of keratinase.

4.1.5 Miscellaneous Penetration Enhancers

4.1.5.1 Inorganic Salts

Some of the inorganic salts were found to enhance the trans-ungual permeation of terbinafine hydrochloride across the nail plate (Nair et al. 2009). The different inorganic salts investigated as potential trans-ungual enhancers were ammonium carbonate, calcium carbonate, potassium carbonate, sodium sulfite, and sodium phosphate. It is hypothesized that the inorganic salts act as trans-ungual penetration enhancers by increasing the hydration of the nail plate. The increased thermodynamic activity of the drug in presence of the salts was also thought to be responsible for the penetration enhancement.

Permeation across human cadaver nail plates mounted on nail adapter was determined in vertical Franz diffusion cells. Terbinafine hydrochloride solution (1 mg/mL, 500 μL) was charged in the donor compartment with or without the salts (0.25–3 M), while the receptor was filled with 5 mL of acidified water (pH 3.0). The permeation of terbinafine hydrochloride was found to be significantly enhanced by threefold to fivefold in presence of the salts (0.5 M) compared to the control that was devoid of any salts. The salts were also found to increase the drug load in the active diffusional area and the peripheral area by 4 to 7 fold and 5 to 10 fold, respectively, compared to the control. Among the five inorganic salts investigated, sodium phosphate was also found to be the most effective trans-ungual penetration enhancer, which increased the drug permeation by approximately fivefold and nail drug load by approximately sevenfold. Sodium phosphate was also found to increase the water uptake of the nail plate from $26.27 \pm 4.55\%$ (for the control) to $36.34 \pm 4.55\%$. It was observed during the studies that the concentration of sodium phosphate was the key in determining the drug transport across the nail plate. Optimum concentration of the salt can be easily incorporated as permeation enhancers in gel formulations intended for topical application to the nail plate.

4.1.5.2 Hydrophobins

Hydrophobins are small amphiphilic fungal proteins made up of 100–250 amino acid residues. They are known to reduce the surface tension and adhere to hydrophilic and hydrophobic structures. Hydrophobins are known to have eight cysteine residues that form four disulfide linkages. They are classified as class I or class II based on the occurrence of the hydrophilic and the hydrophobic amino acid residues on the primary protein structure and the way they are arranged. The hydrophobins identified to date include hydrophobins A, B, and C. Hydrophobin A is a class I hydrophobin, TT1, a fusion protein with glutathione-*S*-transferase got from the fungus *Talaromyces thermophilus*. Hydrophobin B is a chimeric protein composed of N-terminal portion of class I hydrophobin SC3 obtained from *Schizophyllum commune* and C-terminal part of class II hydrophobin HFB2 got from fungus *Trichoderma reesei*. Hydrophobin C is a class I hydrophobin POH3 obtained as a fusion protein with glutathione-*S*-transferase from the fungus *Pleurotus ostreatus*. Hydrophobins have the ability to modify the physical properties of the surface rendering the hydrophilic surface hydrophobic and vice versa. Owing to their ability, hydrophobins are used to modify surface characteristics of medical implants to release the drug in a controlled manner. Hydrophobins are used as stabilizers of foam in the food industry. These proteins find applications as stabilizers of air-filled emulsion used to produce low-fat foods. By virtue of the ability to reduce the surface tension of topical formulations, hydrophobins tend to promote the wetting of the nail surface. Due to their self-assembling property, they are known to coat the drug molecules, thereby improving the drug solubility and stability in the topical formulations. Therefore, hydrophobins are suggested as stabilizers and penetration enhancers in topical trans-ungual formulations.

Hydrophobins were used as permeation enhancers in formulations containing caffeine as a model permeant (Vejnovic et al. 2010a). The test formulations contained 2% w/v caffeine with hydrophobins (0.1%) in water or a binary mixture of 20% v/v ethanol. The reference formulations had the same compositions but were free from hydrophobins. The permeation studies were performed in specially modified Franz diffusion cells having a diffusional area of 0.785 cm^2. The permeability coefficient of caffeine through the human cadaver nail plate was found to increase from 1.56×10^{-8} to 5.6×10^{-8} cm/s when Class II hydrophobins were used as enhancers in hydroalcoholic mixture. The percentages of caffeine remaining in the nail plate following the permeation studies were 1.24-fold and 1.11-fold higher than the reference formulations in aqueous and hydroalcoholic systems, respectively, in the presence of Class II hydrophobins. A 2.8-fold increase in the permeability coefficient was observed in aqueous as well as hydroalcoholic formulations containing Class I hydrophobins compared to the reference formulations. The preliminary studies suggested that hydrophobins are known to act as a surfactant to increase the permeability of a model permeant like caffeine through the nail plate. However, further studies were needed to prove the ability of hydrophobins to enhance the transport of high molecular weight, charged compounds across the nail plate.

The permeation-enhancing ability of hydrophobins was further confirmed in studies involving terbinafine formulations (Vejnovic et al. 2010b). The three hydrophobins that were tested as penetration enhancers in terbinafine formulations were hydrophobin A, hydrophobin B, and hydrophobin C. The hydrophobins were used as penetration enhancers at concentrations of 0.1% w/v in formulations composed of 10% w/v terbinafine in 60% v/v ethanol/water. The reference formulations had the same compositions but were devoid of hydrophobins. The permeation experiments were performed in specially modified Franz diffusion cells having a diffusional area of 0.785 cm^2. The receiver compartment was filled with 60% v/v ethanol that was stirred at 400 rpm, while the donor chamber was loaded with 400 μL of the formulation. Following the permeation studies, the nail plates were milled to detect the amount of caffeine retained. Hydrophobin B was found to be the most effective penetration enhancer of all the hydrophobins investigated, though the improvement in permeability was observed with all the three hydrophobins used. The permeability coefficient of the enhancer-free control formulation was found to be 1.52×10^{-10} cm/s, which increased to 3.52×10^{-10} cm/s, 1.99×10^{-9} cm/s, and 6.78×10^{-10} cm/s for formulations containing hydrophobins A, B, and C, respectively. The corresponding enhancement factors for formulations composed of hydrophobins A, B, and C compared to enhancer-free control formulation were found to be 2.31, 13.05, and 4.45, respectively. The percentage of terbinafine remaining in the nail plate following the permeation studies increased from 0.83 ± 0.02% for the reference to 1.01 ± 1.16% and 1.04 ± 0.36% for hydrophobins B and C, respectively. Hydrophobins, by virtue of their self-assembling property, most likely were able to coat the terbinafine molecule so as to improve the drug solubility and stability in the formulations. The coated terbinafine molecule was assumed to display a higher affinity toward the nail surface compared to the uncoated drug as such. It was also noted during the studies that the hydrophobins did not induce any aggressive structural changes in the nail samples. Since hydrophobins are newer pharmaceutical excipients, limited published data is available on their usage in trans-ungual preparations. This calls for further investigations to evaluate the safety and efficacy issues associated with hydrophobins in trans-ungual application.

4.1.5.3 Dioxolane

The compound 2-*n*-nonyl- 1,3-dioxolane was investigated as a trans-ungual penetration enhancer in an antifungal nail lacquer formulation "EcoNail®" (Hui et al. 2003). The formulation composed of econazole (5% w/w) in a lacquer base of Eudragit® RLPO in ethanol. The ability of 2-*n*-nonyl- 1,3-dioxolane to enhance the drug penetration to the deeper layers of the nail plate was assessed *in vitro*. The test formulation with the enhancer and the control formulation without the enhancer were applied two times a day to human nails for 2 weeks. After the dosing period, the inner ventral section of the nail plate was isolated employing a micrometer-controlled drilling technique and assayed for the absorbed econazole. The drug content in the ventral part of the nail plate was found to be 11.1 ± 2.6 μg/mg for the test formulation and 1.78 ± 0.32 μg/mg for the control formulation. The penetration enhancing effect was attributed to the plasticizing effect of 2-*n*-nonyl- 1,3-dioxolane on the lacquer film, which positively influenced the film adhesion to the nail plate that in turn promoted the permeation of econazole. The amount of econazole that percolated to the support bed under the nail was 47.5 ± 22 mg with the test lacquer and 0.2 ± 0.1 mg for the control lacquer. The surface of the nail was found to contain higher amounts of econazole when the lacquer was devoid of the enhancer. The drug concentration in the deeper layers of the nail plate for the test formulation was about 14,000 times higher than the minimum inhibitory concentration for dermatophytes. Since the subsequent radiolabeling studies ruled out the penetration of 2-*n*-nonyl- 1,3-dioxolane deeper into the nail plate, the enhancer was assumed to exert a plasticizing effect on lacquer film, that would have promoted the film adhesion to the nail surface, which in turn might have facilitated the drug penetration. In spite of the promising results, the efficacy of the 2-*n*-nonyl- 1,3-dioxolane as a trans-ungual penetration enhancer has not been further confirmed with other permeants.

4.1.6 Etchants

Etching agents or surface modifiers were used to disrupt the dorsal surface of the nail plate (Repka et al. 2002). The process of etching would result in development of microporosities within the dorsal surface of the nail plate, increasing the contact surface area and decreasing the contact angle. Disruption of the dorsal surface of the nail plate would promote the penetration of topically applied drug and thereby help to build up and maintain therapeutic drug concentrations in the nail bed. PA and TTA were used as etching agents or nail surface modifiers to disrupt the dorsal surface of the nail plate. The morphology and topography of surface treated nail plates were visualized by atomic force microscopy (AFM), scanning electron microscopy (SEM) and polarized light microscopy (PLM). The image analysis was determined by AFM using NanoScope IIIA v4.23 software to determine the roughness score. Scanning electron micrographs of control and etchant-pretreated nail plates are captured in Figure 4.6. SEM revealed that TTA was found to disrupt and in some cases completely remove the dorsal surface of the nail plate. The PA treatment caused a more disrupted surface morphology resulting in a highly etched surface. The decreased surface integrity and the increased surface area due to etching were thought to facilitate the drug diffusion through the etched nail compared to the untreated counterparts. The atomic force micrographs of a normal nail plate and an etchant-pretreated nail plate are depicted in Figures 4.7 and 4.8. Application of the polymeric dispersion Carbopol 971P was found to render the dorsal surface of the

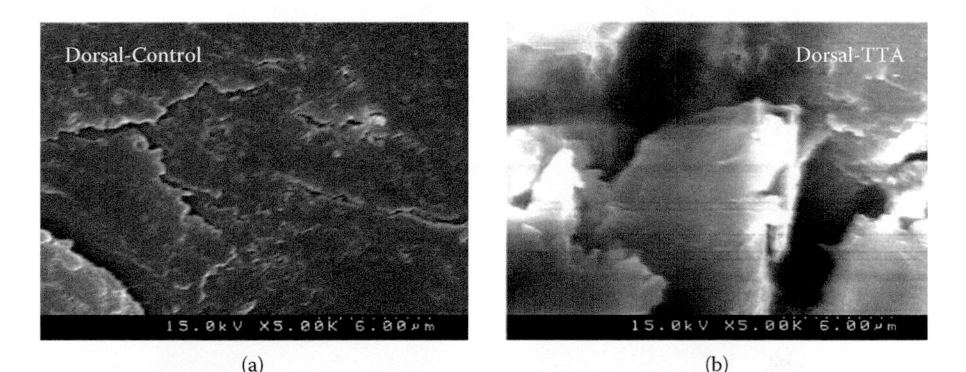

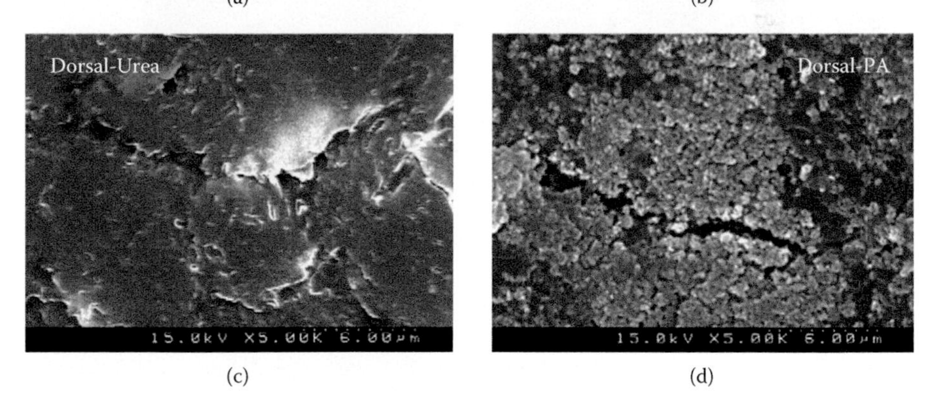

FIGURE 4.6 Scanning electron micrographic pictures of the human dorsal nail plate, (a) control (untreated surface), and those exposed to (b) urea solution, (c) tartaric acid (TTA) solution (90 sec), and (d) phosphoric acid (PA) gel (30 sec). The excess chemicals were removed by washing with approximately 100 mL of distilled water for 3 min. The nails were dried for 30 min before taking the micrographic pictures. (Reproduced from Repka et al., *International Journal of Pharmaceutics*, 245, 25–36, 2002. With kind permission from Elsevier Inc.)

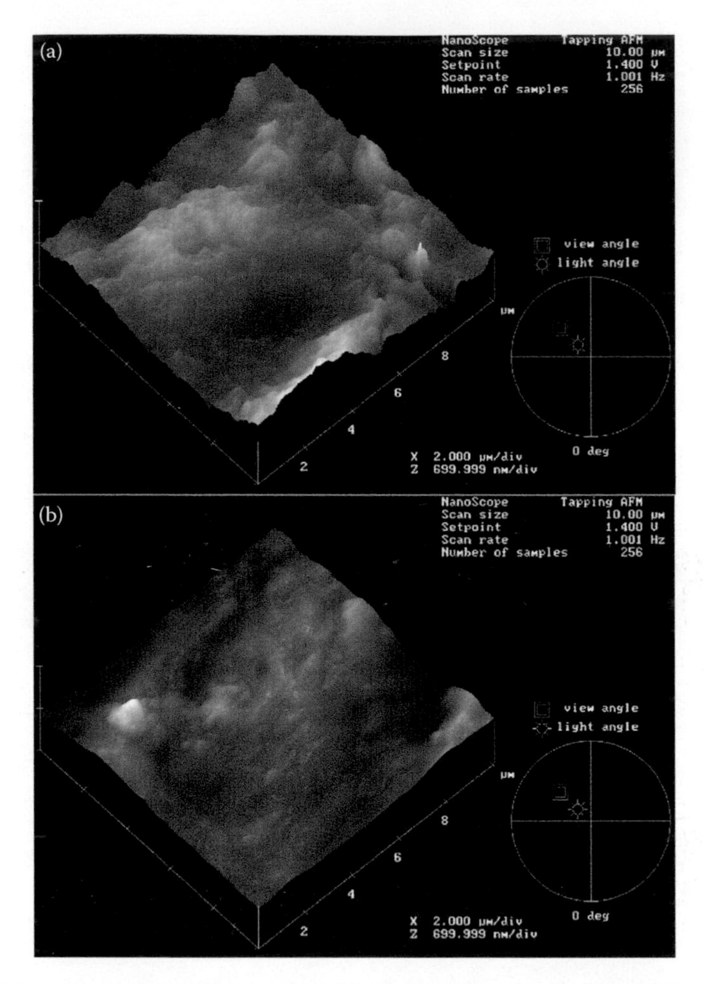

FIGURE 4.7 Atomic force micrographs of the human dorsal nail plate. (a) Untreated nail plate, (b) treated with Carbopol® 971P gel. These served as controls for treated nails shown in Figure 4.8. (Reproduced from Repka et al., *International Journal of Pharmaceutics*, 245, 25–36, 2002. With kind permission from Elsevier Inc.)

nail smooth, while treatment with surface modifiers increased the surface roughness and eventually, the contact surface area. The mean surface scores of the TTA- and the PA-treated nail plates were found to be 112.2 and 147.8 nm, respectively, which were significantly higher ($P < 0.05$) than the untreated nail plate that displayed a roughness score of 85 nm. Though the studies undertaken did not assess the drug permeation through the nail plate, the microscopic studies were suggestive of the fact that the technique could help in effective delivery of topically applied drugs used in the treatment of various disorders of the nails.

The effect of etching on the bioadhesion of the topically applied film and the permeation of ketoconazole from topically applied film or gel was studied (Repka et al. 2004). The dorsal surfaces of the nail plate were treated with the etchant PA gel (10%) for 60 sec. The bioadhesion of the hot-melt extruded films containing the

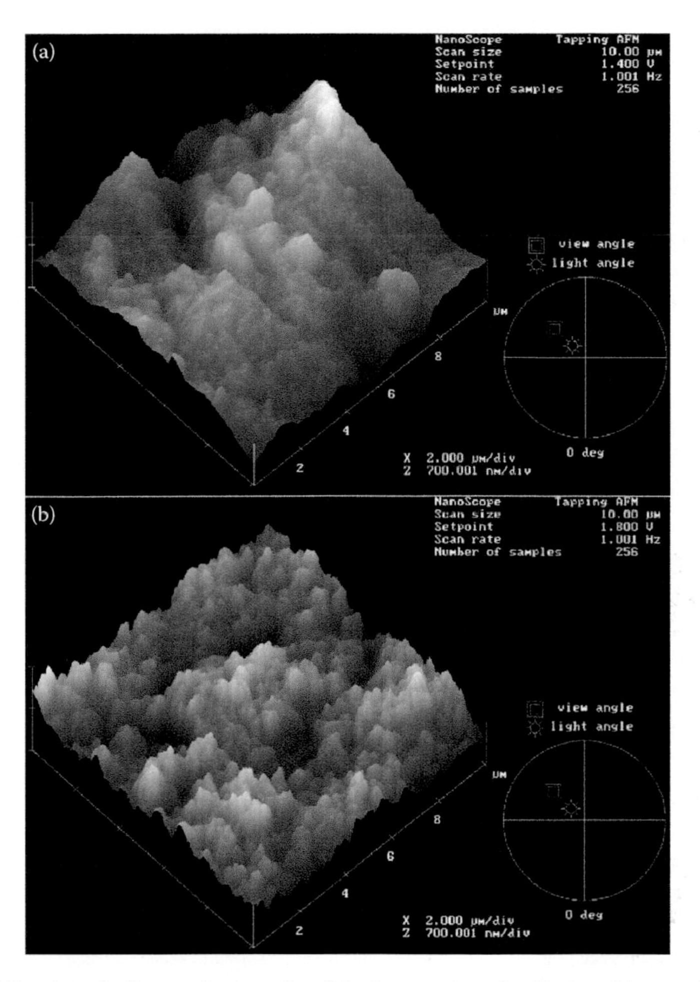

FIGURE 4.8 Atomic force micrographs of the human dorsal nail plate (a) treated with tartaric acid solution (90 sec) and (b) treated with phosphoric acid gel (30 sec). (Reproduced from Repka et al., *International Journal of Pharmaceutics*, 245, 25–36, 2002. With kind permission from Elsevier Inc.)

drug was determined using etchant-treated or untreated nail plates as substrates in a texture analyzer. Tensile strength and peel strength were determined to estimate the bioadhesive profiles of the films. The maximum force termed as "peak adhesion force" that is required to detach the film on the upper probe from the secured nail plate and the "area under the curve" that represents the work of adhesion were recorded. The peak adhesive force for a contact time of 10 sec was approximately 2.5-fold higher for the etchant-treated nail plates compared to the untreated control nail plate. The increase in the bioadhesion of the drug-loaded film to the treated nail plates was attributed to the increase in the surface roughness and contact surface area as a consequence of etching. The treatment with surface modifiers is likely to therefore retain the topically applied drug delivery system on the nail surface for a prolonged period of time.

The permeation of ketoconazole from topically applied film or gels was determined in Franz diffusion cells using etchant-treated or untreated nail plates as barrier. The nail plates were sandwiched between polypropylene adapters and mounted on the cells with the dorsal surface of the nail plate facing the donor chamber. The film or the gel was applied on the dorsal surface of the nail plate, while the receiver was filled with the buffer solution. From the gel formulations, the amount of ketoconazole permeated through the etchant-treated nail plate was 60% higher than through the untreated nail plate. The amount of ketoconazole permeated from the drug-loaded films was found to increase by sixfold through the etchant-treated nail plate compared to the untreated counterparts. The disruption of the dorsal surface eventually reduced the effective thickness of the etched nail plate, thereby increasing the drug permeation. Moreover, the increased contact surface area as a result of etching was also thought to contribute to the increased drug transport through the treated nail plate. The significant increase in the drug transport from the films across the treated nail plate was ascribed to the interpenetration and bonding of the polymeric drug delivery system into the etched nail surface. The studies clearly demonstrated the efficacy of etching in enhancing the drug permeation across the nail plate from drug delivery systems like gels and films. The potential of surface modifiers has made etching an attractive treatment modality to improve the efficacy of the topically applied therapeutic agents. The short pretreatment durations with the etchants prior to topical application of various topical drug delivery systems would suit most of the clinical settings.

4.2 SCREENING METHODS OF UNGUAL PENETRATION ENHANCERS

Several screening methods are proposed to identify the right trans-ungual penetration enhancer for the given drug. Most of the screening procedures put forth are expected to overcome the tedious and laborious *in vitro* permeation studies. In addition, these procedures make use of small samples of nail clippings and are therefore known to cut down the preformulation developmental costs. The screening methods reported to date are based on any one of the following changes noted in the hoof membrane/nail plate, which have been utilized as barriers in the studies:

i. Increase in thickness
ii. Increase in weight
iii. Change in bending stress
iv. Extent of hydration
v. Increase in the drug load or drug uptake

4.2.1 Nail Swelling/Nail Weight/Nail Bending Stress

When immersed in water, the human nail is known to gain weight owing to uptake and retention of water. Nail swelling indicates the rate and extent of drug uptake by the nail, while drug partitioning is considered to be a measure of drug migration into the nail. Nail swelling and drug partitioning studies were used to demonstrate the penetration of the antifungals into the nail when cysteines were used as enhancers (Sun et al. 1997). These studies were carried out by immersing the nail clippings in the test formulations at 32°C for 2 days following which they were examined for weight gain and drug content. The test formulations were composed of itraconazole (1%), cysteine derivatives (5%), urea (10%), and salicylic acid (5%) in a vehicle comprising a mixture of propylene carbonate, propylene glycol, and water. The swelling enhancement factor and the partitioning enhancement factor were calculated as per Equations 4.2 and 4.3, respectively:

$$\text{Swelling enhancement factor} = \frac{\%\ \text{Weight gain of the test nail sample with enhancer}}{\%\ \text{Weight gain of the control nail sample without enhancer}} \tag{4.2}$$

$$\text{Partitioning enhancement factor} = \frac{\text{Drug concentration in the test nail sample with enhancer}}{\text{Drug concentration in the control nail sample without enhancer}} \tag{4.3}$$

The swelling enhancement factors for NAC, l-cysteine, dl-homocysteine, cysteamine, l-cysteine methyl ester, and l-cysteine ethyl ester were found to be 3.18, 4.57, 2.04, 3.03, 2.09, and 1.82, respectively, while the partitioning enhancement factors were found to be 93.6, 105, 23.5, 56.5, 30.7, and 26.2, respectively. Among the six different cysteine derivatives tested, l-cysteine and NAC displayed significant nail swelling and drug partitioning. As a continuation of the screening study, the ability of NAC to increase the *in vitro* permeation of itraconazole was assessed in the presence of urea. The study demonstrated that urea in combination with NAC enhanced the *in vitro* permeation of itraconazole through the nail plate. However, no correlation between the nail swelling and the drug partitioning studies with the *in vitro* permeation of the drug could be drawn at this point, as the results of permeation studies with other enhancers were not available.

The changes in nail weights and bending stresses were used as indices to predict the permeation through the nail plate (Kobayashi et al. 1998). Nail pieces were pretreated with aqueous or lipophilic solutions containing the drug with a permeation enhancer (3%). The aqueous systems had the drug along with either NAC or menthol or MPE as enhancers in 40% ethanol-water, whereas the lipophilic systems contained the drug with the same enhancers in 10% ethanol in isopropyl myristate. The nail weights were recorded using an electronic balance, while the bending stresses were determined using a rheometer with an attachment guide. The weights and the stresses of the nail pieces were recorded before (day 0) and after treatment (day 6) with different solvent systems. The weight/stress ratios were computed from the ratio of the weight/stress after treatment to the weight/stress before treatment. The weight ratio of water that served as control increased by 0.2 point, while the stress ratio decreased by 0.7 point. When compared to control, 8 M urea and 40% sodium salicylate increased the weight ratio by 0.25–0.47 point and decreased the nail stress by 0.06–0.10 point. Significant increase in the weights of the nail pieces and decrease in the bending stresses were noted on immersion of the nail pieces in aqueous vehicles containing NAC. Pretreatment with NAC increased the nail weight by 2.45 points and decreased the nail stress by 0.915 point. Aqueous solvents in general were found to bring about maximum changes in the ratio of the weight/stress on treatment as they were thought to hydrate and swell the nail plate. Good correlation was noticeable between the extent of changes in the weight/stress ratios induced by the enhancers in aqueous systems and the corresponding permeation enhancement obtained with the same system.

The studies indicated that nail swelling and bending stress could be useful predictors of the extent of permeation enhancement, particularly in aqueous systems. However, no significant changes in the weight/stress ratios were noted, in spite of a good enhancement in the flux when NAC or MPE were used as enhancers in lipophilic systems. However, the results clearly indicated that the nail swelling and bending stress cannot be used to predict the extent of permeation in lipophilic systems, which may fail to hydrate or swell the nail plate.

Nail swelling was used as a surrogate marker of penetration enhancement in preformulation screening of putative trans-ungual enhancers (Khengar et al. 2007). The method was based on the principle that the rate and extent of drug penetration into the nail plate depended on the ungual uptake of solvent molecules, which in turn was quantified by the weight gain. To determine the weight gain, clean human nail clippings or horse hoof sections were placed in individual wells of a 24-well plate. Solutions of penetration enhancers or control solutions without any enhancer were added to the wells to completely immerse the nail clippings or the horse hoof sections. The plates were covered with lids and incubated at room temperature for 20 h following which the excess moisture was wiped off and the nail clippings or the horse hoof sections were reweighed to find out the increase in weight. Among 20 different compounds tested, TGA (5% in 20% ethanol) and hydrogen peroxide (35% in water) were found to increase the weight of human nail clippings by 71 ± 5% and 69 ± 7%, respectively, and horse hoof sections by 82 ± 16% and 93 ± 13%, respectively. On the other hand, the increase in the weight of horse hoof sections and human nail clippings was 40 ± 9% and 27 ± 3%, respectively, when immersed in control solutions that contained no enhancers. Attempts were made to correlate the weight gain in the human nail clippings or horse hooves with the penetration of a radiolabeled marker and a model permeant in order to assess the efficacy of potential penetration enhancers. The increase in the weight of the human nail and horse hoof clippings on pretreatment with selected chemicals is represented in Figure 4.9.

For the *in vitro* permeability studies, nails were held between the donor and the receptor compartments of Franz diffusion cells. The nails were pretreated with the enhancer by loading the donor compartment with 0.5 mL of the solution of penetration enhancer. After allowing it to stand for 20 h, the solution was removed and the dorsal surface of the nail was washed with water. The permeability of radiolabeled mannitol was determined by filling the donor compartment with radiolabeled mannitol solution. A 2.8- and 3.2-fold increase in the permeability of the radiolabeled mannitol was seen on pretreatment with TGA and hydrogen peroxide, respectively, when compared to the control. The permeation of radiolabeled mannitol following pretreatment with various chemical penetration enhancers is portrayed in Figure 4.10. TGA, which was found to be one of the promising enhancers as per the prescreen studies, was further assessed for its potential to increase the *in vitro* permeation of caffeine through full-thickness human nail. The steady-state flux of the caffeine was found to be 14.10 ± 3.42 μg/cm²/h on pretreatment with TGA, which was 3.8-fold greater than the pretreatment with enhancer-free solvent (3.64 ± 1.54 μg/cm²/h). The results of the studies indicated that the nail swelling could be useful in screening

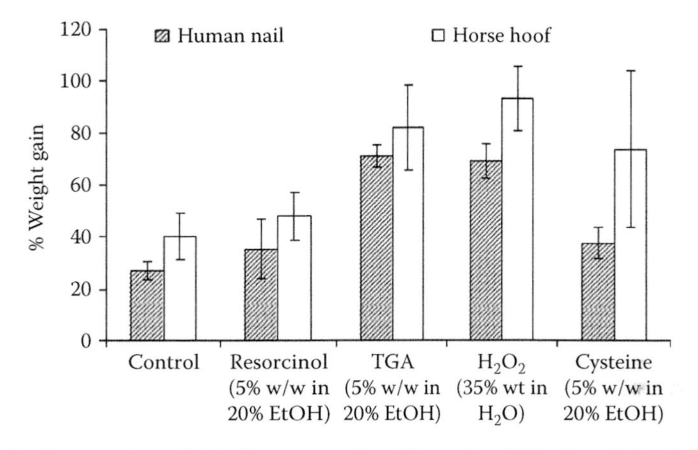

FIGURE 4.9 Increase in weight of human nail and horse hoof clippings following 20-h pretreatment with selected chemical agents. The agents used for pretreatment were resorcinol, thioglycolic acid (TGA), hydrogen peroxide (H_2O_2), and cysteine. All agents were dissolved in 20% ethanol (EtOH) in water apart from hydrogen peroxide, which was used as pre-prepared in water. The control for the study was human nails or horse hooves immersed in water and 20% ethanol in water. (Reproduced from Khenger et al., *Pharmaceutical Research*, 24, 2207–12, 2007. With kind permission from Springer.)

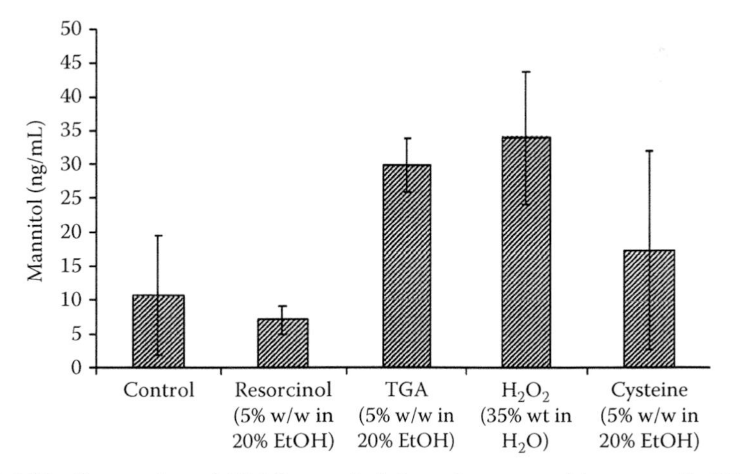

FIGURE 4.10 Permeation of [C-14]-mannitol through pretreated human nail, 120 h after application. Pretreatment time was 20 h. The agents used for pretreatment were resorcinol, thioglycolic acid (TGA), hydrogen peroxide (H_2O_2), and cysteine. All agents were dissolved in 20% ethanol (EtOH) in water apart from hydrogen peroxide, which was used as pre-prepared in water. (Reproduced from Khenger et al., *Pharmaceutical Research*, 24, 2207–12, 2007. With kind permission from Springer.)

large numbers of putative trans-ungual penetration enhancers. The validity of the nail swelling prescreening technique was further confirmed in subsequent studies using model permeants of varied polarity (Brown et al. 2009).

4.2.2 TranScreen-N™

Nail swelling cannot always be considered an indicator for the increased trans-ungual permeation since many trans-ungual enhancers are not likely to promote the swelling of the nail plate. As mentioned earlier, most of the screening methods cannot be used to predict the extent of permeation from lipophilic systems, which may fail to hydrate or swell the nail plate. In this context, TranScreen-N™ was proposed as a high-throughput screening procedure to identify the right trans-ungual penetration enhancer for terbinafine hydrochloride (Murthy et al. 2009). The screening technique that was devised to screen 48 putative enhancers used the high molecular weight (328 Da), charged compound with poor polarity (log $P = 3.3$) as the model permeant. The rapid technique that predicted the permeation through the nail plate, based on the drug uptake or drug loaded into the nail plate, involved two different treatment procedures termed "simultaneous exposure" and "sequential exposure."

In the simultaneous exposure procedure, each preweighed nail piece was placed in an individual well in a microplate containing "drug + enhancer" formulation and incubated at 35°C for 24 h. Control experiments were run in parallel by incubating the nail pieces in "drug-only" formulation at the same temperature for the same time period. The "drug + enhancer" formulation was composed of drug (1 mg/mL) with the putative enhancer in acidified water (pH 3.0), while the "drug-only" formulation was devoid of any enhancer in acidified water. The drug load enhancement factor (L) was calculated as

$$L = \frac{\text{Drug loaded per mg of the nail in ``drug + enhancer'' formulation}}{\text{Drug loaded per mg of the nail in ``drug-only'' formulation}} \quad (4.4)$$

In the sequential exposure procedure, each preweighed nail piece was placed in an individual well in a microplate containing "enhancer-only" formulation and incubated at the same temperature for 24 h. Control experiments were run in parallel by incubating the nail pieces in "drug and enhancer-free" formulation at the same temperature for the same time period. The "enhancer-only" formulation contained only the putative enhancer in acidified water, while the "drug and enhancer-free" formulation was without any drug or the enhancer in acidified water. In the next step, the nail plates were washed and dried before incubating again in microwells filled with "drug-only" formulation for 2 h. The uptake rate enhancement factor (R) was calculated as

$$R = \frac{\text{Amount of drug penetrated per mg of the nail pretreated with ``enhancer-only'' formulation}}{\text{Amount of drug penetrated per mg of the nail pretreated with ``drug and enhancer-free'' formulation}} \quad (4.5)$$

The sequence of procedures employed in TranScreen-N™ is represented in Figure 4.11.

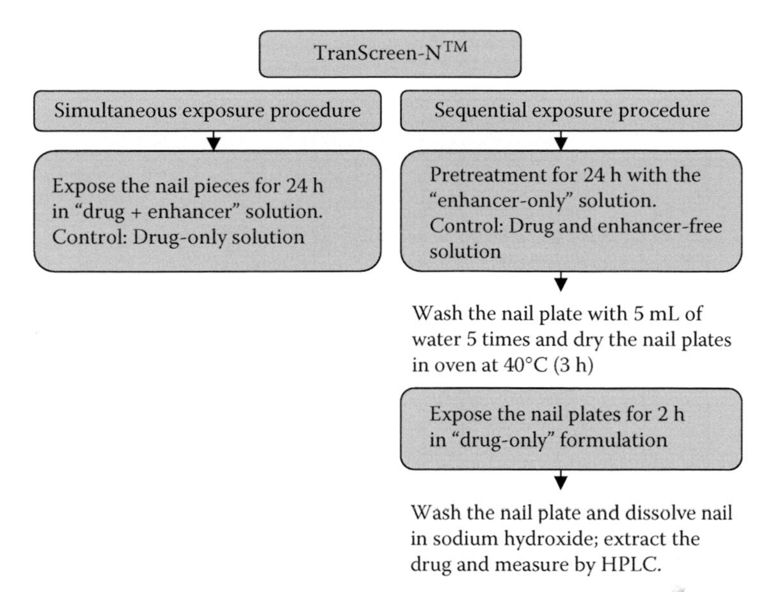

FIGURE 4.11 Schematic representation of sequence of procedures in TranScreen-N™, which is a microwell plate–based method of screening trans-ungual drug delivery enhancers. This involved "simultaneous exposure" and "sequential exposure" procedures run in parallel to determine the drug load enhancement factor (*L*) and the uptake rate enhancement factor (*R*), respectively. (Reproduced from Murthy et al., *Journal of Pharmaceutical Sciences*, 98, 4264–71, 2009. With kind permission from Wiley.)

The *R* and the *L* factors determined were plotted on an *x*–*y* graph to place the 48 different chemicals screened during the study into four quadrants representing four different categories of enhancers.

1. *Non-enhancers (NE):* These chemicals that fell in the first quadrant failed to increase the drug load or the uptake rate, for example, keratolytic agents like salicylic acid and benzoic acid, thiolytic agents like cysteine, and non-ionic surfactants like tweens.
2. *Drug load enhancers (LE):* These chemicals that were placed in the second quadrant were successful in increasing the drug load, for example, resorcinol, propylene glycol, and sodium stearate.
3. *Uptake rate and drug load enhancer (RLE):* These chemicals that fell in the third quadrant were found to enhance the drug load as well as the drug uptake rate, for example, salts such as sodium metabisulfite, sodium sulfite, sodium phosphate, sodium citrate, potassium phosphate and ammonium bicarbonate with moderate to high values of R and L. The other chemicals in this quadrant were sodium dodecyl sulfate, and polyethylene glycol. The *R* and the *L* factors of the polyethylene glycols (PEGs) were found to decrease with increase in the molecular weight.
4. *Uptake rate enhancers (RE):* These chemicals that fall in the fourth quadrant included sodium and potassium hydroxide, sodium carbonate, urea, and acetone.

The four groups of chemical penetration enhancers categorized as per TranScreen-N™ are represented in Figure 4.12.

In vitro trans-ungual diffusion studies were performed in Franz diffusion cells in order to correlate the TranScreen-N™ data with the *in vitro* drug permeation across the human nail plate. The total amount of drug delivered through the nail plate in 24 h was determined in the presence and the absence of enhancers. The drug delivery enhancement factor (EQ_{24}) was calculated as per

$$EQ_{24} = \frac{\text{Drug delivered across the nail plate in 24 h from "drug + enhancer" formulation}}{\text{Drug delivered across the nail plate in 24 h from "drug-only" formulation}} \tag{4.6}$$

An excellent correlation was noted between the L and the EQ_{24} values for the enhancers that fell in the first three quadrants, proving the validity of TranScreen-N™ (Figure 4.13). The R and the EQ_{24} values correlated well for the enhancers of the fourth quadrant.

The application of TranScreen-N™ was further extended to identify the potential trans-ungual etchants for terbinafine hydrochloride and 5-fluorouracil (Vaka et al. 2011). Terbinafine hydrochloride is a cationic lipophilic compound, whereas 5-fluorouracil is an anionic hydrophilic compound. The technique predicted the *in vitro* trans-ungual drug permeation based on the drug uptake or drug loaded into the nail plate following pretreatment with etchants. The etchants screened included PA, lactic acid (LA), glycolic acid, TTA, and citric acid. The nail pieces pretreated with plain gel that was devoid of any etching agent served as a control, whereas the test gel had any one of the etching agents. The nail plates pretreated with the test or control gels were immersed in drug solution for 24 h following which the drug loaded in the nail pieces was determined after extraction. The loading factor (L) was calculated as

$$L = \frac{\text{Drug loaded per mg of the nail pretreated with gels containing potential etchant}}{\text{Drug loaded per mg of the nail pretreated with plain gel devoid of etchant}} \tag{4.7}$$

To assess the drug uptake rate, the pretreated nail plates were kept in contact with the drug solution for 2 h following which the drug penetrated into the nail pieces was determined by analysis. The drug uptake rate enhancement factor (R) was calculated using the formula

$$R = \frac{\text{Amount of drug penetrated per unit weight of the nail pretreated with gels containing potential etchant}}{\text{Amount of drug penetrated per unit weight of the nail pretreated with plain gel devoid of etchant}} \tag{4.8}$$

The drug load and the drug uptake studies demonstrated that PA was a potential surface modifier or etchant for either compound studied. The R and the L values for terbinafine were found to be 2.65 ± 0.24 and 1.67 ± 0.19, respectively, while these values for 5-fluorouracil were 2.14 ± 0.13 and 1.71 ± 0.27, respectively. To elucidate whether these results would be translated into appropriate permeation profiles, the *in vitro* permeation of these compounds was determined across pretreated nail

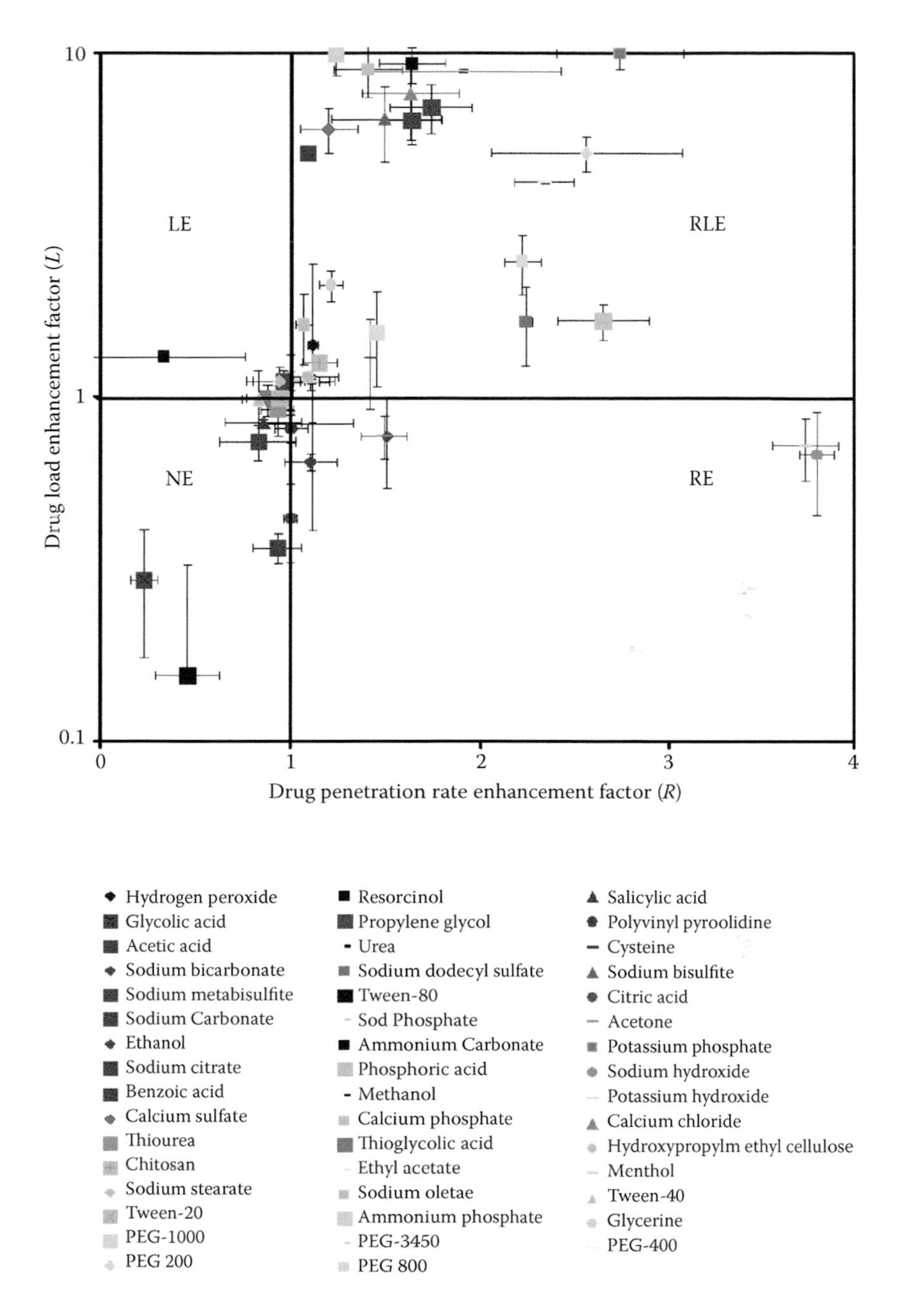

FIGURE 4.12 **(See color insert.)** Categorization of chemical enhancers into four groups based on the R and the L values determined from the TranScreen-N™ studies. L is the load enhancement factor, R is the rate enhancement factor, and EQ24. is the drug delivery enhancement factor. The different quadrants represent non-enhancers (NE) (quadrant 1), drug load enhancers (LE) (quadrant 2), uptake rate and drug load enhancers (RLE) (quadrant 3), and uptake rate enhancers (RE) (quadrant 4). (Reproduced from Murthy et al., *Journal of Pharmaceutical Sciences*, 98, 4264–71, 2009. With kind permission of Wiley.)

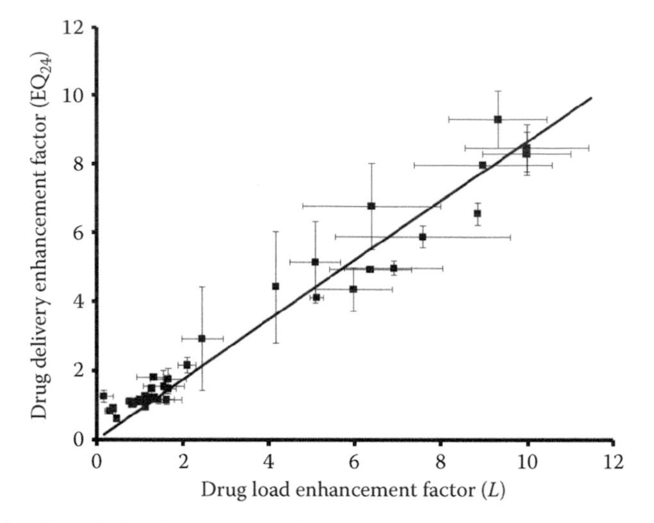

FIGURE 4.13 Correlation between the load enhancement factor (L) determined in the simultaneous exposure procedure of TranScreen-N™ and drug delivery enhancement factor (EQ_{24}) values determined by Franz diffusion cell experiments in case of drug load enhancers (LE), uptake rate and drug load enhancers (RLE), and non-enhancers (NE) groups. The Pearson correlation coefficient, R^2 = 0.95. (Reproduced from Murthy et al., *Journal of Pharmaceutical Sciences*, 98, 4264–71, 2009. With kind permission from Wiley.)

plates. The pretreatment with 1% and 10% PA was found to increase the permeation of terbinafine by twofold and fivefold, respectively, and the drug load by 1.22-fold and 1.64-fold, respectively, compared to the control. The *in vitro* release profiles of terbinafine across etchant-pretreated human nail plate are displayed in Figure 4.14.

Similarly, pretreatment with 1% and 10% PA was found to enhance the permeation of 5-fluorouracil by twofold and threefold compared to the control (1.54 ± 0.06 μg/cm²). Pretreatment with 1% and 10% PA was found to increase the drug load to 277.50 ± 40.24 and 392.65 ± 42.43 ng/mg, respectively, compared to the control (138 ± 32.25 ng/mg). The *in vitro* release profiles of 5-fluorouracil across etchant-pretreated human nail plate are displayed in Figure 4.15.

The studies clearly demonstrated the potential of PA as an etchant to increase the trans-ungual permeation of terbinafine hydrochloride and 5-fluorouracil. It was noted that the drug uptake or drug load data was exactly translated into appropriate *in vitro* permeation profiles, thereby proving the validity of TranScreen-N™. Quantitative topographic analysis by AFM using NanoScope IIIA v4.23 software showed significant difference in the roughness scores of nail plates treated with PA gel when compared to the control. The roughness score of nails pretreated with 1% and 10% PA gels were found to be 118.4 ± 7.8 and 147.8 ± 14.3 nm, respectively, which were significantly higher than that of the control nail plates, which had a score of 85 ± 5.6 nm. These morphological changes were further supported by optical microscopic observations that revealed the reduction in the keratin density of the dorsal surface of the nail plate on pretreatment with etching agents. The decrease in the dorsal surface keratin density due to microstructural alteration as a result of pretreatment with PA was likely to increase the drug permeation.

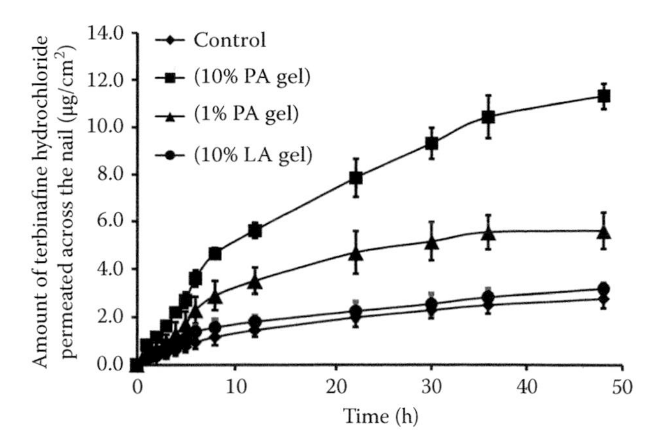

FIGURE 4.14 *In vitro* permeation of terbinafine hydrochloride across the human nail plate pretreated with plain gel (control), phosphoric acid (PA) gels of different strengths and lactic acid (LA) gel for 60 sec. (Reproduced from Vaka et al., *Drug Development and Industrial Pharmacy*, 37, 72–9, 2011. With kind permission of Informa Health.)

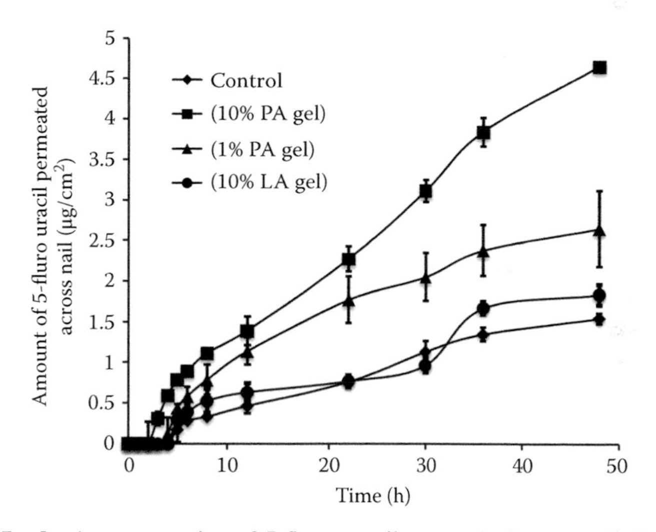

FIGURE 4.15 *In vitro* permeation of 5-fluorouracil across the human nail plate pretreated with plain gel (control), phosphoric acid (PA) gels of different strengths and lactic acid (LA) gel for 60 sec. (Reproduced from Vaka et al., *Drug Development and Industrial Pharmacy*, 37, 72–9, 2011. With kind permission of Informa Health.)

TranScreen-N™ method proved to be a simple, rapid, and less expensive screening procedure that could avoid the need to carry out the *in vitro* permeation studies for several putative trans-ungual enhancers. The technique was found to be extremely useful in identifying not only the trans-ungual enhancers but also the right nail surface modifiers for the given permeant. The procedure also indicates whether the enhancer can be used as a pretreatment or as a coapplication. TranScreen-N™ would most likely decrease the number of experimental trials, reduce the tissue requirement, and eventually cut down the preformulation developmental cost. Though

TranScreen-N™ is a highly prognostic technique, it cannot completely substitute the *in vitro* permeation studies using diffusion cells.

4.3 CONCLUSIONS

Topical management of nail diseases continues to pose a challenge, owing to the poor permeability of the nail plate and the extremely prolonged treatment periods required to maintain therapeutic drug concentrations in the nail bed. The highly dense keratin network of the nail plate makes it difficult for the actives to reach the infected nail bed. The dorsal layer of the nail plate is found to be the major barrier to permeation of drugs. Though the factors that influence the drug penetration through the nail plate have been identified, there are quite a few ambiguous reports in the literature in view of the incomplete understanding of the mechanisms involved. As topical nail formulations are attractive options in treatment of nail disorders, nail penetration enhancers are going to play a pivotal role in the development of trans-ungual topical formulations. Chemical enhancers and etching agents are advantageous over physical methods in that they prove to be cost-effective, noninvasive, and do not require any skilled medical intervention. However, development of an optimal formulation with the right penetration enhancer is a key in delivering therapeutic concentrations of drug across the nail plate into the nail bed. Though the number of trans-ungual penetration enhancers and etching agents identified to date are scarce, many of them are awaiting identification and regulatory approvals.

REFERENCES

Baden, H. 1970. "The Physical Properties of the Nail." *Journal of Investigative Dermatology* 55: 115–22.

Brown, M. B., R. H. Khengar, R. B. Turner, B. Forbes, M. J. Traynor, C. R. G. Evans, and S. A. Jones. 2009. "Overcoming the Nail Barrier: A Systematic Investigation of Ungual Chemical Penetration Enhancers." *International Journal of Pharmaceutics* 370: 61–7.

Franz, T. J. 1992. "Absorption of Amorolfine through Human Nail." *Dermatology* 184 (suppl. 1): 18–20.

Gunt, H. B., and G. B. Kasting. 2007. "Effect of Hydration on the Permeation of Ketoconazole through Human Nail Plate In Vitro." *European Journal of Pharmaceutical Sciences* 32: 254–60.

Hao, J., K. A. Smith, and S. K. Li. 2008. "Chemical Method to Enhance Transungual Transport and Iontophoresis Efficiency." *International Journal of Pharmaceutics* 357: 61–9.

Hui, X., T. C. K. Chan, S. Barbadillo, C. Lee, H. I. Maibach, and R. C. Wester. 2003. "Enhanced Econazole Penetration into Human Nail by 2-n-Nonyl-1,3 Dioxolane." *Journal of Pharmaceutical Sciences* 92 (1): 142–8.

Hui, X., Z. Shainhouse, H. Tanojo, A. Anigbogu, G. E. Markus, H. I. Maibach, and R. C. Wester. 2002. "Enhanced Human Nail Drug Delivery: Nail Inner Drug Content Assayed by New Unique Method." *Journal of Pharmaceutical Sciences* 91 (1): 189–95.

Khengar, R. H., S. A. Jones, R. B. Turner, B. Forbes, and M. B. Brown. 2007 "Nail Swelling as a Preformulation Screen for the Selection and Optimization of Ungual Penetration Enhancers." *Pharmaceutical Research* 24: 2207–12.

Kligman, A. M. 1965. "Topical Pharmacology and Toxicology of Dimethyl Sulfoxide—Part 1." Journal of the *American Medical Association* 193: 796–804.

Kobayashi, Y., M. Miyamoto, K. Sugibayashi, and Y. Morimoto. 1998. "Enhancing Effect of N-Acetyl Cysteine or 2-Mercaptoethanol on the In Vitro Permeation of 5 Fluorouracil or Tolnaftate through Human Nail Plate." *Chemical Pharmaceutical Bulletin* 46: 1797–802.

Malhotra, G. G., and J. L. Zatz. 2002. "Investigation of Nail Permeation Enhancement by Chemical Modification using Water as a Probe." *Journal of Pharmaceutical Sciences* 91: 312–23.

Mohorcic, M., A. Torkar, J. Friedrich, J. Kristl, and S. Murdan. 2007. "An Investigation into Keratolytic Enzymes to Enhance Ungual Drug Delivery." *International Journal of Pharmaceutics* 332: 196–201.

Murdan, S. 2002. "Drug Delivery to the Nail Following Topical Application." *International Journal of Pharmaceutics* 236: 1–26.

Murdan, S. 2008. "Enhancing the Permeability of Topically Applied Drugs." *Expert Opinion on Drug Delivery* 5 (11): 1–16.

Murthy, S. N., D. E. Wiskirchen, and P. C. Bowers. 2007. "Iontophoretic Delivery across Human Nail." *Journal of Pharmaceutical Sciences* 96: 305–11.

Murthy, S. N., S. R. K. Vaka, S. M. Sammeta, and A. B. Nair. 2009. "Tran-Screen-N™: Method for Rapid Screening of Trans-Ungual Drug Delivery Enhancers." *Journal of Pharmaceutical Sciences* 98: 4264–71.

Nair, A. B., S. M. Sammeta, S. K. Vakka, and S. N. Murthy. 2009. "A Study of the Effect of Inorganic Salts on Transungual Delivery of Terbinafine." *Journal of Pharmacy and Pharmacology* 61: 431–7.

Ng, Y., M. Mohorcic, A. Torkar, J. Friedrich, J. Kristl, and S. Murdan. 2007. "Sodium Sulfite – A Potential Onycheal Enhancer to Increase the Topical Drug Delivery to the Nail." *The AAPS Journal* 9 (S2), Abstract T3181.

Quintanar-Guerrero, D., A. Ganem-Quintanar, P. Tapia-Olguin, Y. N. Kalia, and P. Buri. 1998. "The Effect of Keratolytic Agents on the Permeability of Three Imidazole Antimycotics Drugs through the Human Nail." *Drug Development and Industrial Pharmacy* 24: 685–90.

Repka, M. A., P. K. Mididoddi, and S. P. Stodghill. 2004. "Influence of Human Nail Etching for the Assessment of Topical Onychomycosis Therapies." *International Journal of Pharmaceutics* 282: 95–106.

Repka, M. A., J. O'Haver, C. H. See, K. Gutta, and M. Munjal. 2002. "Nail Morphology Studies as Assessments for Onychomycosis Treatment Modalities." *International Journal of Pharmaceutics* 245: 25–36.

Stuttgen, G., and E. Bauer. 1982. "Bioavailability, Skin and Nail Penetration of Topically Applied Antimycotics." *Mykosen* 25: 74–80.

Sun, Y., J. C. Liu, E. S. Kimbleton, and J. C. T Wang. 1997. Antifungal Treatment of Nails. United States Patents US005696164A.

Sun, Y., J. C. Liu, J. C. T. Wang, and P. De Doncker. 1999. "Nail Penetration. Focus on Topical Delivery of Antifungal Drugs for Onychomycosis Treatment." In *Percutaneous Absorption: Drugs-Cosmetics-Mechanisms-Methodology*, 3rd ed., edited by R. L. Bronaugh, and H. I. Maibach, 759–87. New York: Marcel Dekker, Inc.

Torres-Rodriguez, J. M., N. Madrenys, and M. C. Nicolas. 1991. "Non-Traumatic Topical Treatment of Onychomycosis with Urea Associated with Bifonazole." *Mycoses* 34: 499–504.

Vaka, S. R. K., S. N. Murthy, J. H. O'Haver, and M. A. Repka. 2011. "A Platform for Predicting and Enhancing Model Drug Delivery across Human Nail Plate." *Drug Development and Industrial Pharmacy* 37 (1): 72–9.

Vejnovic, I., L. Simmler, and G. Betz. 2010a. "Investigation of Different Formulations for Drug Delivery through the Nail Plate." *International Journal of Pharmaceutics* 386: 185–94.

Vejnovic, I., L. Simmler, and G. Betz. 2010b. "Investigation of Different Formulations for Drug Delivery through the Nail Plate." *International Journal of Pharmaceutics*, 386: 67–76.

Walters, K. A., and G. L. Flynn. 1983. "Permeability Characteristics of Human Nail Plate." *International Journal of Cosmetic Science* 5: 231–46.

Walters, K. A., G. L. Flynn, and J. R. Marvel. 1985. "Physicochemical Characterization of the Human Nail: Solvent Effects on the Permeation of Homologous Alcohols." *Journal of Pharmacy and Pharmacology* 37: 771–5.

Wang, J. C. T., and Y. Sun. 1999. "Human Nail and its Topical Treatment: Brief Review of Current Research and Development of Topical Antifungal Drug Delivery for Onychomycosis Treatment." *Journal of Cosmetic Science* 50: 71–6.

5 *In Vitro* and *In Vivo* Models to Evaluate Topical Nail Formulations

Anroop B. Nair, Xiaoying Hui, Majella E. Lane, and S. Narasimha Murthy

CONTENTS

The success of any medicated topical nail formulation depends on its ability to deliver the medicament into and across the nail plate. Hence, preclinical evaluation of ungual and trans-ungual penetration of antifungal agents is an essential part of formulation development. Different experimental models have been used to evaluate the ungual penetration (drug transported into the nail plate) and permeation (drug transported across the nail plate) of drugs. This chapter outlines several experimental models or approaches, which have been used to characterize nail barrier properties, to evaluate perungual formulations, and to optimize drug delivery technologies.

5.1 *IN VITRO* NAIL DIFFUSION CELLS

A typical diffusion cell used to investigate the drug transport in nail models consists of a donor compartment and a receiver compartment similar to the diffusion cells used in drug permeation studies in skin. In addition, provision is made to secure the nail plate in between the two chambers. The donor compartment holds the drug reservoir or the formulation. The receiver or receptor phase is generally filled with a buffer or a suitable solvent system to provide sink conditions.

5.1.1 Stainless Steel Diffusion Cells

A stainless steel diffusion cell was designed for *in vitro* evaluation of ungual formulations by Walters et al. (1981). The internal and external diameters of the cell were 0.7 and 1.3 cm, respectively, while the exposed area of the nail plate was 0.38 cm^2. The length of each half cell was 3.9 cm and the capacity was 1.4–1.6 mL. The apparatus essentially provided a tight seal between the nail and the half cell surface. The whole cell was placed in a clamping device operating as a "c-clamp" and was immersed in a constant temperature water bath to maintain the temperature (Figure 5.1).

5.1.2 Vertical and Horizontal Diffusion Cells

Two compartment static diffusion cells of upright (vertical) or side-by-side (horizontal) design are widely used to evaluate perungual drug delivery formulations. These cells generally use a *nail adapter* to hold the nail plate between the donor and the

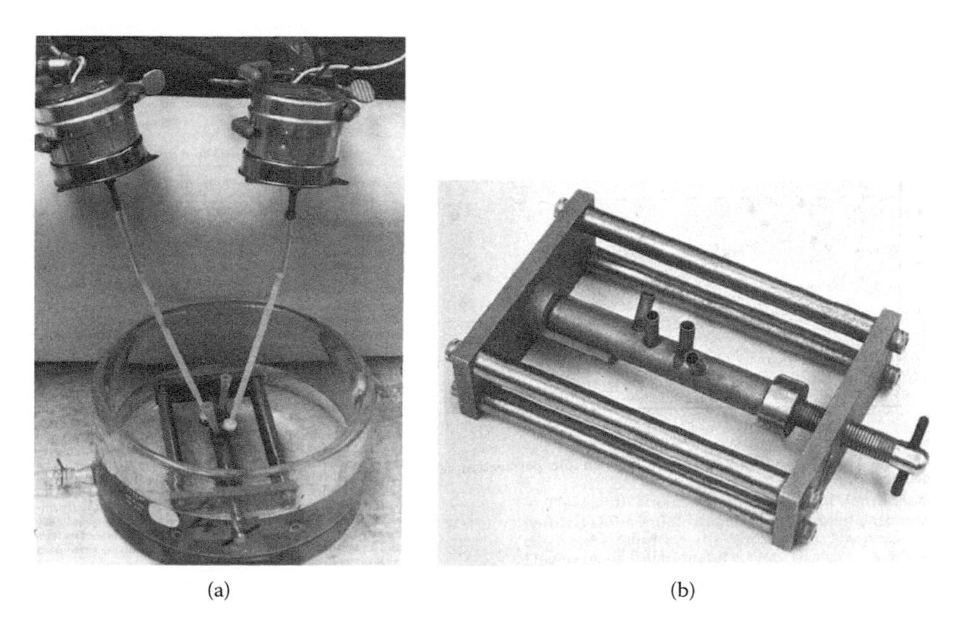

(a) (b)

FIGURE 5.1 (a) The diffusion cell, water bath, and stirring motors. (b) The diffusion cell and clamping device. (Reproduced from Walters et al., *Journal of Investigative Dermatology*, 76, 76–9, 1981. With kind permission of Williams and Wilkins Co.)

receiver compartments. The temperature is maintained by water circulation or by placing the whole setup in a constant temperature oven. The receptor fluid is generally agitated using a Teflon-coated magnetic stir bar at 300–600 rpm.

A side-by side (two-chamber) diffusion cell in which the nail plate was secured in an adapter fabricated using polypropylene with an O-shaped ring (diffusion area 0.049 cm^2) and mounted on the diffusion cell with a water jacket connected to a water bath was reported by Kobayashi et al. (1998). The drug solution was placed on the dorsal side of nail plate. A modified design of this diffusion cell is currently being used in some of the studies (Figure 5.2).

A modified form of the Franz diffusion cell for real-time measurement of drug permeation across the nail plate was developed by Kierstan et al. (2001). This cell consists of a stainless steel receptor compartment in the shape of a UV cuvette. This diffusion cell design provides continuous measurement of the amount of drug transported into the receiver compartment, across the membrane. However, the relatively small diffusion area means that very low amounts of drug will permeate, and the technique will not be applicable for actives that do not have a UV chromophore.

Because of the nail topography, differences in size and shape, roughness, and hard surface mean, it is difficult to fix the nail plate between the donor and the receiver chambers during the *in vitro* studies. Hence, nail adapters are typically used to ensure there are no leaks at the nail–donor interface. Moreover, the material used for fabrication of nail adapters should avoid any binding with the drug molecules. Hao and Li (2008a) used a nail adapter constructed with silicone elastomer (MED-6033, NuSil Silicone Technology, Carpinteria, CA) to accommodate the nail plate curvature and found no noticeable permeant-to-adapter binding. Murthy et al. (2007b) used a custom-made Teflon nail holder to demonstrate the influence of iontophoresis on the trans-ungual delivery. Most of the recent studies have used a nail adapter fabricated by PermeGear Inc, which can be used in both vertical and

FIGURE 5.2 (See color insert.) Picture of horizontal Franz diffusion cell setup. (A) Chambers of the diffusion cell and (B) nail adapter.

horizontal diffusion cells having flat ground or flat flange joints. The nail adapters are fabricated using plastic (acrylonitrile butadiene styrene/Neoflon) or stainless steel (Figure 5.3).

Figure 5.4 illustrates a modified Franz diffusion setup described by Malhotra and Zatz (2002a), in which the donor chamber has two portions. The lower component of the chamber is attached to the portal of the receptor compartment. The nail is then sandwiched between the lower and the upper components of the donor chamber with the help of a clamp.

FIGURE 5.3 **(See color insert.)** Picture of vertical Franz diffusion cell with (a) nail adapter and (b) different components of a nail adapter. (By kind permission of PermeGear, Inc., Hellertown, PA.)

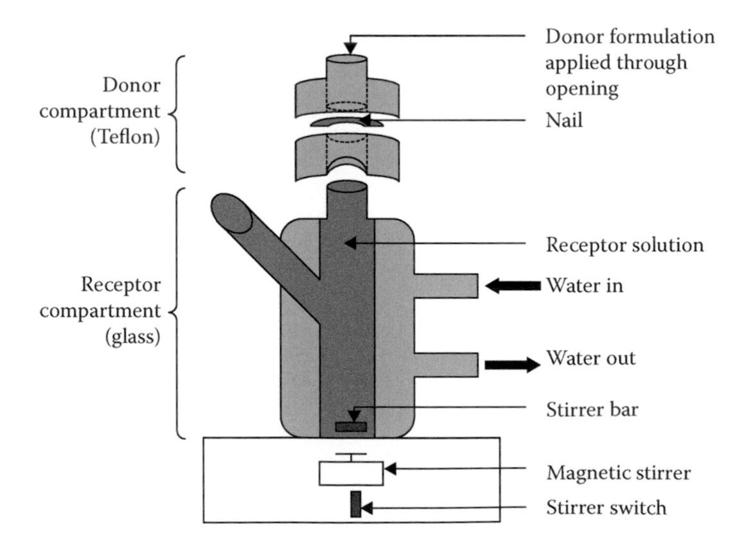

FIGURE 5.4 Schematic of the diffusion cell used for *in vitro* permeation studies. (Reproduced from Malhotra and Zatz, *Journal of Cosmetic Science*, 51, 367–77, 2002a. With kind permission of The Society of Cosmetic Chemists.)

Donnelly et al. (2005) developed a customized diffusion cell as illustrated in Figure 5.5. Hoof membrane was inserted into a purpose-built stainless steel washer designed with a circular hole of identical dimensions to the actual tissue. The hoof sample was supported on a Cuprophan® membrane resting on a stainless steel filter support grid.

A one-chamber diffusion cell was used by Gunt and Kasting (2007) to investigate drug delivery from lacquer or gel-type ungual formulations (Figure 5.6). This cell is unusual in that it is essentially simply a receptor compartment, with a curved top edge designed to accommodate the natural curvature of the nail. Individual fingernails were fixed to this curved edge with a silicone adhesive, and permeation studies were performed by placing the drug formulation over the dorsal surface of the nail plate and sampling the receptor phase through a side port.

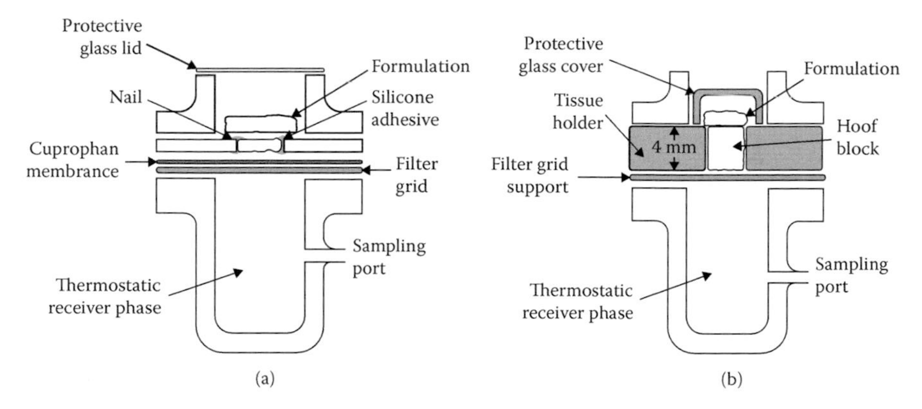

FIGURE 5.5 Schematic illustration of the apparatus used to study 5-aminolevulinic acid penetration across nail (a) and into hoof (b). (Reproduced from Donnelly et al., *Journal of Controlled Release*, 103, 381–92, 2005. With kind permission from Elsevier Inc.)

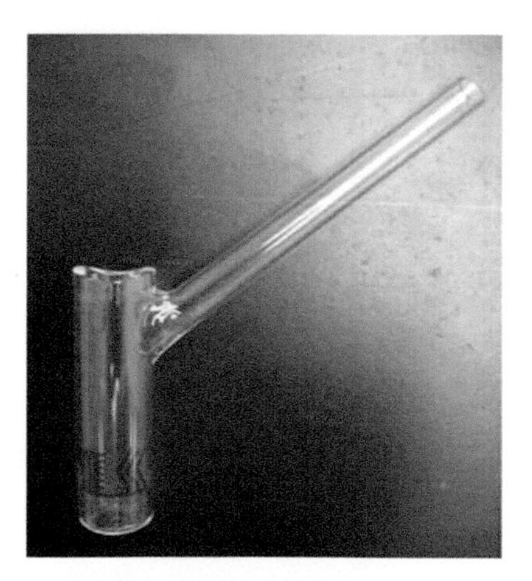

FIGURE 5.6 **(See color insert.)** Picture of the diffusion cell for nail permeation studies. (Reproduced from Gunt and Kasting, *European Journal of Pharmaceutical Sciences*, 32, 254–60, 2007. With kind permission from Elsevier.)

5.1.3 Hydration-Controlled Nail Incubation Systems

Hui et al. (2002) reported the use of a one-chamber diffusion cell device with a Teflon holder for nail plate samples. At the end of the study period, the amount of drug in the inner nail strata is measured by drilling the ventral side of nail, using a micrometer-controlled nail sampling instrument. A detailed description of the nail hydration system and the nail sampling device is provided in Chapter 6.

An understanding of the ability of ungual formulations to modulate nail water properties and thus nail barrier properties is critical to the development of efficacious therapies for nail conditions. A stand-alone method for evaluation of nail hydration and solvent modulation of nail barrier properties has been investigated by Abdalghafor et al. (2011). Nail clippings collected from human subjects are mounted in a dynamic vapor sorption (DVS) chamber coupled with a near-Infrared (NIR) spectroscopic probe (Figure 5.7). The DVS setup allows precise control of relative

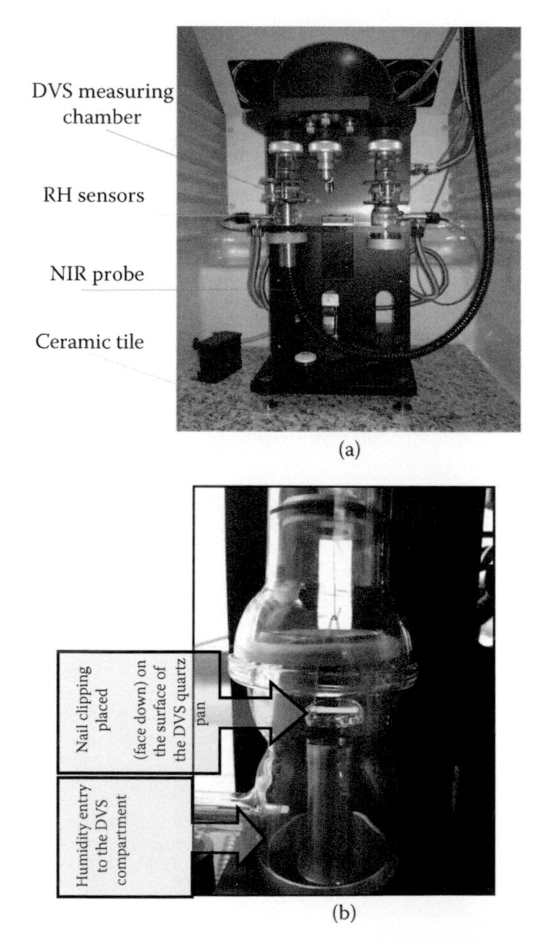

FIGURE 5.7 **(See color insert.)** (a) Dynamic vapor sorption (DVS)/near-infrared (NIR) device. RH, relative humidity. (b) Mounting of nail sample in dynamic vapor sorption (DVS) device.

humidity (RH) with simultaneous measurement of gravimetric changes. At the same time, the water content of the nail is monitored at the molecular level through assessment of changes in free and bound water that have distinct bands at 1400 and 1900 nm, respectively. When nail clippings from five volunteers were subject to a fixed temperature of 32°C and relative humidities (RH%) ranging from 40% to 90%, a good correlation was observed between weight increase of nail samples and signal intensity for free water ($r^2 = 0.95$) and bound water ($r^2 = 0.97$).

In addition to investigations on hydration of the nail, it is also possible to study solvent modulation of nail water content under controlled conditions. Using this approach, pretreatment of nail clippings with isopropyl myristate (IPM) was shown to increase total water uptake by nails. NIR analysis revealed that this change was associated solely with increased free water uptake without any significant change in the bound water content.

5.2 *IN VITRO* TISSUE MODELS

Selection of a suitable *in vitro* model is of great importance in the evaluation of perungual formulations. *In vitro* data are often used to predict the performance of a product during an *in vivo* situation. This also signifies that the evaluation of newly designed formulation(s) or new approaches or techniques with an inappropriate model may eventually lead to incorrect data interpretation. Table 5.1 summarizes the trans-ungual permeation studies reported in the literature where various biological membranes are employed as models of the human nail plate.

TABLE 5.1
Particulars of the *In Vitro* Models, Formulations and Drugs/Permeants Used in Trans-Ungual Delivery

Membrane/Model	Method/Diffusion Cell	Drug/Penetrant	Formulation Type	References
Porcine hoof	Glass penetration chamber on dampened gauze	Amorolfine	Nail lacquer	Pittrof et al. (1992)
Porcine hoof/human nail plate	Modified Franz diffusion cell	5-Aminolevulinic acid	Patch	Donnelly et al. (2005)
Bovine hoof/human nail plate	Modified Franz diffusion cell	Amorolfine/bifonazole/ciclopirox/clotrimazole/econazole/griseofulvin/ketoconazole/naftifine/nystatin/tolnaftate	Solution	Mertin and Lippold (1997a)

(*Continued*)

TABLE 5.1 (*Continued*)
Particulars of the *In Vitro* Models, Formulations and Drugs/Permeants Used in Trans-Ungual Delivery

Membrane/ Model	Method/ Diffusion Cell	Drug/Penetrant	Formulation Type	References
Bovine hoof/ human nail plate	Modified Franz diffusion cell	Nicotinic acid esters/ paracetamol/ phenacetin/benzoic acid/pyridine	Solution	Mertin and Lippold (1997b)
Bovine hoof/ human nail plate	Modified Franz diffusion cell	Chloramphenicol	Solution/nail lacquer	Mertin and Lippold (1997c)
Bovine hoof	Modified Franz diffusion cell	Metformin	Solution	Mohorcic et al. (2007)
Horse hoof/ human nail plate	ChubTur® permeation cell	Mannitol/caffeine	Solution	Khengar et al. (2007)
Pig skin/ human *in vivo*	*In vivo*	Ciclopirox	Nail lacquer	Ceschin-Roques et al. (1991)
Human nail plate	Side-by-side diffusion cell	5-Fluorouracil/ tolnaftate	Suspension	Kobayashi et al. (1998)
Human nail plate	Side-by-side diffusion cell	Miconazole nitrate/ ketoconazole/ itraconazole	Solution	Quintanar-Guerrero (1998)
Human nail plate	Side-by-side diffusion cell	5-Fluorouracil/ ibuprofen	Suspension	Kobayashi et al. (1999)
Human nail plate	Hydration-controlled nail incubation system	Urea/ketoconazole/ salicylic acid	Solution	Hui et al. (2002)
Human nail plate	Hydration-controlled nail incubation system	Econazole	Solution	Hui et al. (2003)
Human nail plate	Side-by-side diffusion cell	Several drugs	Suspension/ solution	Kobayashi et al. (2004)
Human nail plate	Hydration-controlled nail incubation system	Ciclopirox	Gel/lacquer	Hui et al. (2004)
Human nail plate	Vertical Franz diffusion cell	Ketoconazole	Film	Repka et al. (2004)
Human nail plate	Hydration-controlled nail incubation system	Oxaborole	Solution	Hui et al. (2007a)
Human nail plate	Hydration-controlled nail incubation system	Panthenol	Solution	Hui et al. (2007b)

TABLE 5.1 (*Continued*)
Particulars of the *In Vitro* Models, Formulations and Drugs/Permeants Used in Trans-Ungual Delivery

Membrane/ Model	Method/ Diffusion Cell	Drug/Penetrant	Formulation Type	References
Human nail plate	Modified Franz diffusion cell	Glucose/mannitol	Solution	Murthy et al. (2007a)
Human nail plate	Modified Franz diffusion cell	Salicylic acid	Solution	Murthy et al. (2007b)
Human nail plate	ChubTur permeation cell	Caffeine/ methylparaben, terbinafine	Solution	Brown et al. (2009)
Human nail plate	ChubTur permeation cell	Thioglycolic acid/urea/ hydrogen peroxide	Solution	Khengar et al. (2010)
Human nail plate	Franz diffusion cell	Mannitol	Solution	Dutet and Delgado-Charro (2010a)
Human nail plate	Franz diffusion cell	Terbinafine	Solution	Amichai et al. (2010)
Human nail plate	Franz diffusion cell	Sodium/lithium	Solution	Dutet and Delgado-Charro (2010b)
Excised cadaver nail	Modified Franz diffusion cell	Ketoconazole	Solution	Gunt and Kasting (2007)
Excised cadaver nail	Modified Franz diffusion cell	Tritiated water (^3H_2O)	Solution	Gunt et al. (2007)
Excised cadaver fingernail/ toenail	Modified Franz diffusion cell	Tritiated water	Gel	Malhotra and Zatz (2002a, b)
Excised cadaver nail	Side-by-side diffusion cell	Mannitol/urea/ tetraethylammonium	Solution	Hao and Li (2008a)
Excised cadaver nail	Side-by-side diffusion cell	Mannitol/urea	Solution	Hao and Li (2008b)
Excised cadaver nail	Franz diffusion cell	Terbinafine	Solution/gel	Nair et al. (2009a, c, e)
Excised cadaver nail	Side-by-side diffusion cell	Sodium/chloride	Solution	Smith et al. (2010)
Excised cadaver nail	Side-by-side diffusion cell	Tetraethylammonium ion	Solution	Smith et al. (2009)
Excised cadaver nail	Side-by-side diffusion cell	Ciclopirox	Solution	Hao et al. (2009)
Excised cadaver nail	Side-by-side diffusion cell	Phosphate buffered saline	Solution	Hao et al. (2010)
Excised cadaver nail	Franz diffusion cell	Terbinafine	Nail lacquer	Shivakumar et al. (2010)
Excised cadaver nail	Franz diffusion cell	Terbinafine/5- flurouracil	Solution	Vaka et al. (2010)

(*Continued*)

TABLE 5.1 (*Continued*)
Particulars of the *In Vitro* Models, Formulations and Drugs/Permeants Used in Trans-Ungual Delivery

Membrane/ Model	Method/ Diffusion Cell	Drug/Penetrant	Formulation Type	References
Excised cadaver nail	Franz diffusion cell	Caffeine	Solution	Vejnovic et al. (2010a)
Excised cadaver nail	Franz diffusion cell	Terbinafine	Solution	Vejnovic et al. (2010b)
Infected nail plate	Agar incubation	Bifonazole/sodium pyrithione/amorolfine/ terbinafine	Solution	Nakashima et al. (2002)
Infected cow hoof plate	Agar incubation	Ciclopirox	Nail lacquer	Bayerl (1999)
Infected bovine hoof	Agar incubation	Piroctone olamine	Solution	Dubini et al. (2005)
Diseased nail	Side-by-side diffusion cell	Several drugs	Suspension/ solution	Kobayashi et al. (2004)
Intact toe	Ex vivo	Terbinafine	Gel	Nair et al. (2009b)

5.2.1 Hoof Membrane

The difficulty in obtaining intact human nails for *in vitro* studies has led to a search for more readily available models (Mertin and Lippold 1997a). In some instances, animal hooves have been shown to be a useful alternative to human nails, particularly for *in vitro* permeation studies. Porcine, sheep, and bovine hooves have been adopted for many studies to provide an acceptable *in vitro* prediction of drug permeation. Some authors have reviewed the similarities between the hoof membrane and the human nail structure. Both are rich in keratin and release soluble protein on incubation with keratinase (Vignardet et al. 2001). However, hoof membranes are much thicker than human nail plates (Khengar et al. 2007; Mertin and Lippold 1997a). Notwithstanding this, extensive work has been done by several research groups using the porcine hoof and bovine hoof membranes to assess their suitability as a model for human nail.

5.2.2 Porcine Hoof

Porcine hoof membranes are widely used to evaluate *in vitro* trans-ungual drug delivery. Pittorf et al. (1992) have evaluated various amorolfine nail lacquers using porcine hoof using a glass penetration chamber on dampened gauze, where the moistened gauze simulates the nail bed. A novel *in vitro* nail apparatus simulating both the nail plate and the nail bed was developed by Kim et al. (2001) to investigate the permeability of a commercial ciclopirox lacquer (8%) using porcine hoof. The nail penetration capacity of the formulation was assessed by loading the lacquer on the dorsal surface of porcine hooves. Subsequently, the nail was placed on a plastic

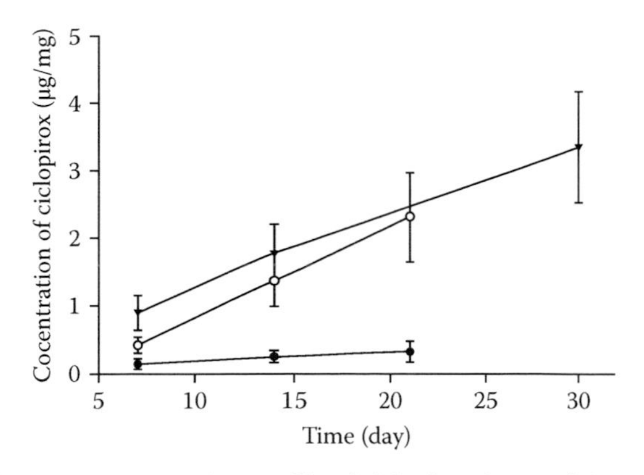

FIGURE 5.8 The *in vitro* penetration profile of ciclopirox from nail lacquer containing 8% drug, in a porcine hoof model ($n = 8$) and the *in vivo* profile in human. Data are shown as mean ± SD. (●, amount in the gel; ○, amount in the porcine hoof membrane; ▼, amount in the human nail). (Reproduced from Kim et al., *Pharmaceutical Research*, 18, 1468–71, 2001. With kind permission of the Plenum Publishing Corporation.)

penetration chamber prepared using a poloxamer gel. The system was sealed to prevent evaporation of water from the gel and to avoid any variation in temperature during the experiment. The penetration chamber was then dissolved and the solution (2 mL) was then transferred into a plastic chamber of diameter 2.3 cm and maintained at room temperature until solidification. The amount of drug permeated was assessed by separating the gel from the hooves, and the amount of drug in the gel was analyzed after extraction. Further, the authors compared the *in vitro* profile (amount of drug permeated through the porcine hoof membrane) with *in vivo* data for the same formulation. Figure 5.8 illustrates the *in vitro* and *in vivo* profiles for a commercial ciclopirox lacquer formulation (8%). The penetration profiles of ciclopirox across the porcine hoof and human nail *in vivo* were similar and suggest that porcine hoof is a useful model of the human nail. Approaches to limit intersample variability of porcine hooves could also have been considered by the authors, for example, by using only freshly excised hooves, using hooves of comparable weight and thickness, and by maintaining a constant prehydration treatment time.

5.2.3 Bovine Hoof

The suitability of bovine hoof membranes as an alternative to human nail plate for the trans-ungual permeation was assessed by Hemidy et al. (1994). The relationship between the permeability coefficients of bovine hoof membrane and of human nail plate was reported (Figure 5.9) as follows:

$$\log P_N = 3.723 + 1.751 \log P_H$$

where P_N is the permeability coefficient of the nail plate and P_H is the permeability coefficient of the hoof membrane.

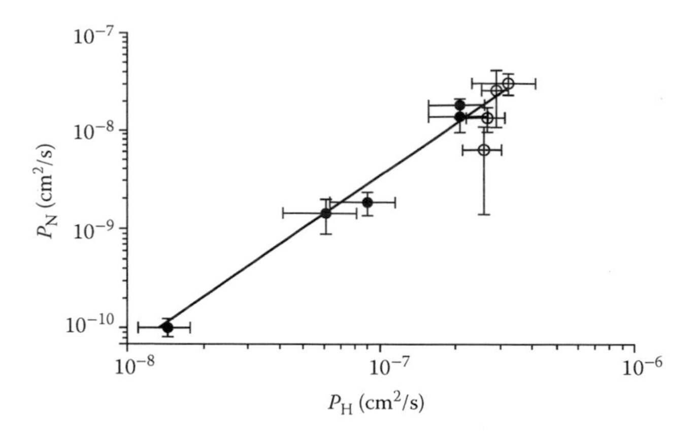

FIGURE 5.9 Relationship between the permeability coefficient of the nail plate (P_N) and the permeability coefficient of the hoof membrane P_H at 32°C (n = 3–8, means ± SD). ○, nicotinic acid esters; ●, remaining substances. Log P_N = 3.723 ± 1.751; log P_H, r = 0.971. (Reproduced from Mertin and Lippold, *Journal of Pharmacy and Pharmacology*, 49, 866–72, 1997a. With kind permission of Wiley.)

The findings of the study suggested that both barriers behave as hydrophilic gel membranes rather than lipophilic membranes and that bovine hoof membranes may be useful models of the human nail.

A systematic investigation was carried out by Mertin and Lippold (1997b) on the permeability properties of bovine hoof compared with the human nail plate. The permeability of 10 antimycotics with different molecular weights was assessed through human nail plates and bovine hooves. The data indicated a linear relationship with a negative slope between permeability coefficient and molecular weight for both the nail plates (Figure 5.10). The permeability of the bovine hooves was found to be approximately twice that of the human nail plate (Mertin and Lippold 1997a) (Figure 5.10).

In a later study, the effect of lipophilic vehicles and nail lacquers on the *in vitro* permeability of chloramphenicol was assessed in both human nail plate and bovine hooves (Mertin and Lippold 1997c). The authors found that the drug flux was independent of the nature of the vehicle, and penetration of the drug was membrane-controlled in the initial stages of the experiment, which eventually changed to matrix control from the formulation as the drug content of lacquer decreased. Bovine hooves have also been used to evaluate a water-soluble nail lacquer of ciclopirox using vertical diffusion cells (Monti et al. 2005). These authors investigated the trans-ungual permeation of ciclopirox from a water-soluble polymer-based formulation compared with water-insoluble nail lacquer of ciclopirox (Penlac®) using vertical diffusion cells (Monti et al. 2010). In a further study, a formulation of a new antifungal agent (Myfungar®) containing 0.5% of piroctone olamine was also evaluated using the bovine hoof model. The effect of keratinolytic enzymes on ungual permeation enhancement was also assessed using bovine hoof membranes of ~150–200 μm thick (Mohorcic et al. 2007). Recently, bovine hoof membrane was used in the evaluation of a ciclopirox nail lacquer formulation (Togni and Mailland 2010).

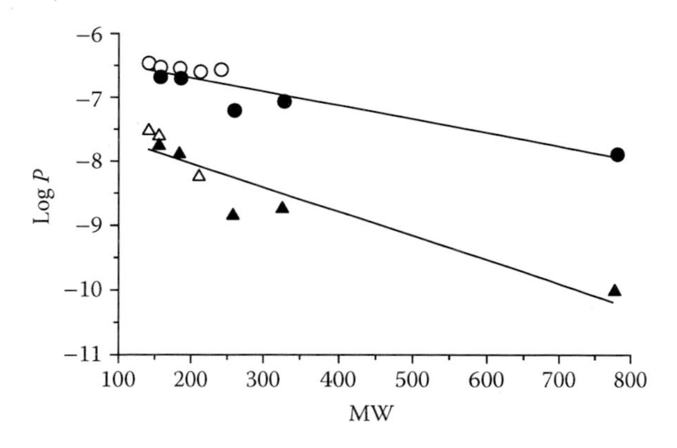

FIGURE 5.10 Relationship between the logarithm of the permeability coefficient *P* for the nail plate (△▲) or the hoof membrane (○●) and the molecular weight (MW) (n = 3–8, means ± SD) Medium: phosphate buffer pH 7.4. *P* is expressed in cm²/s. (○), △, nicotinic acid esters; ●▲, remaining substances (paracetamol, phenacetin diprophylline, chloramphenicol, and iopamidol). (Reproduced from Mertin and Lippold, *Journal of Pharmacy and Pharmacology*, 49, 866–72, 1997a. With kind permission of Wiley.)

5.2.4 Other Membranes

A limited number of investigations have probed the potential use of membranes from other sources as a model of the human nail plate. The permeation of ciclopirox nail lacquer was assessed using cow horn plate as a surrogate model of the fingernail (Bayerl 1999). Sheep hooves have been used to assess the trans-ungual permeation of antifungal agents (Yang et al. 1997). A high-throughput technique has been developed to assess a range of perungual permeation enhancers using horse hoof as a membrane, and the data were compared with results from human nail plates (Khengar et al. 2007).

5.2.5 Keratin Film Model

Keratin from human hair was investigated as a potential model of the nail plate (Reichl and Müller-Goymann 2009). Keratin was extracted from hair and any toxic components were removed by dialysis. To the isolated keratin solution, 2% of diluted glycerol was added as a plasticizer, and the resulting preparation was molded into Teflon rings ($d = 2$ cm). Permeation studies were carried out using sodium fluorescein (water-soluble), rhodamine B (lipid-soluble), and fluorescein isothiocyanate-dextran MW 4000 Da (large-molecule drug model). Similar permeation profiles were obtained for both 100- and 120-μm hoof membranes for sodium fluorescein and fluorescein isothiocyanate-dextran MW 4000 Da. In addition, the same rank order for marker permeability coefficients was observed for both bovine hooves and keratin films.

5.3 HUMAN NAIL PLATE

The human nail plate is a specialized epithelial structure that gives rise to a fully keratinized, multilayered, flattened sheet of horny cells cemented together and filled with keratin. Typically, it is a hardened epidermal tissue, is thin, slightly elastic, translucent, and possesses a convex structure. The chemical composition of human nail plate differs significantly from other membranes in the human body. The nail plate itself is composed of keratin fibers with many disulfide linkages, globular proteins, and relatively low lipid levels (0.1–1%). The nail plate may be divided into a hard/brittle dorsal layer and a "plastic" ventral layer. The upper dorsal layer is a relatively more compact layer of keratin filled with keratinocytes and it is considered to be the main barrier for drug penetration and permeation (Gupchup and Zatz 1999). A micrograph of a cross-section of nail plate is shown in Figure 5.11.

Intrasubject and intersubject variability is likely to be a critical factor when evaluating *in vitro* studies using human nail plates. It is well known that nail plate thickness differs between fingers and toes. Hence, it is likely that permeation of drug molecules may vary greatly (Malhotra and Zatz 2002a). Usually, using nail plates from the index, middle, and ring fingers is preferred as they are reported to be more

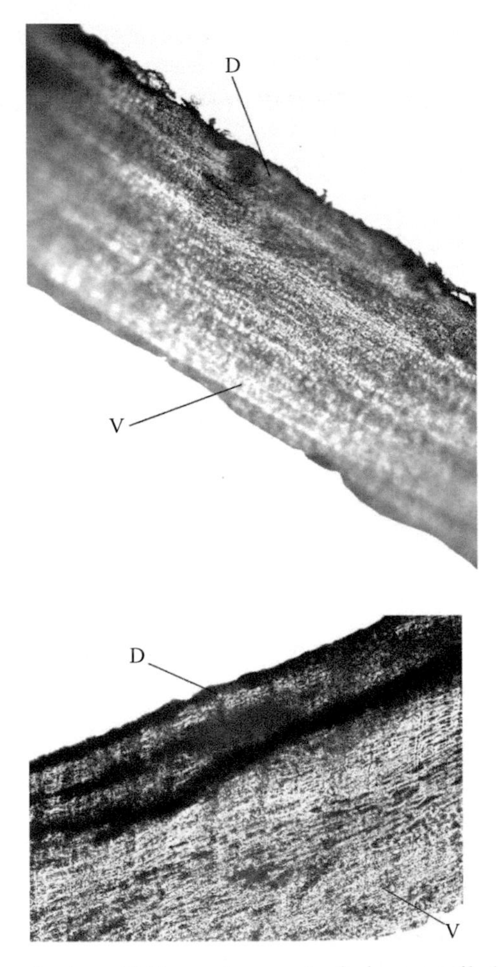

FIGURE 5.11 (See color insert.) Photomicrograph of a human nail plate. D, dorsal layer of the nail plate; V, ventral layer of the nail plate.

comparable in size, weight, and thickness, and more reproducible. A comprehensive statistical evaluation study on the intervariation and the intravariation is still lacking. Factors such as age, sex, and racial difference are likely to alter the ungual drug permeability, but to date, these issues have not been studied in any detail.

Typically, the nail plate can be divided into two distinct layers, dorsal and ventral, based on differential ultrasound transmission and microscopy (Jemec and Serup 1989; Nair et al. 2009e; Wortsman and Jemec 2006). However, there are reports that consider the nail to be a three-layer structure, namely, dorsal, intermediate, and ventral, with a respective weight ratio of 3:5:2 (Kobayashi et al. 1999). The characteristics and properties of the three nail layers were found to be different, for example, the dorsal layer is very hard, while the intermediate plate is soft and flexible, and the ventral layer is a thin lamina. Although the nail plate is a hard membrane, hydration results in considerable softening of this tissue. Moreover, hydration may allow the passage of hydrophilic drugs through the nail plate, even though this permeation route is not very specific. However, there are contradictory reports in the literature in relation to the permeability of hydrophobic molecules across nail plate (Mertin and Lippold 1997a; Walters et al. 1983, 1985). In addition, the physicochemical properties of drug molecules (size, shape, charge, partition coefficient, keratin binding, and hydrophobicity), formulation characteristics (nature of the vehicle, pH, and drug concentration), presence of chemical permeation enhancers, and nail properties (extent of hydration and disease condition) will influence the permeation, retention, and efficacy of antifungal drugs in nail layers to varying extents.

Several investigators have utilized the nail plate as a model for the evaluation of topical nail formulations, screening of chemical enhancers, and in mechanistic studies. Kobayashi et al. (1998, 1999) have assessed the enhancing effect of various vehicles on the trans-ungual delivery of a hydrophilic model drug (5-fluorouracil) and a lipophilic drug (tolnaftate) using a modified side-by-side diffusion cell. Results indicate that the nail behaves as a hydrophilic gel, upon hydration, rather than a lipophilic membrane. Moreover, the authors suggested that the dorsal (upper) layer is the main barrier to drug permeation (Kobayashi et al. 1999).

The permeability of miconazole nitrate, ketoconazole, and itraconazole in the presence of keratolytics (papain, urea, and salicylic acid) was investigated using human nail plates in side-by-side diffusion cells (Quintanar-Guerrero et al. 1998). Hui et al. (2002) have used nail plates to assess the inner nail drug content following topical delivery of radiolabeled urea, ketoconazole, and salicylic acid after dosing twice daily for 7 days, using a hydration-controlled nail incubation system. The lower inner section (ventral) of the nail plate was assayed for absorbed drug content using a unique drilling/removal system. In a later study, the same authors also determined the effects of 2-*n*-nonyl-1,3-dioxolane on the penetration of econazole, an antifungal drug, into the deeper layers of the human nail (Hui et al. 2003).

Kobayashi et al. (2004) have studied the relationship between drug physicochemical properties and the permeability in human nail plate. The authors have also assessed the effect of drug octanol/water partition coefficients on nail permeability coefficients using a homologous series of *p*-hydroxybenzoic acid esters. The effects of molecular weight and drug dissociation on nail permeability were also investigated using the same model. The results suggest that drug flux through nail plates with fungal infection was very similar to that through healthy nail plates (nail plate samples had a thickness range of 800–1200 μm). Hui et al. (2004) have compared the nail penetration of ciclopirox from a marketed gel containing 0.77% of ciclopirox, with an experimental gel containing 2% of ciclopirox, and a marketed lacquer containing 8% of ciclopirox.

5.4 EXCISED CADAVER NAILS AND NAIL CLIPPINGS

Several investigators have utilized excised cadaver nail plates to screen trans-ungual drug delivery enhancers. Evaluation of ungual formulations and physical enhancement techniques has also been conducted using these samples. Testing of topical antifungal agents on excised cadaver nail plates rather than animal tissues is more likely to mimic the *in vivo* situation. The barrier properties (structure) and biochemical components of nail plate (differentiated dead keratinocytes) are unlikely to alter following separation from the body if used relatively soon after collection. Moreover, *in vitro* investigations using excised cadaver nail plates for up to 2 months have indicated that the membrane is stable, and the nail integrity is not affected (Hao and Li 2008a).

Using excised cadaver fingernail and toenail plates, Malhotra and Zatz (2002b) screened a pool of chemicals to identify novel trans-ungual penetration enhancers. The authors found that greater enhancement effects were observed with *N*-(2-mercaptopropionyl) glycine, a mercaptan derivative of an amino acid, in combination with urea, among the chemical agents studied. Using a similar model, the authors also investigated variations in nail permeability as a function of location (toenails versus fingernails). The results showed that there is a large individual variability in nail permeation and that the water flux in the thumbnail is equivalent to that of the toenails. The structure of the thumbnail was also found to differ from other fingernails (Malhotra and Zatz 2002a).

The diffusivity of water in human nail has been determined in excised cadaver fingernails (Gunt et al. 2007). Results suggested that the diffusivity increased with increasing relative humidity. In a later study, the impact of hydration on the permeation of the antifungal drug, ketoconazole, through excised cadaver nails *in vitro* was evaluated. The authors suggested that the formulations or treatments, which increase nail hydration, could have the potential to improve topical therapy for nail diseases if an optimal balance between drug delivery and growth conditions for the dermatophytes could be achieved (Gunt and Kasting 2007).

A series of experiments has been conducted to augment the permeation of antimycotic agents and to determine the mechanism of selected perungual enhancers on trans-ungual drug delivery, using excised cadaver nail plate. The rapid iontophoretic delivery of a potent antifungal agent, terbinafine from a Tween®-water system and a gel formulation, was studied by Nair et al. (2009a, e) using this approach. The results were promising and substantiated earlier findings, suggesting that the use of the trans-ungual iontophoresis approach in nail disease therapy could result in safer and more efficacious outcomes with less frequent treatments (Murthy et al. 2007b). Moreover, the potential of inorganic salts and polyethylene glycols in the trans-ungual delivery of terbinafine was also assessed using excised cadaver nail

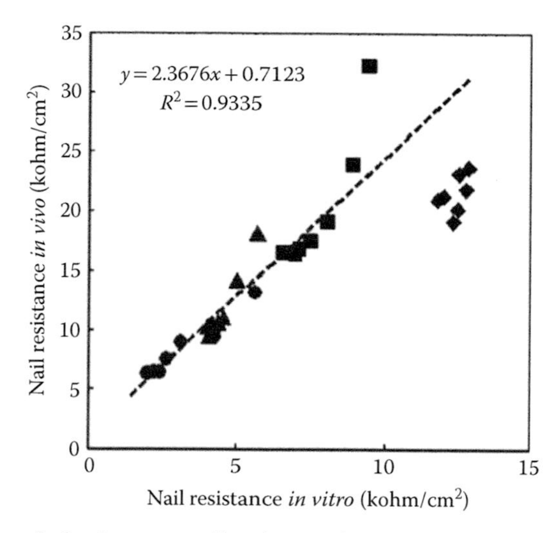

FIGURE 5.12 Correlation between nail resistance *in vivo* and *in vitro* upon 0.5–6-h hydration in phosphate buffered saline (PBS) of 0.01 (◆), 0.06 (■), 0.15 (▲), and 0.6 M (●). Nail resistance was measured by the application of 0.1 mA constant current for 1 min. Each data point represents the mean of four samples. The trend line was calculated using the 0.5–6-h resistance data but excluding the nail resistance in 0.01 M PBS and at 0.5 h in 0.06 M PBS. (Reproduced from Hao et al., *Journal of Pharmaceutical Sciences*, 99, 107–18, 2010. With kind permission of Wiley.)

plates (Nair et al. 2009d, 2010). In another approach, the Murthy group has also evaluated how modulation of nail barrier properties may influence trans-ungual delivery using excised cadaver nail plates (Nair et al. 2009c). Abraded nails (dorsal layer was partially removed by filing) were used and the findings suggested that the barrier properties of the nail are largely associated with the dorsal layer, which is in agreement with earlier reports (Kobayashi et al. 1999).

The Murthy group has also developed a high-throughput technique for screening potential trans-ungual enhancers using cadaver nail plates. These results are discussed in detail in Chapter 4.

It is interesting to note that the effects of hydration and solution ion concentration on the electrical properties of human nail compare well with those for excised cadaver nail plate (Martinsen et al. 2008; Hao et al. 2010) (Figure 5.12).

5.5 DISEASED NAILS

Onychomycosis causes the nail plate to become thick, deform, hyperkeratize, crack, discolor, and to undergo onycholysis (Scher 1996). Disease may also increase nail thickness and lead to alterations in nail structure (Jillson and Piper 1957; Sagher 1948). The presence of dermatophytoma in onychomycosis may also hamper drug permeation. The partial or complete detachment of the nail plate from the nail bed may also hamper drug delivery to the nail bed following topical therapy. Overall, the difference in anatomical and biological features of diseased nails should be expected to influence nail permeability and effective trans-ungual delivery.

The differences in transonychial water loss in diseased nails compared with healthy nails have been measured *in vivo*. The results suggest that the healthy nail is more permeable than diseased human nail plates, likely due to less intimate apposition between nail and nail bed in diseased nails (Kronauer et al. 2001). However, there are clinical situations in which nail plate porosity is increased, which may increase permeability. Very few studies have used diseased nails as an *in vitro* model, likely due to poor availability of such tissues.

Attempts to develop models of diseased nail plates have been reported. One study attempted to infect excised healthy nail plate. Initially, the nail plate was placed on an agar plate previously inoculated with conidia of *Trichophyton mentagrophytes*, and it was then incubated at 28°C for 5 days, which allowed the organism to grow in the nail plate. The antifungal effects of various drugs were then assessed (Nakashima et al. 2002).

In another study, Kobayashi et al. (2004) investigated and compared the permeability of 5-fluorouracil in healthy nails and in nails with mild fungal infection. In this study, the permeability of the nail did not differ significantly between healthy and diseased tissues, most likely due to the fact that the mild infection did not alter the nail plate structure. Similar findings were reported by Nair et al. (2011) where the authors found that terbinafine permeability did not differ between diseased and healthy nail plates obtained from cadaver fingernails and toenail.

5.6 INTACT TOE MODEL

Nair and coworkers have utilized an excised intact cadaver toe (hallux) model to evaluate prototype iontophoretic patch systems. This model is regarded as an *in vitro* model, which closely resembles the nail apparatus *in vivo*, except for the normal physiological clearance component. Figure 5.13 illustrates the prototype iontophoretic

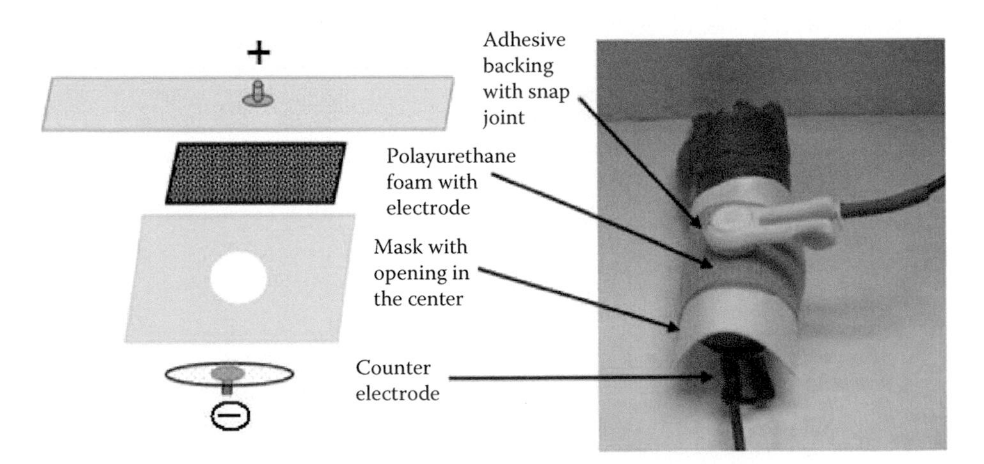

FIGURE 5.13 **(See color insert.)** Schematic representation of the nail-only applicator (left) and a picture demonstrating the application in the ex vivo intact toe model for ungual and transungual iontophoretic delivery of terbinafine (Reproduced from Nair et al., *Pharmaceutical Research*, 26, 2194–201, 2009b. With kind permission of Springer.)

applicator tested using an intact cadaver toe. The drug was loaded in the patch and secured onto the toe. After the treatment, the toenail, lateral nail fold, distal nail fold, proximal nail fold, nail bed, and nail under cuticle were separated, and the amount of drug in various components of the nail apparatus was determined (Nair et al. 2009b).

5.7 DRUG RELEASE STUDIES

Drug retention due to irreversible binding generally hampers the permeation of drug into deeper tissues in topical drug delivery. The drug retention is a factor of its solubility in the membrane, affinity for the membrane, pH-pK_a conditions of the experiment, and drug concentration. Reports suggest that drug accumulated in the nail following prolonged oral therapy was found to persist even after the termination of therapy. Thus, it appears that long-term systemic therapy creates a drug depot in the nail plate, and this may be the reason for the success of long-term oral therapy.

During topical nail drug delivery, it is possible that the binding of the drug to keratin may be reversible with eventual releases of drug into the nail bed over a period of time. This is likely to improve the clinical efficacy of nail therapies. Hence, the maintenance of effective drug levels in nail apparatus will be critical for successful management of nail disease. Thus, the core approach in topical nail therapy should be to produce a drug depot in the nail plate, in addition to actual drug delivery across nail plate.

A quantitative method to assess drug discharge over time was reported by Nair et al. (2009a). Essentially, this model consists of nail tissue mounted on an agar plate. The nail plates were loaded with drug by passive and iontophoresis processes, and the drug-loaded nail plates (~3 × 3 mm) were then placed on Sabouraud dextrose agar (SDA) plates, such that the dorsal side was facing up, and incubated. The amount of drug released into agar after 4-day intervals was determined by extracting the drug from the agar. Nail samples were then transferred to new SDA plates and the process was repeated until the release reached a plateau. The authors observed that the drug loaded by iontophoresis exhibited a biphasic release profile, consisting of an initial rapid release (may be due to the free drug) phase, which was followed by a slower release phase. This type of reversible binding of drug to the nail plate (forming a depot) is likely to prolong the therapeutic effect, owing to continuous drug release during the treatment period.

The same group extended these release studies by using *Trichophyton rubrum* inoculated agar plates. Nail plates loaded with terbinafine were subjected to antifungal activity studies by placing the nail SDA plates inoculated with the test organism (Nair et al. 2009a). The zone of inhibition was measured on each plate after 4 days and the nail was transferred into a new agar plate, and this procedure was continued until there was no measurable zone of inhibition. This study clearly showed that the drug released from the depot is released slowly in a sustained manner, and this may provide prolonged antifungal activity in the target tissue.

5.8 *IN VITRO* PHARMACODYNAMIC STUDIES

Although *in vitro* drug permeation studies provide some support for *in vivo* drug delivery, more specific tests are needed to assess the efficacy of the drug or formulation. Very few clinically relevant experimental models are available for evaluating existing or novel antifungal drugs and formulations in nails. The fungicidal activity of ciclopirox nail lacquer was evaluated in a pigskin model. Stratum corneum was inoculated with *T. mentagrophytes* and the organism was allowed to grow. The nail lacquer was then applied, and at certain time intervals, the epidermis, which contained the fungal elements, was separated from the cutis and the viable cell counts, measured as colony forming units (CFUs), were determined in epidermal homogenates, using a conventional agar plating technique (Markus 1999).

The efficacy of terbinafine in nail plate was tested against the nail invasion by arthroconidia of *T. mentagrophytes* in a model to represent the parasitic infection of dermatophyte fungi (Rashid et al. 1995). Briefly, nail fragments were inoculated with a suspension of arthroconidia and incubated for a period of time. The nail plates were then saturated with terbinafine and the nail colonization was measured by light and electron microscopy. The authors suggested that this *in vitro* model provided an alternative system for studying the activity of antifungal agents in nail.

A nail powder–based *in vitro* culture model for testing of antifungal compounds was developed. It was presumed that the proposed model mimics the course of a natural nail fungal infection, with antifungal treatment starting only after mycelial growth is established. Human nail powder was inoculated with *T. rubrum* and incubated in liquid RPMI 1640 salt medium for extensive and invasive mycelial growth. Antifungal drugs were added and cultures were incubated for 1–4 weeks. Fungal survival was determined by spreading cultures on potato dextrose agar (PDA) plates without drug and measuring colony forming units after 1–4 weeks incubation. Drug activity was expressed as the nail minimum fungicidal concentration required for 99.9% elimination of viable fungus (Osborne et al. 2004).

An *in vitro* tool for assessing antifungal activity using human nail has also been reported. Figure 5.14 illustrates a model of experimental tinea unguium infection. The O-ring was placed on the dorsal surface of the nail fragment and fixed with silicone bonding. The agar base was seeded with a conidial suspension of *T. mentagrophtes* and nail specimens were infected with the fungi by placing nail plate on agar and incubating at 28°C for 5 days. The antifungal formulations were applied inside the O-ring and the whole system was incubated at 28°C for 14 days.

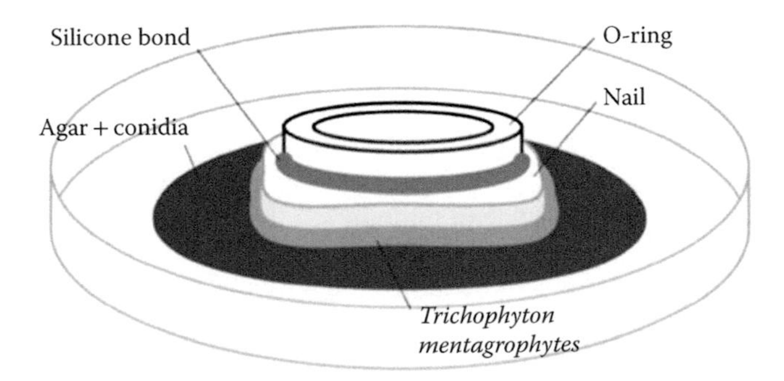

FIGURE 5.14 Experimental tinea unguium by *in vitro Trichophyton mentagrophytes* to the human nail tissue: evaluation system for antifungal activity. (Reproduced from Nakashima et al. (2002) by kind permission of The Japanese Society of Chemotherapy and The Japanese Association for Infectious Diseases.)

The area of distribution of fungal colony with time on the agar surrounding nail tissues and the amount of *T. mentagrophytes* recovered from the nail plate were assessed (Nakashima et al. 2002).

Recently, a novel diffusion cell model has been reported for assessment of antifungal drug delivery to the nail namely the TurChub® model. Sections of human nail were placed over an agar-filled receptor chamber that was previously inoculated with dermatophytes. The test formulation was placed over the nail barrier and then incubated. The drug penetration was approximated by measuring the zone of inhibition in the agar medium present in the receiver compartment.

5.9 *IN VIVO* MODELS

Evaluation of pharmaceutical formulations in animal models is frequently performed to establish safety and efficacy of formulations prior to progressing to clinical investigations. The animal model should be selected to simulate human conditions, and any data gathered should be relevant for scaling up from animals to humans. For regional drug delivery, to make the best use of animal models, it is necessary to understand the similarity and variation between the anatomy and physiology of the target tissue, in this case the nail apparatus. Guinea pigs have been used as an *in vivo* model to investigate the delivery and efficacy of antifungal drugs. The major difference between guinea pig nail and human nail plate is the size of the nail plate, which is too small in the former.

5.9.1 Studies in Animal Models

In one of the studies, the therapeutic efficiency of antifungal agents against tinea unguium infection was evaluated using a guinea pig model. Animals were infected with *T. mentagrophytes* by applying the paper discs (immersed in fungal suspension) between the toes of the hind paw with a foam pad, fixed with an adhesive elastic tape adhered at the site for 21 days. The presence of infection was observed in nails by viewing the nail sections under light microscopy. Guinea pigs with tinea pedis were treated by applying formulations to the nails and whole sole of the foot, once daily. Five days after the last treatment, cultures were taken from the left and right feet by both conventional and modified recovery culture methods. The therapeutic efficacy of formulations was evaluated by evaluation of nail specimens and number of fungal colonies (Tatsumi et al. 2002).

In a later study, guinea pig skin was infected to evaluate the clinical and mycological efficacy of nail formulations. A specific area of the back skin of the guinea pig was clipped and shaved and abraded with sandpaper. The skin was infected by applying a cell suspension of *T. mentagrophytes* conidia. Formulations were applied topically to the infected area, beginning 72-h postinfection, once a day, and the treatment was continued for 7 days. Evaluation of clinical and mycological efficacy was performed 72 h after completion of a 7-day treatment regimen. Clinical evaluation was carried out by visual examination for changes in redness, ulceration, scaling, or hair loss at the site of inoculation. Mycological evaluation was carried out by hair root invasion, that is, a number of hairs were removed from each animal and then implanted on the surface of potato dextrose agar petri dishes and incubated. Hairs showing fungal growth at the hair root were counted. Skin biopsy samples were processed for histopathological examination of the tissues for the presence of fungal elements; inflammation and tissue destruction were evaluated using light microscopy (Ghannoum et al. 2009).

An appropriate animal model for evaluation of perungual formulations has not been identified to date. More extensive research in this area is required to determine whether animal models can realistically be used to simulate the anatomic and physiological features of the nail apparatus.

5.10 CONCLUSIONS

Diffusion cells of different types are used to evaluate the perungual formulations. Diffusion cells conventionally used for skin permeation studies may be modified with a nail adapter to secure the various tissue types used as models of the human nail. Several *in vitro* models such as hoof membranes (cow, pig, sheep, and horse), keratin films, human nail plate, excised human nails, diseased nails, keratin discs, and intact toe models have been reported in the literature. Human cadaver nails and intact toes present the closest simulation of the real-life situation. Currently, there are no suitable animal models available for performing pharmacokinetic and pharmacodynamic investigations of medicated perungual products.

REFERENCES

Abdalghafor, H. M., J. Hadgraft, and M. E. Lane. 2011. Novel Techniques in Probing the Nail Water Content Using Dynamic Vapour Sorption-Near Infrared Spectroscopy. *Proceedings of the Second UKPharmSci Meeting*. Nottingham. August 31–September 2. Abstract 28.

Amichai, B., B. Nitzan, R. Mosckovitz, and A. Shemer. 2010. "Iontophoretic Delivery of Terbinafine in Onychomycosis: A Preliminary Study." *British Journal of Dermatology* 162(1): 46–50.

Bayerl, T. M. 1999. "Measurement of Ciclopirox Permeation through Finger Nail Models by Novel Spectroscopic Techniques." In *Hydroxy-Pyridones As Antifungal Agents with Special Emphasis on Onychomycosis*, edited by S. Shuster, 36–8. Berlin: Springer-Verlag.

Ceschin-Roques, C. G., H. Hanel, S. M. Pruja-Bougaret, J. Luc, J. Vandermander, and G. Michel. 1991. "Ciclopirox Nail Lacquer 8%: In Vivo Penetration into and through Nails and In Vitro Effect on Pig Skin." *Skin Pharmacology and Physiology* 4(2): 89–94.

Donnelly, R. F., P. A. McCarron, J. M. Lightowler, and A. D. Woolfson. 2005. "Bioadhesive Patch-based Delivery of 5-Aminolevulinic Acid to the Nail for Photodynamic Therapy of Onychomycosis." *Journal of Controlled Release* 103: 381–92.

Dubini, F., M. G. Bellotti, A. Frangi, D. Monti, and L. Saccomani. 2005. "In Vitro Antimycotic Activity and Nail Permeation Models of a Piroctone Olamine (Octopirox) Containing Transungual Water Soluble Technology." *Arzneimittelforschung/Drug Research* 55(8): 478–83.

Dutet, J. and M. B. Delgado-Charro. 2010a. "Electroosmotic Transport of Mannitol across Human Nail During Constant Current Iontophoresis." *Journal of Pharmacy and Pharmacology* 62(6): 721–29.

Dutet, J. and M. B. Delgado-Charro. 2010b. "Transungual Iontophoresis of Lithium and Sodium: Effect of pH and Co-ion Competition on Cationic Transport Numbers." *Journal of Controlled Release* 144(2): 168–74.

Ghannoum, M. A, L. Long, and W. R. Pfister. 2009. "Determination of the Efficacy of Terbinafine Hydrochloride Nail Solution in the Topical Treatment of Dermatophytosis in a Guinea Pig Model." *Mycoses* 52: 35–43.

Gunt, H. B., and G. B. Kasting. 2007. "Effect of Hydration on the Permeation of Ketoconazole through Human Nail plate In Vitro." *European Journal of Pharmaceutical Sciences* 32: 254–60.

Gunt, H. B., M. A. Miller, and G. B. Kasting. 2007. "Water Diffusivity in Human Nail Plate." *Journal of Pharmaceutical Sciences* 96: 3352–62.

Gupchup, G. V. and J. L. Zatz. 1999. "Structural Characteristics and Permeability Properties of the Human Nail: A Review." *Journal of Cosmetic Science* 50: 363–85.

Hao, J., and S. K. Li. 2008a. "Trans-ungual Iontophoretic Transport of Polar Neutral and Positively Charged Model Permeants: Effects of Electrophoresis and Electroosmosis." *Journal of Pharmaceutical Sciences* 97: 893–905.

Hao, J., and S. K. Li. 2008b. "Mechanistic study of electroosmotic transport across hydrated nail plates: Effects of pH and ionic strength." *Journal of Pharmaceutical Sciences* 97: 5186–97.

Hao, J., K. A. Smith, and S. K. Li. 2009. "Iontophoretically Enhanced Ciclopirox Delivery into and across Human Nail Plate." *Journal of Pharmaceutical Sciences* 98: 3608–16.

Hao, J., K. A. Smith, and S. K. Li. 2010. "Time-Dependent Electrical Properties of Human Nail upon Hydration In Vivo." *Journal of Pharmaceutical Sciences* 99: 107–18.

Hemidy P. Y., S. Makki, P. Muret, J. P. Chaumont, and J. Millet. 1994. "The Use of Sheep Hoof Plates for Substituting Human Nails in Trans-ungual Absorption Studies." *Journal of Applied Cosmetology* 12: 73–84.

Hui, X., T. C. K. Chan, S. Barbadillo, C. Lee, H. I. Maibach, and R. C. Wester. 2003. "Enhanced Econazole Penetration into Human Nail by 2-n-Nonyl-1,3-Dioxolane." *Journal of Pharmaceutical Sciences* 92: 142–8.

Hui, X., Z. Shainhouse, H. Tanojo, A. Anigbogu, G. E. Markus, H. I. Maibach, and R. C. Wester. 2002. "Enhanced Human Nail Drug Delivery: Nail Inner Drug Content Assayed by New Unique Method." *Journal of Pharmaceutical Sciences* 91: 189–95.

Hui X., R. C. Wester, S. Barbadillo, C. Lee, B. Patel, M. Wortzmann, E. H. Gans, and H. I. Maibach. 2004. "Ciclopirox Delivery into the Human Nail Plate." *Journal of Pharmaceutical Sciences* 93: 2545–8.

Jemec G. B. E., and J. Serup. 1989. "Ultrasound Structure of the Human Nail Plate." *Archives of Dermatology* 125: 643–6.

Jillson O. F., and E. L. Piper. 1957. "The Role of the Saprophytic Fungi in the Production of Eczematous Dermatitis". *Journal of Investigative Dermatology* 28: 137–45.

Khengar, R. H., M. B. Brown, R. B. Turner, M. J. Traynor, K. B. Holt, and S. A. Jones. 2010. "Free Radical Facilitated Damage of Ungual Keratin." *Free Radical Biology and Medicine* 49(5): 865–71.

Khengar R. H., S. A. Jones, R. B. Turner, B. Forbes, and M. B. Brown. 2007. "Nail Swelling As a Pre-Formulation Screen for the Selection and Optimisation of Ungual Penetration Enhancers." *Pharmaceutical Research* 24: 2207–12.

Kierstan K. T. E., A. E. Beezer, J. C. Mitchell, J. Hadgraft, S. L. Raghavan, and A. F. Davis. 2001. "UV Spectrophotometry Study of Membrane Transport Processes with a Novel Diffusion Cell." *International Journal of Pharmaceutics* 229: 87–94.

Kim J., C. H. Lee, and H. Choi. 2001. "A Method to Measure the Amount of Drug Penetrated across the Nail Plate." *Pharmaceutical Research* 18: 1468–71.

Kobayashi, Y., T. Komatsu, M. Sumi, S. Numajiri, M. Miyamoto, D. Kobayashi, K. Sugibayashi, and Y. Morimoto. 2004. "In Vitro Permeation of Several Drugs through the Human Nail Plate: Relationship between Physicochemical Properties and Nail Permeability of Drugs." *European Journal of Pharmaceutical Science* 21: 471–7.

Kobayashi Y., M. Miyamoto, K. Sugibayashi, and Y. Morimoto. 1998. "Enhancing Effect of N-acetyl-L-cysteine or 2-Mercaptoethanol on the In Vitro Permeation of 5-Fluorouracil or Tolnaftate through the Human Nail Plate." *Chemical and Pharmaceutical Bulletin* 46: 1797–802.

Kobayashi Y., M. Miyamoto, K. Sugibayashi, and Y. Morimoto. 1999. "Drug Permeation through the Three Layers of the Human Nail Plate." *Journal of Pharmacy and Pharmacology* 51: 271–8.

Kronauer, C., M. Gfesser, J. Ring, and D. Abeck. 2001. "Transonychial Water Loss in Healthy and Diseased Nails." *Acta Dermato-Venereologica*: 81: 175–7.

Malhotra G. G., and J. L. Zatz. 2002a. "Characterization of the Physical Factors Affecting Nail Permeation Using Water As a Probe." *Journal of Cosmetic Science*: 51: 367–77.

Malhotra G. G., and J. L. Zatz. 2002b. "Investigation of Nail Permeation Enhancement by Chemical Modification Using Water As a Probe." *Journal of Pharmaceutical Sciences* 91: 312–23.

Markus, A. 1999. "Hydroxy-Pyridones. Outstanding Biological Properties." In *Hydroxy-Pyridones As Antifungal Agents with Special Emphasis on Onychomycosis*, edited by S. Shuster, 1–10. Berlin: Springer-Verlag.

Martinsen O. G., S. Grimnes, and S. H. Nilsen. 2008. "Water Sorption and Electrical Properties of a Human Nail." *Skin Research and Technology* 14: 142–6.

Mertin, D., and B. C. Lippold. 1997a. "In-Vitro Permeability of the Human Nail and of a Keratin Membrane from Bovine Hooves: Prediction of the Penetration Rate of Antimycotics through the Nail Plate and Their Efficacy." *Journal of Pharmacy and Pharmacology* 49: 866–72.

Mertin D., and B. C. Lippold. 1997b. "In-Vitro Permeability of the Human Nail and of a Keratin Membrane from Bovine Hooves: Influence of the Partition Coefficient Octanol/Water and the Water Solubility of Drugs on Their Permeability and Maximum Flux." *Journal of Pharmacy and Pharmacology* 49: 30–4.

Mertin, D., and B. C. Lippold. 1997c. "In-Vitro Permeability of the Human Nail and of a Keratin Membrane from Bovine Hooves: Penetration of Chloramphenicol from Lipophilic Vehicles and a Nail Lacquer." *Journal of Pharmacy and Pharmacology* 49: 241–5.

Mohorcic, M., A. Torkar, J. Friedrich, J. Kristl, and S. Murdan. 2007. "An Investigation into Keratinolytic Enzymes to Enhance Ungual Drug Delivery." *International Journal of Pharmaceutics* 332: 196–201.

Monti, D., L. Saccomani, P. Chetoni, S. Burgalassi, M. F. Saettone, and F. Mailland. 2005. "In Vitro Trans-ungual Permeation of Ciclopirox from a Hydroxypropyl Chitosan-Based, Water-Soluble Nail Lacquer." *Drug Development and Industrial Pharmacy* 31: 11–7.

Monti, D., L. Saccomani, P. Chetoni, S. Burgalassi, S. Senesi, E. Ghelardi, and F. Mailland. 2010. "Hydrosoluble Medicated Nail Lacquers: In Vitro Drug Permeation and Corresponding Antimycotic Activity." *British Journal of Dermatology* 162: 311–7.

Murthy, S. N., D. C. Waddell, H. N. Shivakumar, A. Balaji, and C. P. Bowers. 2007a. "Iontophoretic Permselective Property of Human Nail." *Journal of Dermatology Science*: 46: 150–2.

Murthy, S. N., D. E. Wiskirchen, and C. P. Bowers. 2007b. "Iontophoretic Drug Delivery across Human Nail." *Journal of Pharmaceutical Sciences* 96: 305–11.

Nair, A. B., C. Bireswar, and S. N. Murthy. 2010. "Effect of Polyethylene Glycols on the Trans-ungual Delivery of Terbinafine." *Current Drug Delivery* 7: 407–14.

Nair, A. B., H. D. Kim, B. Chakraborty, J. Singh, M. Zaman, A. Gupta, P. M. Friden, and S. N. Murthy. 2009a. "Ungual and Trans-Ungual Iontophoretic Delivery of Terbinafine for the Treatment of Onychomycosis." *Journal of Pharmaceutical Sciences* 98: 4130–40.

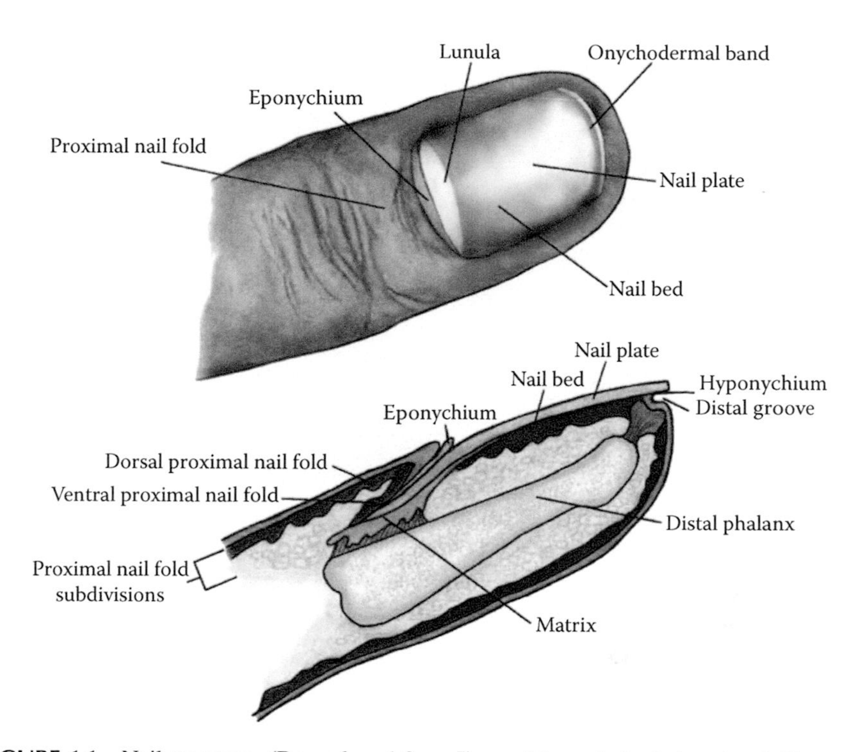

FIGURE 1.1 Nail structure. (Reproduced from Jiaravuthisan et al., *J. Am. Acad. Dermatol.* 57, 1, 2007. With kind permission from the American Academy of Dermatology, Inc.)

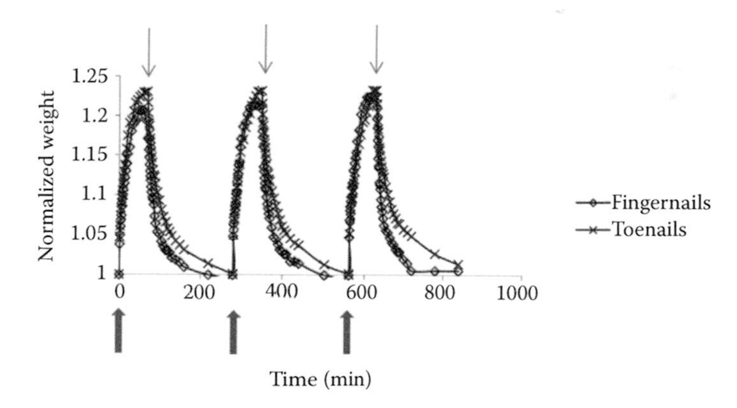

FIGURE 1.7 Rapid uptake and loss of water when a nail clipping is sequentially placed in (up arrows) and removed from water (down arrow) to air-dry. The curves show averages (±sd) of three clippings each for fingernail and toenail. Fingernail and toenail clippings absorb similar amounts of water, with consequent increase in mass of between 20% and 23%. When placed in water, nail clippings' mass increases rapidly and saturation is reached after about 70 min. Once removed from water, most of the absorbed water is lost rapidly although the initial nail mass is not regained for 3 h. Toenail clippings seem to lose the last of the absorbed water at a slower rate compared to fingernail clippings.

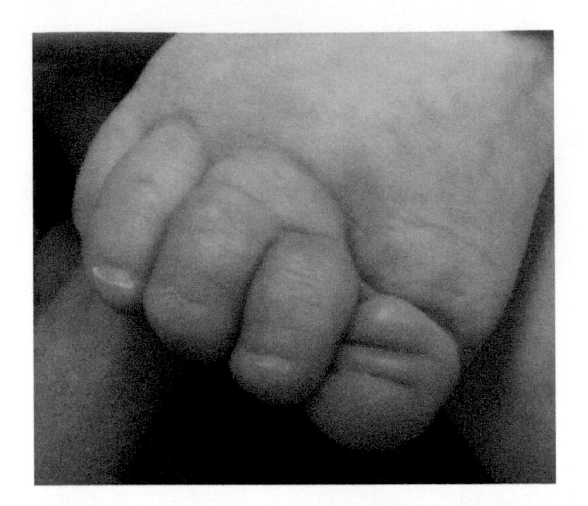

FIGURE 1.12 Partially and totally absent toenails in a newborn on digits 3 and 1, respectively.

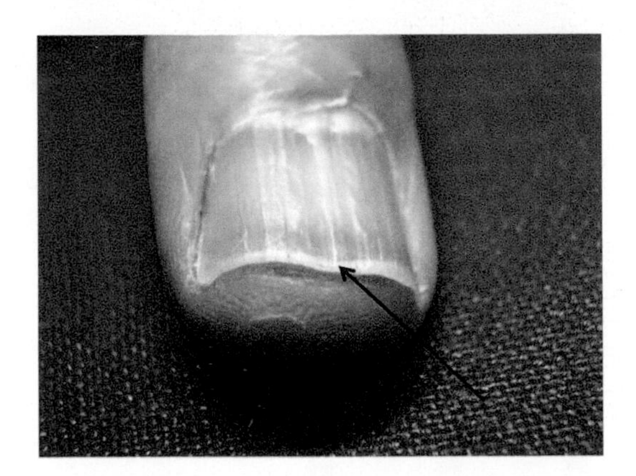

FIGURE 1.13 Longitudinal grooves in a fingernail.

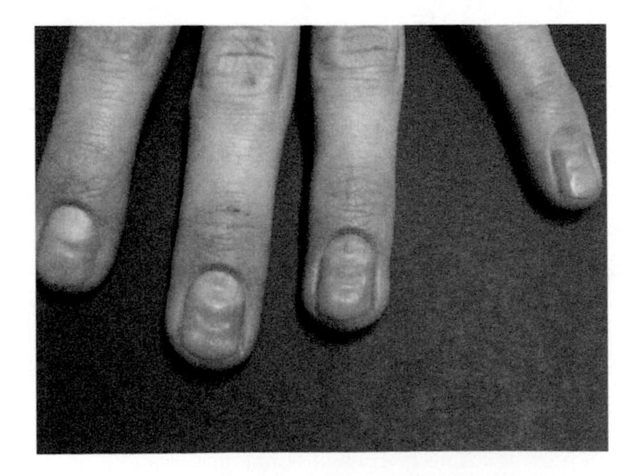

FIGURE 1.14 Transverse ridges and grooves on fingernails.

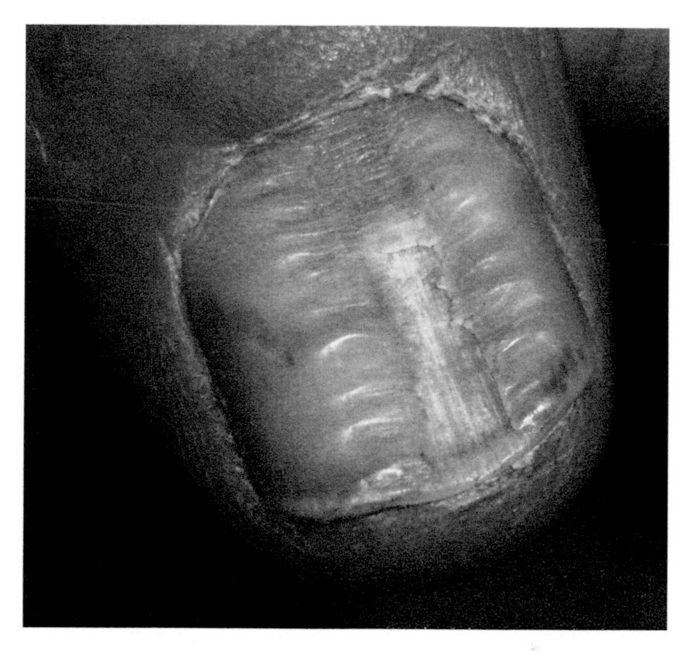

FIGURE 1.15 Transverse grooves in the nail plate; part of the nail plate has been lost due to the damage.

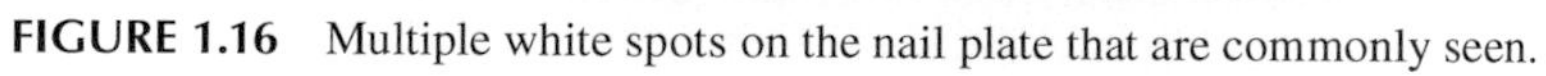

FIGURE 1.16 Multiple white spots on the nail plate that are commonly seen.

FIGURE 1.17 Damaged to the nail folds without any overt abnormality on the nail plate.

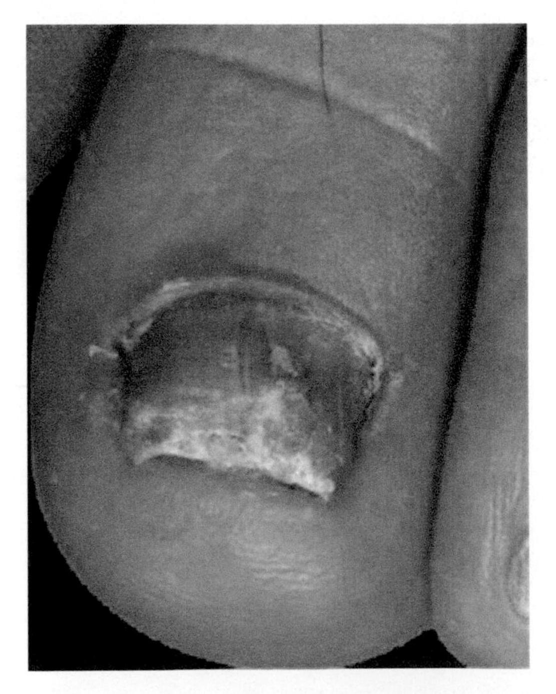

FIGURE 1.18 Distal lateral onychomycosis. (Reproduced from Welsh et al., *Clin. Dermatol.*, 28, 151, 2010. With kind permission from Elsevier Inc.)

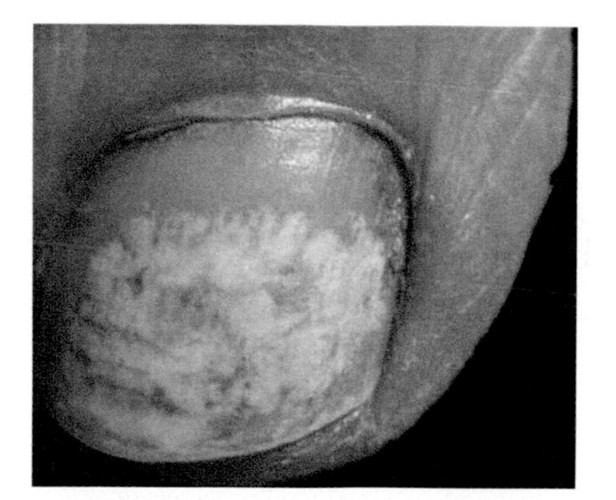

FIGURE 1.19 Superficial white onychomycosis. (Reproduced from Welsh et al., *Clin. Dermatol.*, 28, 151, 2010. With kind permission from Elsevier Inc.)

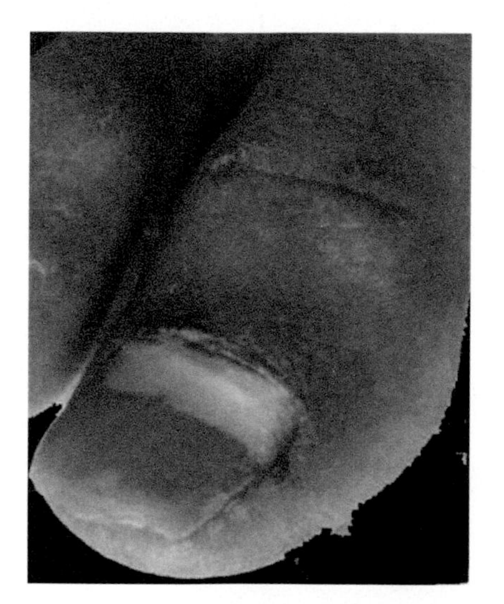

FIGURE 1.20 Proximal white onychomycosis. (Reproduced from Welsh et al., *Clin. Dermatol.*, 28, 151, 2010. With kind permission from Elsevier Inc.)

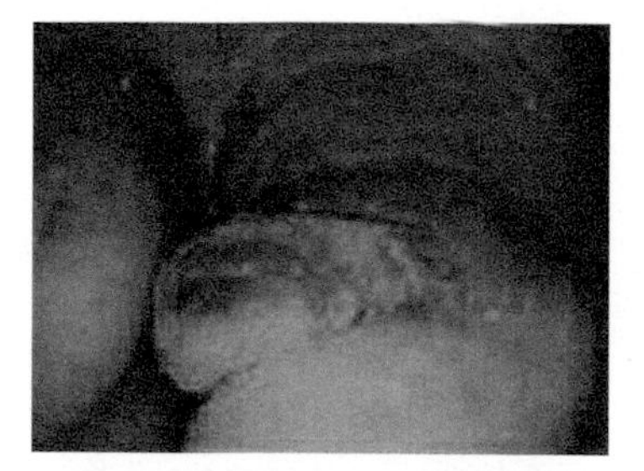

FIGURE 1.21 Total dystrophic onychomycosis. (Reproduced from Welsh et al., *Clin. Dermatol.*, 28, 151, 2010. With kind permission from Elsevier Inc.)

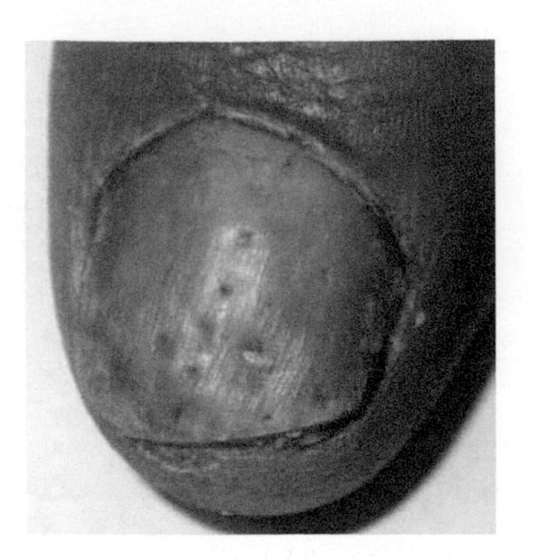

FIGURE 1.22 Nail pitting. (Reproduced from Jiaravuthisan et al., *J. Am. Acad. Dermatol.* 57, 1, 2007. With kind permission from the American Academy of Dermatology, Inc.)

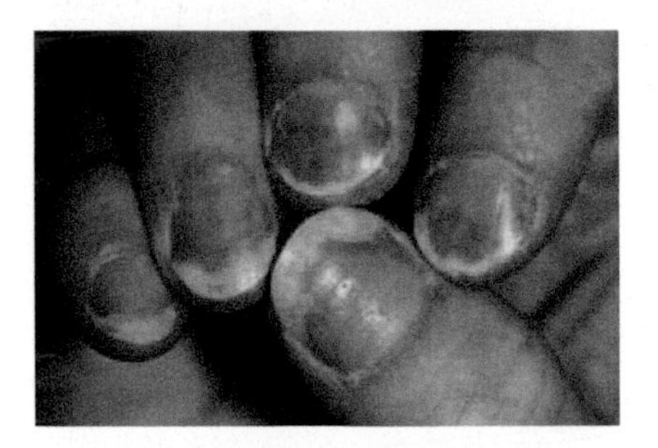

FIGURE 1.23 Onycholysis at the distal nail plate. (Reproduced from Jiaravuthisan et al., *J. Am. Acad. Dermatol.* 57, 1, 2007. With kind permission from the American Academy of Dermatology, Inc.)

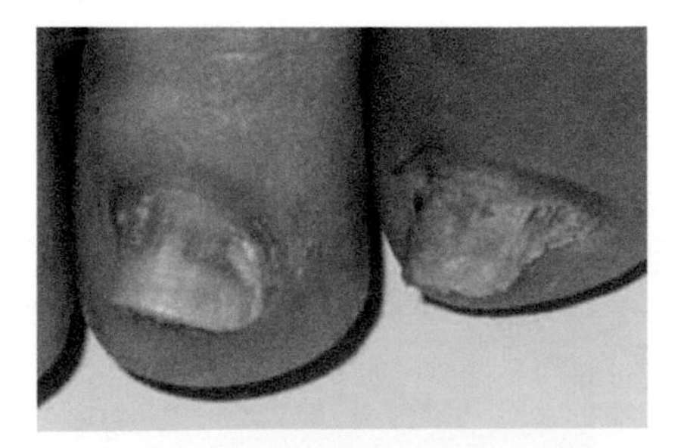

FIGURE 1.24 Subungual hyperkeratosis. (Reproduced from Jiaravuthisan et al., *J. Am. Acad. Dermatol.* 57, 1, 2007. With kind permission from the American Academy of Dermatology, Inc.)

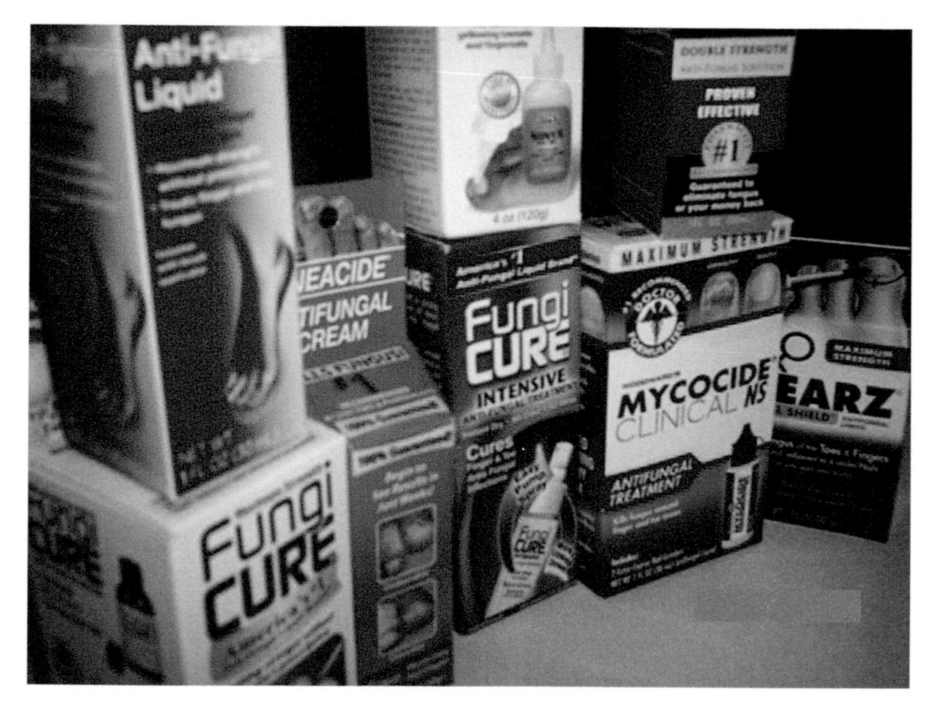

FIGURE 3.1 Topical nail formulations.

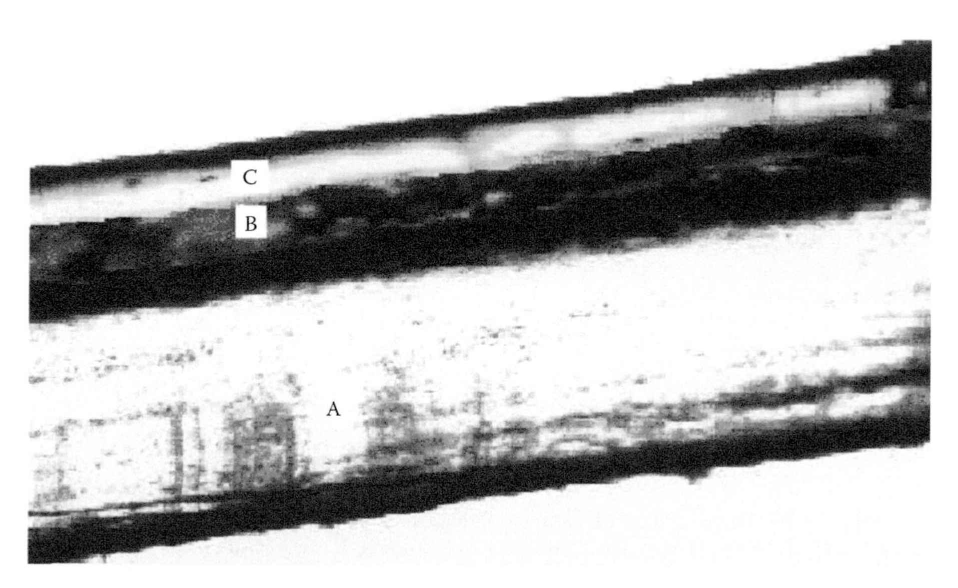

FIGURE 3.2 Cross-section of the nail (A) applied with the bilayered nail lacquer showing the drug-loaded hydrophilic lacquer (B) over laid with the hydrophobic vinyl lacquer (C). Reproduced from Shivakumar et al., *Journal of Pharmaceutical Sciences* 99 (10): 4267–76, 2010. With kind permission from Wiley Interscience®.

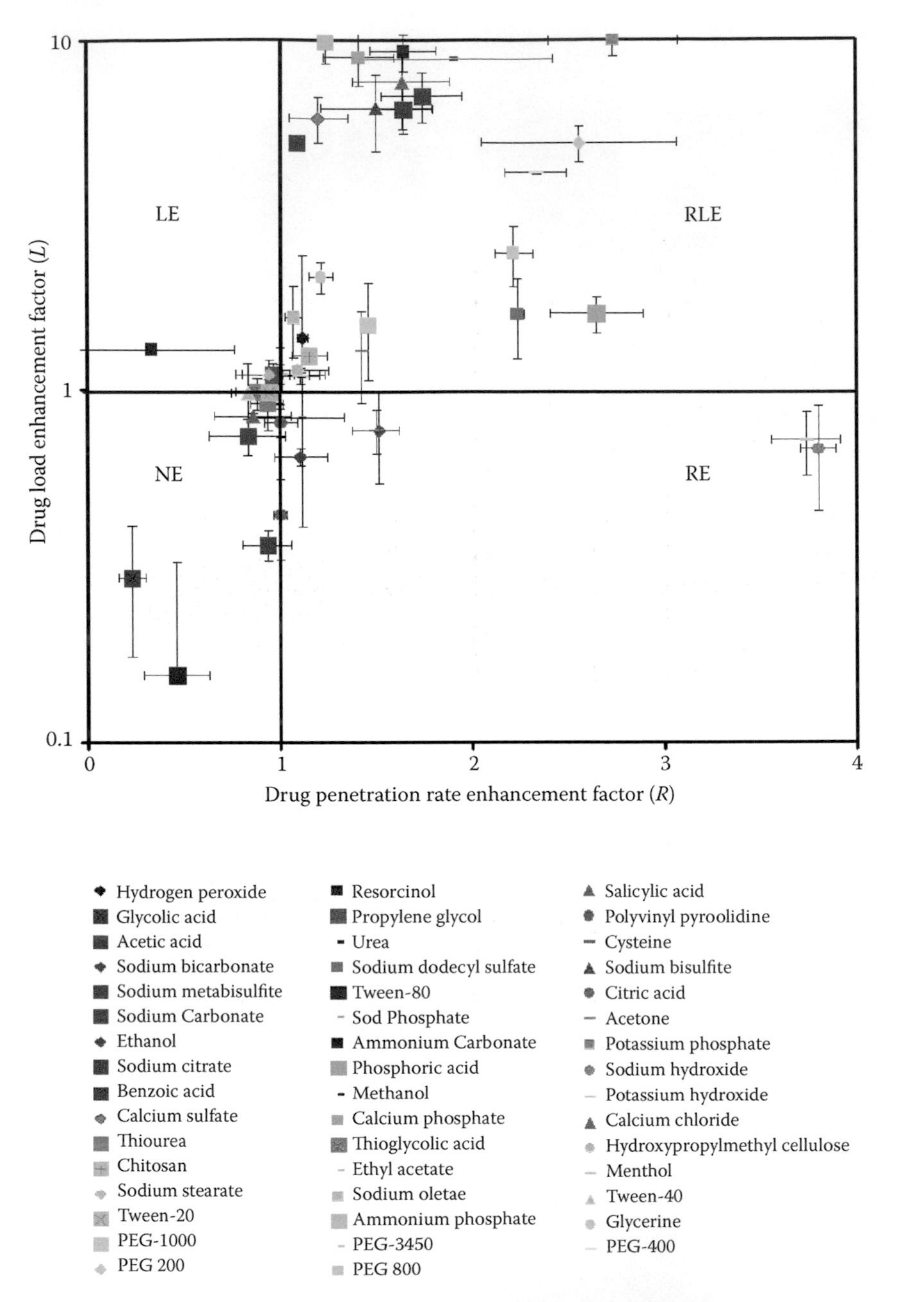

FIGURE 4.12 Categorization of chemical enhancers into four groups based on the *R* and the *L* values determined from the TranScreen-N™ studies. *L* is the load enhancement factor, *R* is the rate enhancement factor, and EQ_{24} is the drug delivery enhancement factor. The different quadrants represent non-enhancers (NE) (quadrant 1), drug load enhancers (LE) (quadrant 2), uptake rate and drug load enhancers (RLE) (quadrant 3), and uptake rate enhancers (RE) (quadrant 4). (Reproduced from Murthy et al., *Journal of Pharmaceutical Sciences*, 98, 4264–71, 2009. With kind permission of Wiley.)

FIGURE 5.2 Picture of horizontal Franz diffusion cell setup. (A) Chambers of the diffusion cell and (B) nail adapter.

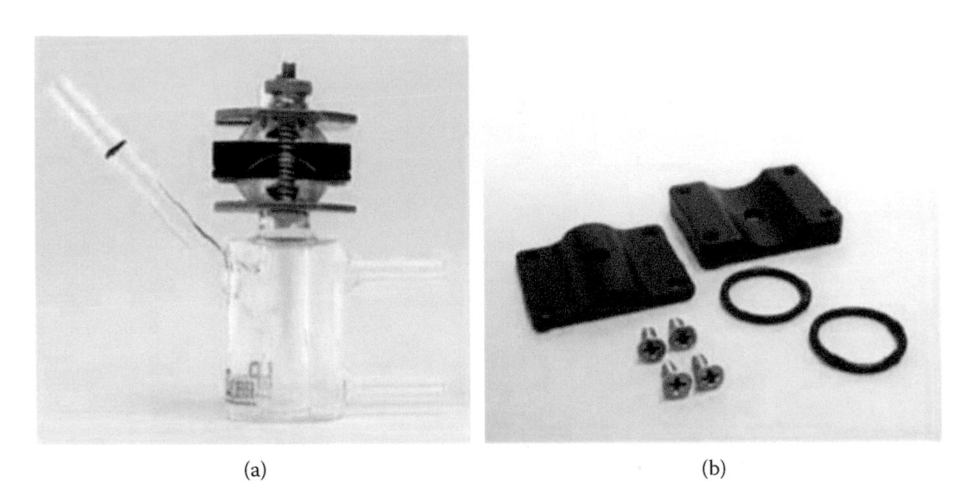

(a) (b)

FIGURE 5.3 Picture of vertical Franz diffusion cell with (a) nail adapter and (b) different components of a nail adapter. (By kind permission of PermeGear, Inc., Hellertown, PA.)

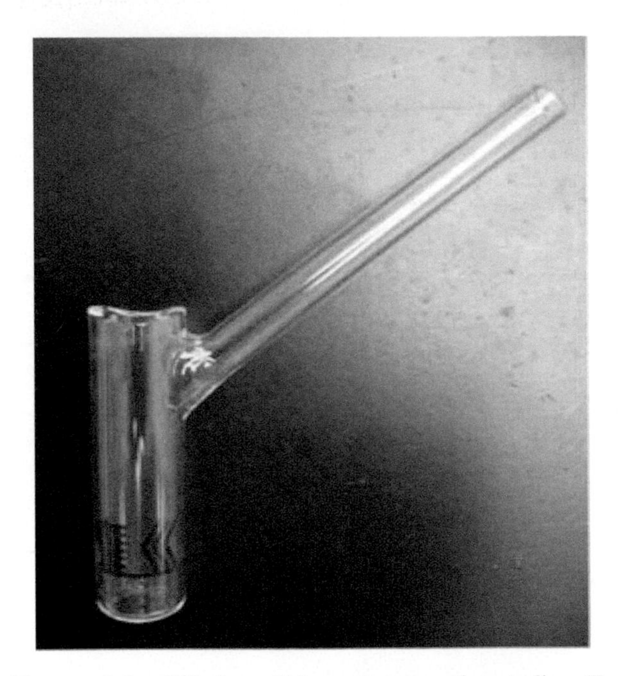

FIGURE 5.6 Picture of the diffusion cell for nail permeation studies. (Reproduced from Gunt and Kasting, *European Journal of Pharmaceutical Sciences*, 32, 254–60, 2007. With kind permission from Elsevier.)

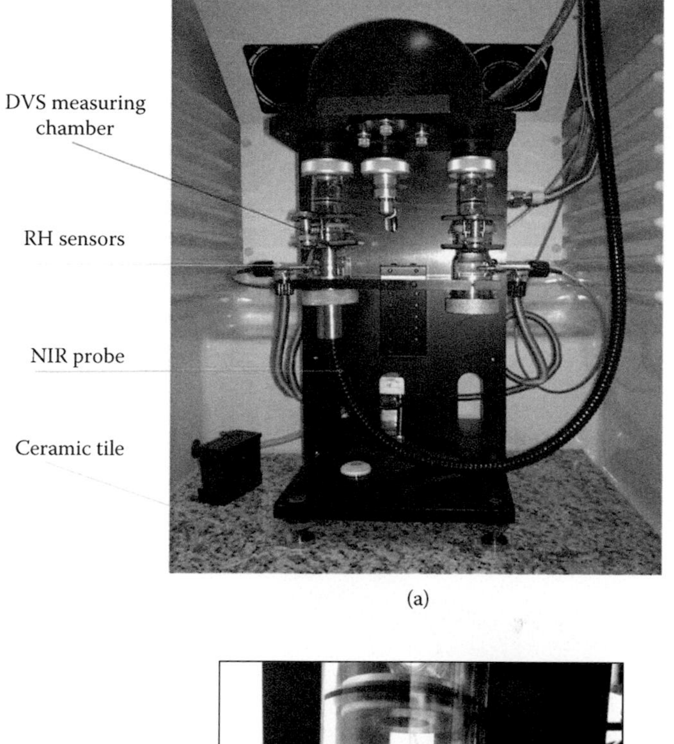

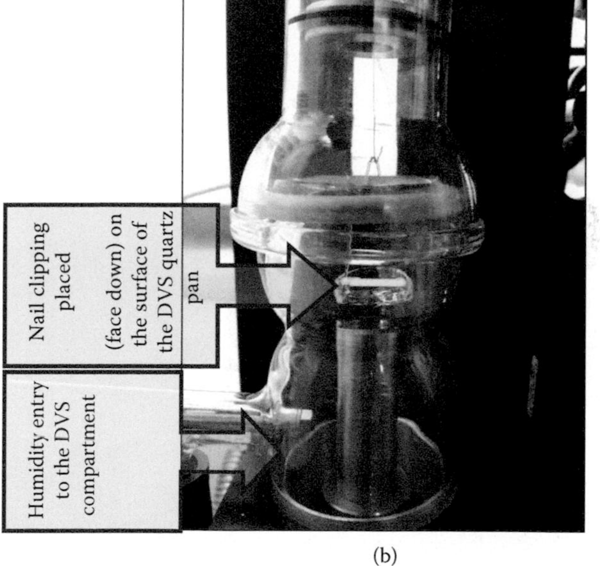

FIGURE 5.7 (a) Dynamic vapor sorption (DVS)/near-infrared (NIR) device. RH, relative humidity. (b) Mounting of nail sample in dynamic vapor sorption (DVS) device.

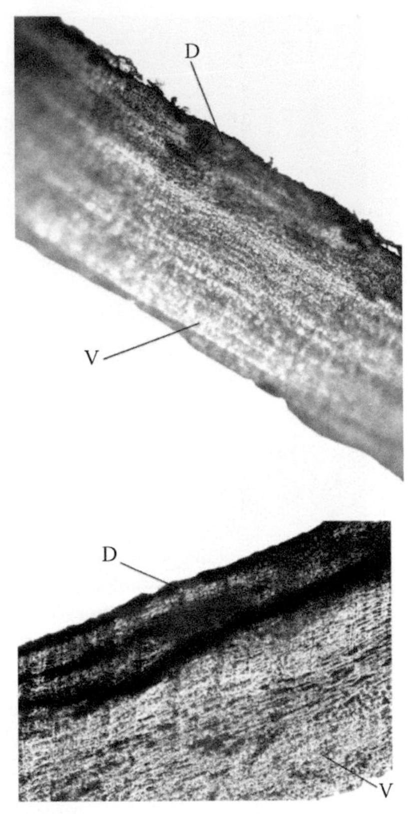

FIGURE 5.11 Photomicrograph of a human nail plate. D, dorsal layer of the nail plate; V, ventral layer of the nail plate.

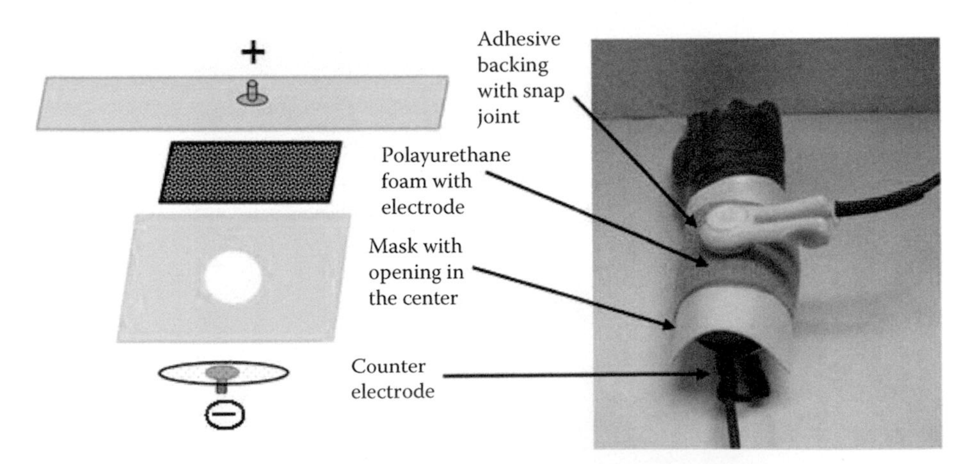

FIGURE 5.13 Schematic representation of the nail-only applicator (left) and a picture demonstrating the application in the ex vivo intact toe model for ungual and trans-ungual iontophoretic delivery of terbinafine. (Reproduced from Nair et al., *Pharmaceutical Research*, 26, 2194–201, 2009b. With kind permission of Springer.)

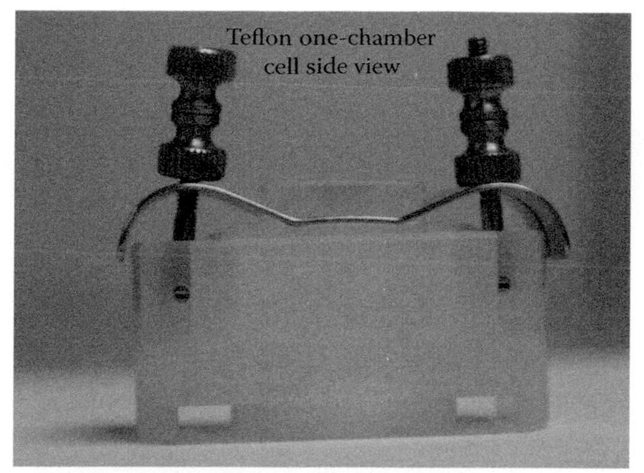

(a)

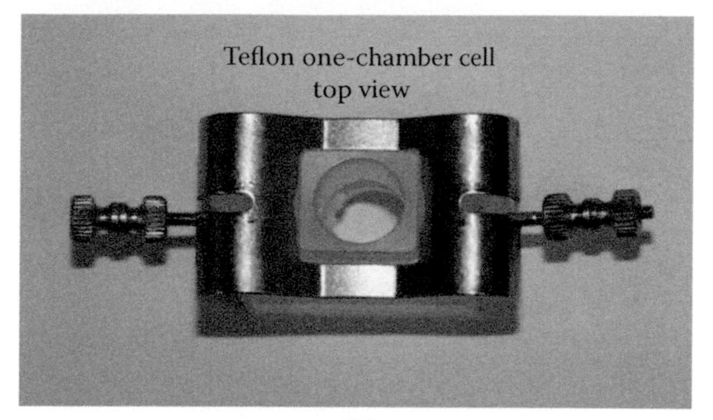

(b)

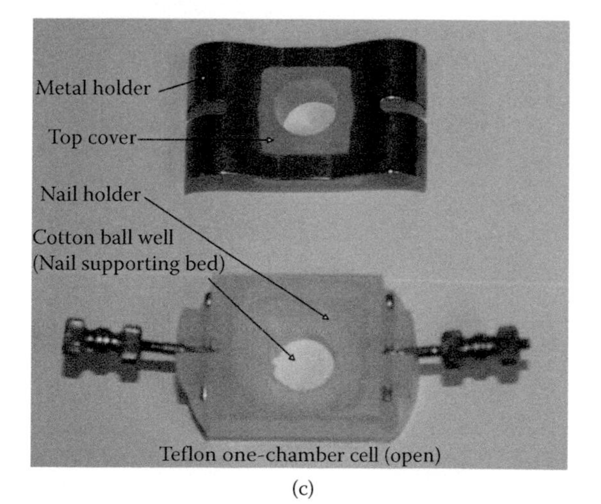

(c)

FIGURE 6.1 Hydration-controlled nail incubation system: (a) side view of a Teflon one-chamber diffusion cell side; (b) top view of Teflon one-chamber diffusion cell; (c) Teflon one-chamber diffusion cell components. (From Elkeeb et al., *International Journal of Pharmaceutics*, 384, 1–8, 2010.)

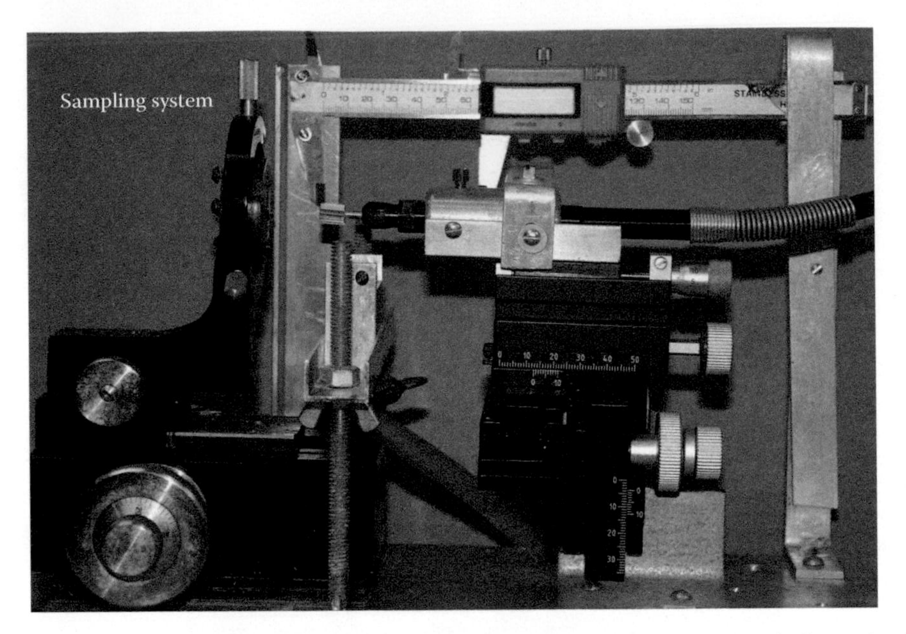

FIGURE 6.2 Nail sampling device. (From Elkeeb et al., *International Journal of Pharmaceutics*, 384, 1–8, 2010.)

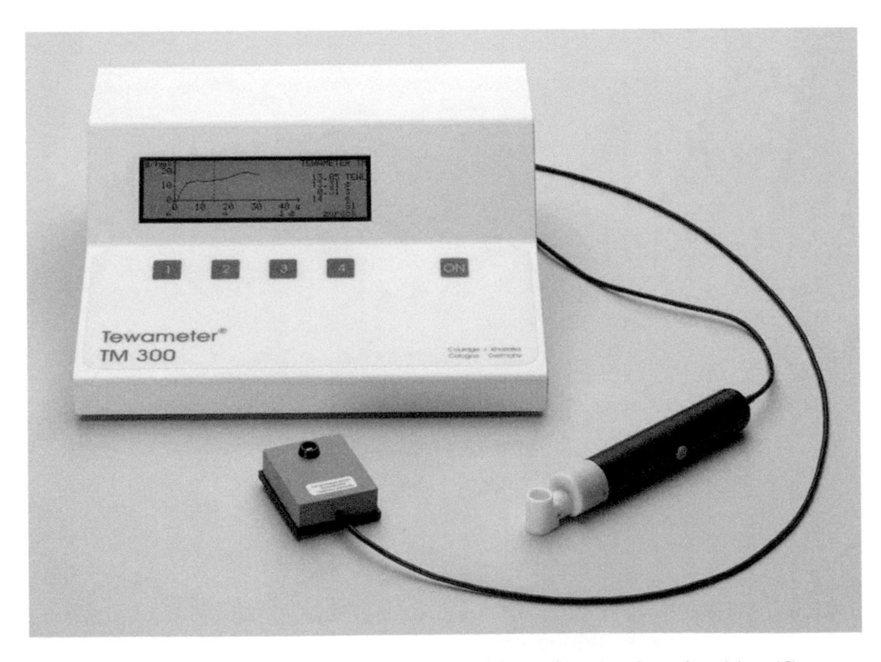

FIGURE 7.1 Tewameter for measuring transepidermal water loss in skin. (Courtesy of Courage + Khazaka electronic GmbH, Köln, Germany.)

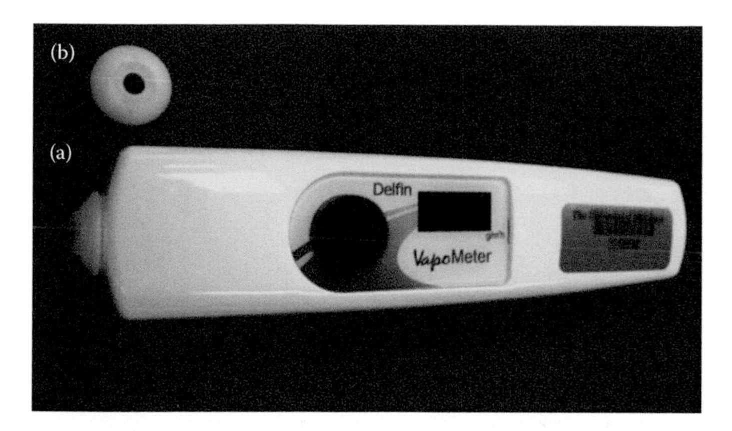

FIGURE 7.2 (a) VapoMeter™ used to measure transepidermal water loss or transonychial water loss. (b) Nail adapter. (Courtesy of Delfin Technologies Ltd, Kuopio, Finland.)

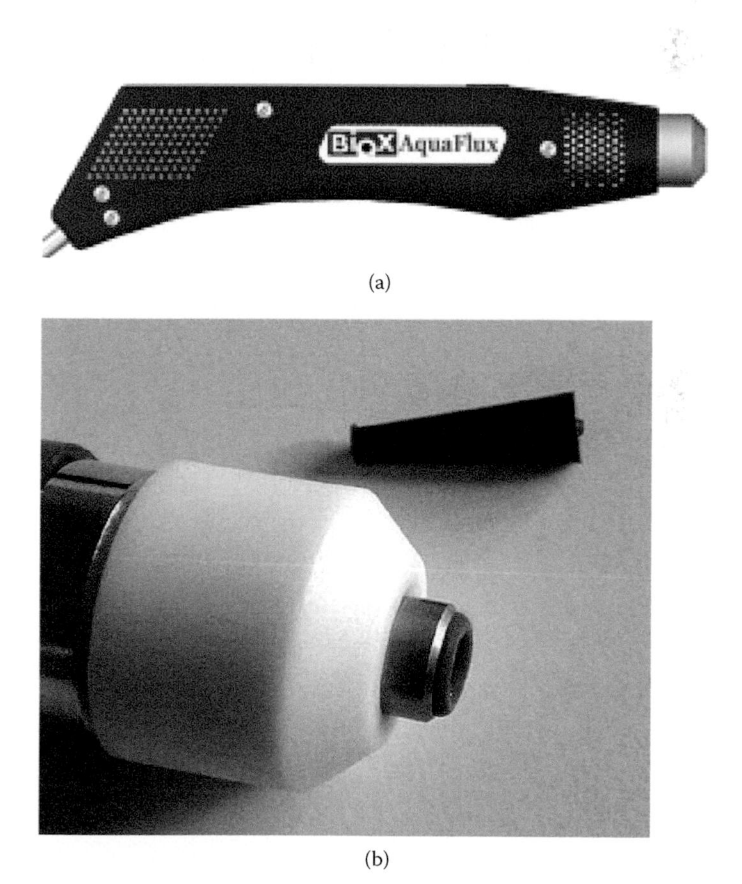

FIGURE 7.3 (a) AquaFlux AF200; (b) TOWL measurement adaptor. (Courtesy of Biox Systems Ltd, London, UK.)

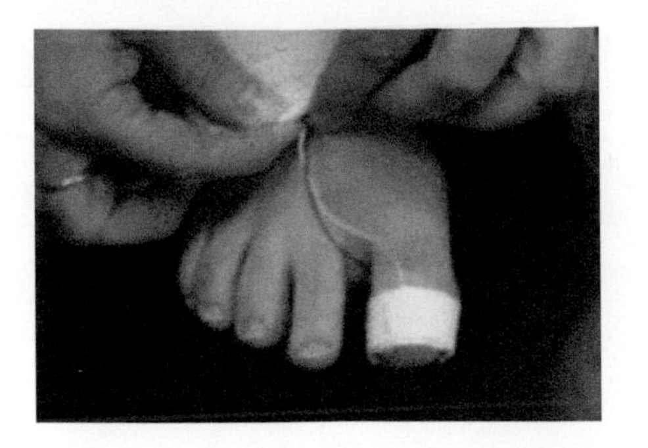

(a)

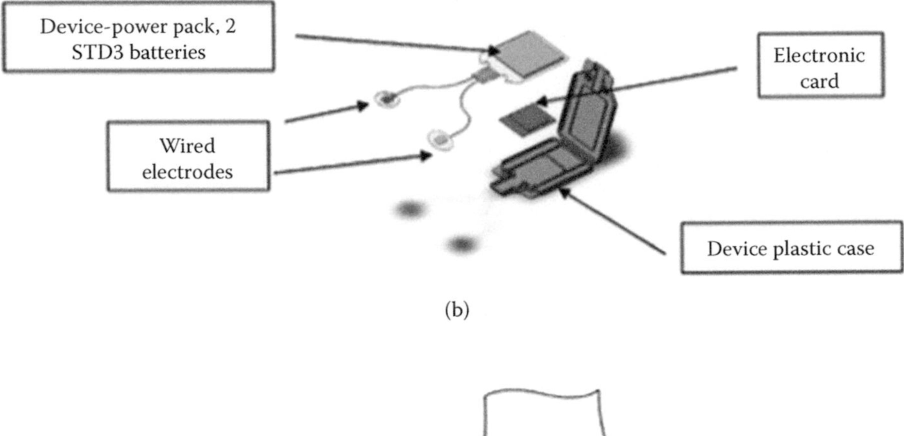

(b)

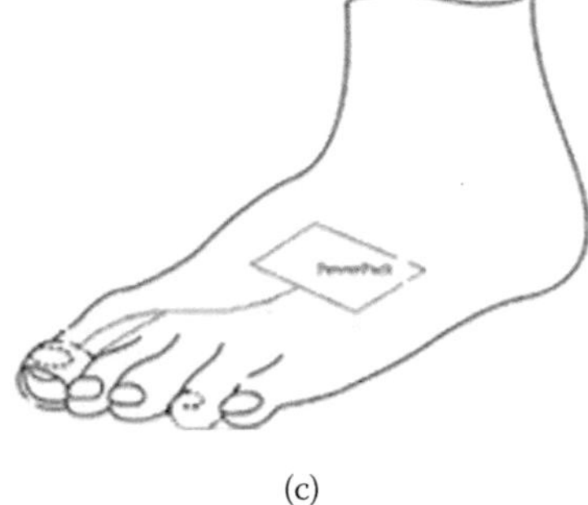

(c)

FIGURE 8.1 (a) Appearance of the iontophoretic device. (b) Device components. (c) Device application: the plastic case containing an electronic card and power pack is situated on the upper foot; the treatment patch is placed on the infected nail and two-wired electrodes applied above (active) and beneath (counter) the toe. (Reprinted from Amichai et al., *British Journal of Dermatology*, 162, 46–50, 2010. With permission of Wiley and Blackwell publisher.)

Nair, A. B., H. D. Kim, S. P. Davis, R. Etheredge, M. Barsness, P. M. Friden, and S. N. Murthy. 2009b. "An Ex Vivo Toe Model Used to Assess Applicators for the Iontophoretic Ungual Delivery of Terbinafine." *Pharmaceutical Research* 26: 2194–201.

Nair, A. B., S. M. Sammeta, H. D. Kim, B. Chakraborty, P. M. Friden, and S. N. Murthy. 2009c. "Alteration of the Diffusional Barrier Property of the Nail Leads to Greater Terbinafine Drug Loading and Permeation." *International Journal of Pharmaceutics* 375: 22–7.

Nair, A. B., S. M. Sammeta, S. R. K. Vaka, and S. N. Murthy. 2009d. "A Study on the Effect of Inorganic Salts in Trans-ungual Drug Delivery of Terbinafine." *Journal of Pharmacy and Pharmacology* 61: 431–7.

Nair, A. B., S. R. K. Vaka, and S. N. Murthy. 2011. "Trans-ungual Delivery of Terbinafine in Onychomycotic Nails." *Drug Development and Industrial Pharmacy* 37 (10): 1253–8.

Nair, A. B., S. R. K. Vaka, S. M. Sammeta, H. D. Kim, P. M. Friden, C. Bireswar, and S. N. Murthy. 2009e. "Trans-Ungual Iontophoretic Delivery of Terbinafine." *Journal of Pharmaceutical Sciences* 98: 1788–96.

Nakashima, T., A. Nozawa, T. Ito, and T. Majima. 2002. "Experimental Tinea Unguium Model to Assess Topical Antifungal Agents Using the Infected Human Nail with Dermatophyte In Vitro." *Journal of Infection and Chemotherapy* 8: 331–5.

Osborne, C. S., I. Leitner, B. Favre, and N. S. Ryder. 2004. "Antifungal Drug Response in an In Vitro Model of Dermatophyte Nail Infection." *Medical Mycology* 42: 159–63.

Pittrof, F., J. Gerhards, W. Erni, and G. Klecak. 1992. "Loceryl Nail Lacquer—Realization of a New Galenical Approach to Onychomycosis Therapy." *Clinical and Experimental Dermatology* 1: 26–8.

Quintanar-Guerrero, D., A. Ganem-Quintanar, P. Tapia-Olguin, Y. N. Kalia, and P. Buri. 1998. "The Effect of Keratolytic Agents on the Permeability of Three Imidazole Antimycotic Drugs through the Human Nail." *Drug Development and Industrial Pharmacy* 24: 685–90.

Rashid, A., E. M. Scott, and M. D. Richardson. 1995. "Inhibitory Effect of Terbinafine on the Invasion of Nails by *Trichophyton mentagrophytes*." *Journal of the American Academy of Dermatology* 33: 718–23.

Reichl, L. S., and C. C. Müller-Goymann. 2009. Keratin Film from Human Hair As a Novel Model in Replacing Human Nail Plate for Studying Drug Permeation. *Proceedings of the 2nd PharmSciFair*. Nice, June 8–12.

Sagher, F. 1948. "Histologic Examinations of Fungus Infections of the Nails." *Journal of Investigative Dermatology* 11: 337–54.

Scher, R. K. 1996. "Onychomycosis: A Significant Medical Disorder." *Journal of the American Academy of Dermatology* 35: S2–5.

Shivakumar, H. N., S. R. K. Vaka, N. V. S. Madhav, H. Chandra, and S. N. Murthy. 2010. "Bilayered Nail Lacquer of Terbinafine for Treatment of Onychomycosis." *Journal of Pharmaceutical Sciences* 99(10): 4267–76.

Smith, K. A., J. Hao, and S. K. Li. 2009. "Effects of Ionic Strength on Passive and Iontophoretic Transport of Cationic Permeant across Human Nail." *Pharmaceutical Research* 26(6): 1446–55.

Smith, K. A., J. Hao, and S. K. Li. 2010. "Influence of pH on Transungual Passive and Iontophoretic Transport." *Journal of Pharmceutical Science* 99(4): 1955–67.

Tatsumi, Y. M., Y. T. Arika, and H. Yamaguchi. 2002. "In Vivo Fungicidal Effect of KP-103 in a Guinea Pig Model of Interdigital Tinea Pedis Determined by Using a New Method for Removing the Antimycotic Carryover Effect." *Microbiology and Immunology* 46: 433–9.

Togni, G., and F. Mailland. 2010. "Antifungal Activity, Experimental Infections and Nail Permeation of an Innovative Ciclopirox Nail Lacquer Based on a Water Soluble Biopolymer." *Journal of Drugs in Dermatology* 9: 525–30.

Vaka, S. R. K., S. N. Murthy, J. H. O'Haver, and M. A. Repka. 2011. "A Platform for Predicting and Enhancing Model Drug Delivery across Human Nail Plate." *Drug Development and Industrial Pharmacy* 37(1): 72–9.

Vejnovic, I., L. Simmler, and G. Betz. 2010a. "Investigation of Different Formulations for Drug Delivery through the Nail Plate." *International Journal of Pharmaceutics* 386: 185–94.

Vejnovic, I., L. Simmler, and G. Betz. 2010b. "Investigation of Different Formulations for Drug Delivery through the Nail Plate." *International Journal of Pharmaceutics*, 386: 67–76.

Vignardet, C., Y. C. Guillaume, L. Michel, J. Friedrich, and J. Millet. 2001. "Comparison of Two Hard Keratinous Substrates Submitted to the Action of a Keratinase Using an Experimental Design." *International Journal of Pharmacy* 224: 115–22.

Walters, K. A., G. L. Flynn, and J. R. Marvel. 1981. "Physicochemical Characterization of the Human Nail: I. Pressure Sealed Apparatus for Measuring Nail Plate Permeabilities." *Journal of Investigative Dermatology* 76: 76–9.

Walters, K. A., G. L. Flynn, and J. R. Marvel. 1983. "Physicochemical Characterization of the Human Nail Permeation Pattern for Water and the Homologous Alcohols and Differences With Respect To the Stratum Corneum." *Journal of Pharmacy and Pharmacology* 35: 28–33.

Walters, K. A., G. L. Flynn, and J. R. Marvel. 1985. "Physicochemical Characterization of the Human Nail Solvent Effects on the Permeation of Homologous Alcohols." *Journal of Pharmacy and Pharmacology* 37: 771–5.

Wortsman, X., and G. B. E. Jemec. 2006. "Ultrasound Imaging of Nails." *Dermatology Clinical* 24: 323–38.

Yang, D., J. P. Chaumont, S. Makki, and J. Millet-Clerc. 1997. "A New Simulation Model for Studying In Vitro Topical Penetration of Antifungal Drugs into Hard Keratin." *Journal de Mycologie Medicale* 7: 195–8.

6 Hydration-Controlled Nail System for the Evaluation of Topical Formulations and a Novel Nail Sampling Device

Rania Elkeeb, Xiaoying Hui, and Howard I. Maibach

CONTENTS

6.1 INTRODUCTION

Due to the impermeable nature of the nail plate, topical therapy of the nail in onychomycosis has been a real challenge. Current research on nail permeation focuses on overcoming the nail barrier properties by means of chemical treatments (Kobayashi et al. 1998), penetration enhancers, as well as physical and mechanical approaches. Efficacy of these methods has been mainly carried out by the most common *in vitro* screening, which is a conventional long-term diffusion study using conventional diffusion cells, which has limitations. As a result of failure or poor performance of topical treatment, a lot of interest has been directed toward finding a more efficient and robust method for screening.

6.2 CURRENT *IN VITRO* UNGUAL DRUG DELIVERY TECHNIQUES

6.2.1 Conventional Permeation Studies with Diffusion Cells

Horizontal and vertical cells discussed in Chapter 5 are used for flux determination. Drug is initially applied to the nail dorsal surface. Permeation is measured by sampling the solution on the ventral nail plate at successive points in time and by calculating the drug flux through the nail (Mohorcic et al. 2007). This method bears similarities to skin penetration studies. However, skin penetration studies are not limited simply to the determination of flux but also include the separation of skin layers to quantify drug concentration in each layer.

6.2.2 Hydration-Controlled Nail Incubation System

Hui et al. (2005) has modified the conventional diffusion cell by using a Teflon one-chamber diffusion cell (PermeGear, Inc., Hellertown, PA) along with a cotton ball soaked in normal saline in order to maintain the nail at physiological levels of temperature and humidity.

6.2.3 TranScreen-N™-Based Delivery

TranScreen-N™ is a simple and rapid technique developed by Murthy and co-workers for screening trans-ungual drug permeation enhancers by utilizing a simple microwell plate-based high-throughput method (Murthy et al. 2009).

These three available drug permeation assay methods cannot be directly compared, but we can make possible extrapolations of what can be learned from the derived data (Table 6.1).

TABLE 6.1
Current *In Vitro* Ungual Drug Delivery Techniques

In Vitro Method	Attributes	Limitations	References
Conventional diffusion method	• Measures diffusion flux	• Time consuming • Does not provide drug measurement at site of disease	Mohorcic et al. (2007)
Hydration-controlled system and nail drilling method	• Unique in sampling the nail core. • Provides drug measurement at site of disease • Use of cotton ball provides moisture	• Measured by radioactivity	Hui et al. (2002)
TransScreen method	• Simpler • Rapid • Less expensive • A preformulation tool	• No diffusion flux • Not providing drug measurement at site of disease	Murthy et al. (2009)

6.3 METHODOLOGY

6.3.1 Hydration-Controlled Nail Incubation System

Nail samples were incubated in a Teflon one-chamber diffusion cell (PermeGear, Inc.) to keep the nail at physiological levels of temperature and humidity. The setup is as follows: the nail surface (top center) was open to air and the inner surface was in contact with a small cotton ball acting as a nail supporting bed (Figure 6.1a through c). This small cotton ball was kept wet by normal saline and was placed in

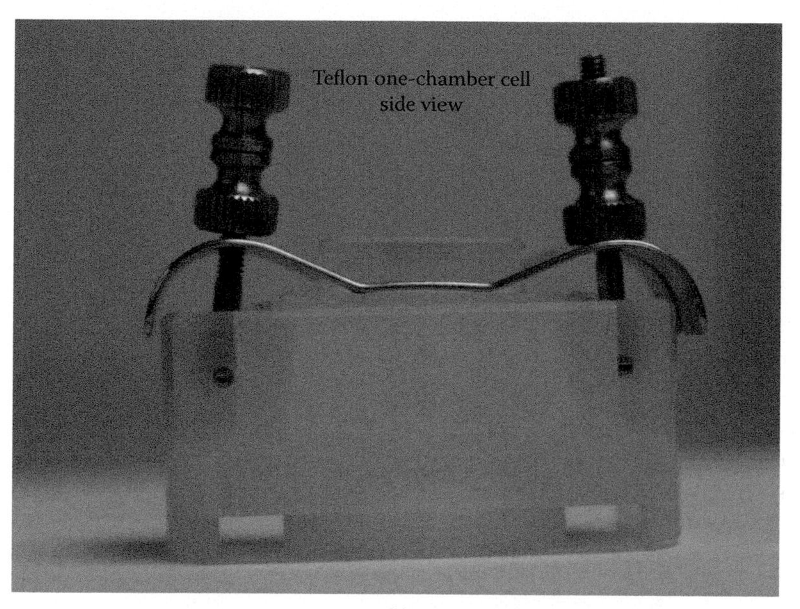

(a)

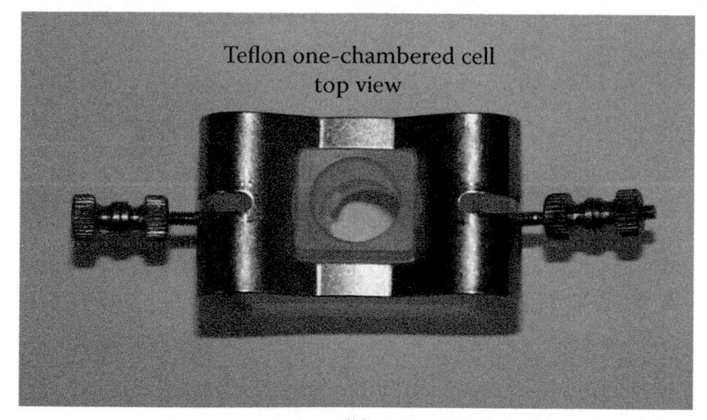

(b)

FIGURE 6.1 **(See color insert.)** Hydration-controlled nail incubation system: (a) side view of a Teflon one-chamber diffusion cell side; (b) top view of Teflon one-chamber diffusion cell; (c) Teflon one-chamber diffusion cell components. (From Elkeeb et al., *International Journal of Pharmaceutics*, 384, 1–8, 2010.)

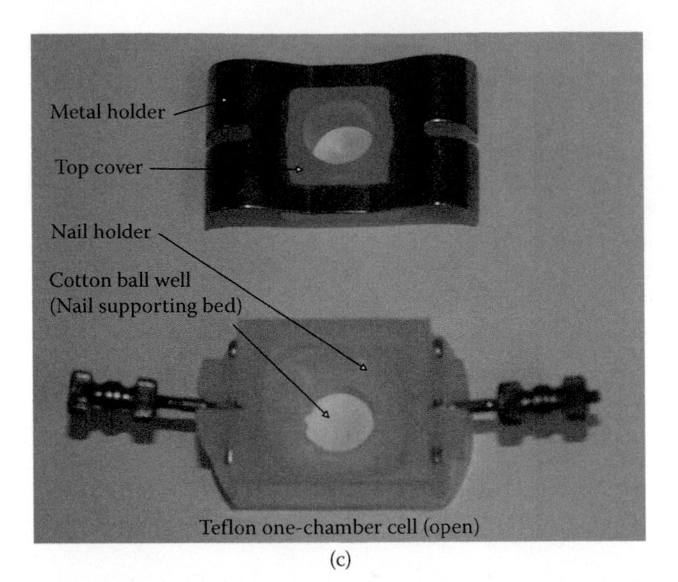

(c)

FIGURE 6.1 (*Continued*)

TABLE 6.2
Hydration of the Nail Plate and Nail Bed (Hui et al. 2005)

Measurement			Hydration (AU)	
Source	***n***	**Time**	**Nail Plate**	**Nail Bed**
Human cadaver	6	24-hour postmortem	7.6 ± 0.9	99.9 ± 8.9
Diffusion cell	8	Twice a day for 7 days	8.5 ± 2.4	109.9 ± 6.2

the chamber beneath the nail plate, and the degree of hydration was monitored and controlled during the experiment (Hui et al. 2003). Hydration of the nail plate and the supporting cotton bed was measured with Corneometer CM 820 (Courage & Khazaka, Cologne, Germany).

Table 6.2 shows the average hydration of the wet cotton balls (109 ± 6.2 AU), which resembles the average hydration of the human nail bed (99.9 ± 8.9 AU) measured from fresh human cadavers, where AU is an arbitrary unit, a digital expression of capacitance (Hui et al. 2002).

During the experiment, the holding tank temperature was 25 ± 2°C, and the relative humidity was 44 ± 8%. The advantage of this incubation system is that it is nonocclusive and hydration-controlled, where normal physical condition is attained.

6.3.2 Test Formulations

Human cadaver nails were used in an experiment where a radiolabeled compound (3H or 14C) was added to the test formulation, and the radioactivity of the samples was measured by a liquid scintillation counting. Test formulations were applied to the surface of the nails at a predetermined time interval for a certain number of days, and prior to daily dosing, the surface of the nails was washed by cotton tips in a cycle to mimic daily bathing as follows: a dry tip, a tip wet with 50% Ivory liquid soap (Proctor & Gamble, Cincinnati, OH) followed by a tip wet with distilled water and then by another one wet with distilled water and finally a dry tip. Those samples from each cycle are pooled and collected by breaking the cotton tip off and placing it into scintillation glass vials. A predetermined aliquot amount of methanol is added to the vials to extract the test material. Then, the radioactivity of each sample is measured with a liquid scintillation counter (Hui et al. 2005).

6.3.3 Nail Sampling Procedure

This unique technique was developed by Hui et al. (2005) in order to determine the drug concentration within the nail where the disease resides. This drilling system makes it possible to sample the inner core of the nail without disturbing the nail surface (Figure 6.2), thus determining drug penetration into the nail. This is made possible by the use of a micrometer-precision nail sampling instrument that enables finely controlled drilling into the nail and collection of the powder generated by the drilling process.

FIGURE 6.2 (See color insert.) Nail sampling device. (From Elkeeb et al., *International Journal of Pharmaceutics*, 384, 1–8, 2010.)

This nail sampling instrument is composed of the following two parts (Hui et al. 2003):

1. A nail sampling stage
2. A drill

The nail sampling stage is composed of the following parts:

1. Copper nail holder
2. Three adjustments control vertical movement
   a. The first coarse adjustment (on the top) is used to change the copper cell and for taking powder samples from the capture device.
   b. The sec coarse adjustment (lower) is used in sampling and allows movement of 25 mm.
   c. The third fine adjustment (lower) is also used in sampling and provides movement of 0.20 mm.
3. Device for nail powder capture is located between the copper cell and the cutter.

The inner shape of the capture device is an inverted funnel with the end connected to a vacuum pump. A filter paper is placed inside the funnel in order to capture the nail powder samples while sampling. The nail is fastened in a cutting holder below the cutter and surrounded by the funnel containing the filter paper. Upon drilling, the vacuum draws the powder debris onto the filter paper to collect it and measure it.

Drilling of the nail occurs through the ventral surface, and thus, the dorsal surface and the ventrally accessed nail core can be assayed separately. The dorsal surface sample contains residual drug while the core from the ventral side provides drug measurement at the site of the disease. This method permits drug measurement in the intermediate nail plate, which was previously impossible (Elkeeb et al. 2010).

Table 6.3 shows drug penetration into human nail into dorsal and ventral layers of the nail plate in microgram equivalent per milligram of nail.

TABLE 6.3
Drug Ungual Absorption Parameters

		Nail Absorption (µg eq/mg nail)			
Drug	**MW**	**Dorsal**	**Ventral**	**Remainder**	**References**
Ciclopirox A	207.27	71.7 ± 16.79	0.55 ± 0.28	51.97 ± 62.44	Hui et al. (2004)
Econazole	381.68	0.25 ± 0.03	11.15 ± 2.56	0.16 ± 0.14	Hui et al. (2003)
Terbinafine	291.43	0.81 ± 0.13	0.74 ± 0.38	0.003 ± 0.001[b]	Unpublished
AN2690	152.00	25.6 ± 8.80	20.46 ± 4.72	26.06 ± 12.41	Hui et al. (2007a)
Ketoconazole	531.43	0.38 ± 0.12	0.81 ± 0.39	0.18 ± 0.09[b]	Unpublished
Panthenol	205.25	10.05 ± 3.52	4.97 ± 0.90	0.80 ± 0.33[b]	Hui et al. (2007b)
Drug X[a]	350.00	187 ± 90	349 ± 158	188 ± 64	Unpublished
Drug Y[a]	200.00	8.5 ± 4.71	1.92 ± 0.30	61.16 ± 30.48[b]	Unpublished

[a] Proprietary compound.
[b] Cotton Ball (reservoir data).

6.4 CONCLUSION

The effectiveness of topical therapy in treating nail disorders has been limited by the limited permeability nature of the nail. Hui and coworkers developed an assay system for drug content in the inner nail bed where the infection resides using a micrometer-controlled drilling instrument that collects powder samples of the permeated radiolabled drug (Hui et al. 2005). The intermediate nail plate has finally been breached, and it is now available for serious attention.

Additional experiments comparing *in vivo* to *in vitro* results should verify or deny the veracity of the model.

REFERENCES

Elkeeb, R., A. Alikhan, L. Elkeeb, X. Hui, and H. I. Maibach. 2010. "Trans-ungual Drug Delivery: Current Status." *International Journal of Pharmaceutics* 384: 1–8.

Hui, X., S. J. Baker, R. C. Wester, S. Barbadillo, A. K. Cashmore, V. Sanders, K. M. Hold et al. 2007a. "In Vitro Penetration of a Novel Oxaborole Antifungal (AN2690) into the Human Nail Plate." *Journal of Pharmaceutical Sciences* 96: 2622–31.

Hui, X., T. C. Chan, S. Barbadillo, C. Lee, H. I. Maibach, and R. C. Wester. 2003. "Enhanced Econazole Penetration into Human Nail by 2-n-Nonyl-1,3-Dioxolane." *Journal of Pharmaceutical Sciences* 92: 142–8.

Hui, X., S. B. Hornby, R. C. Wester, S. Barbadillo, Y. Appa, and H. Maibach. 2007b. "In Vitro Human Nail Penetration and Kinetics of Panthenol." *International Journal of Cosmetic Science* 29: 277–82.

Hui, X., Z. Shainhouse, H. Tanojo, A. Anigbogu, G. E. Markus, H. I. Maibach, and R. C. Wester. 2002 "Enhanced Human Nail Drug Delivery: Nail Inner Drug Content Assayed by New Unique Method." *Journal of Pharmaceutical Sciences* 91: 189–95.

Hui, X., R. C. Wester, S. Barbadillo, C. Lee, B. Patel, M. Wortzmman, E. H. Gans, and H. I. Maibach. 2004. "Ciclopirox Delivery into the Human Nail Plate." *Journal of Pharmaceutical Sciences* 93: 2545–8.

Hui, X., R. C. Wester, S. Barbadillo, and H. Maibach. 2005. "Nail Penetration: Enhancement of Topical Delivery of Antifungal Drugs by Chemical Modification of the Human Nail." In *Textbook of Cosmetic Dermatology*, edited by R. Baran and H. I. Maibach, 57–63. New York, NY: Taylor and Francis.

Kobayashi, Y., M. Miyamoto, K. Sugibayashi, and Y. Morimoto. 1998. "Enhancing Effect of N-Acetyl-L-Cysteine or 2-Mercaptoethanol on the In Vitro Permeation of 5-Fluorouracil or Tolnaftate through the Human Nail Plate." *Chemical and Pharmaceutical Bulletin (Tokyo)* 46: 1797–802.

Mohorcic, M., A. Torkar, J. Friedrich, J. Kristl, and S. Murdan. 2007. "An Investigation into Keratolytic Enzymes to Enhance Ungual Drug Delivery." *International Journal of Pharmaceutics* 332: 196–201.

Murthy, S. N., S. R. K. Vaka, S. M. Sammeta, and A. B. Nair. 2009. "Tran-Screen-N™: Method for Rapid Screening of Trans-Ungual Drug Delivery Enhancers." *Journal of Pharmaceutical Sciences* 98: 4264–71.

7 Bioengineering of the Nail

Transonychial Water Loss and Imaging Techniques, an Overview

Rania Elkeeb and Howard I. Maibach

CONTENTS

7.1 INTRODUCTION

Despite the importance of biophysical assessment of nail physiology, minimal literature analyzes biophysical and biochemical processes. This chapter reflects the progress in the field. Based on recent focus on the understanding of nail physiology, this chapter explains the mechanisms involved in nail barrier function and hydration and summarizes the publications on the noninvasive biophysical assessment of nail physiology parameters. This chapter will be useful for scientists in the bioengineering field by presenting methods that offer reliable and reproducible approaches for product testing.

7.2 THE NAIL AND THE STRATUM CORNEUM: SIMILARITIES AND DIFFERENCES

The nail differs significantly in chemical composition from the skin. Skin has a significant amount of lipids in the form of bilayers (~15%); nail, on the other hand, is composed of low lipid levels (<1%) and keratin molecules with many disulfide linkages (Gupchup and Zatz 1999). Drug transport into the nail plate is influenced by the chemical properties of the drug molecule (charge and size), nature of the vehicle, and nail properties (hydration and thickness). There is experimental evidence indicating that ungual drug penetration is influenced by the aqueous pathway (Gunt and Kasting 2006). The permeability nature of the nail plate is discussed in detail in Chapter 3.

7.3 TRANSONYCHIAL WATER LOSS

Transonychial water loss (TOWL) refers to the total water loss flux ($g/m^2/h$) through the nail plate to the outside environment, that is, water passing the layers of the nail by passive diffusion. Extensive studies document transepidermal water loss (TEWL) in skin; on the other hand, there is less extensive research involving TOWL (Murdan et al. 2008; Sattler et al. 2011). This might be attributed to the difficulty in ensuring contact between the instrument probe and the rigid nail plate. Recently, there has been interest in the treatment of nail diseases topically, and consequently, research in this field is on the rise; thus, TOWL will be an important tool in ungual diagnosis as well as ungual drug delivery research (Murdan et al. 2008).

Interestingly, there are no methods for measuring TEWL or TOWL directly. In fact, TEWL/TOWL is inferred from water vapor flux evaporating from skin/nail. If TEWL is the only source of water reaching the skin surface and the skin surface remains dry, then the measured vapor flux equals TEWL (Elkeeb et al. 2010; Farahmand et al. 2009). *In vivo*, TEWL can be measured by various approaches, as follows: (i) closed-chamber method, which measures the increase of relative humidity (RH) in a closed air chamber; (ii) the ventilated chamber method, which measures water picked up by a gas passing through the chamber; and (iii) the open-chamber method, which utilizes an open capsule and estimates the vapor pressure gradient from the difference in vapor pressure at the two fixed heights of measurement. All these methods have inherent drawbacks as they interfere with the microclimate overlying the surface of the skin. The main problem with open-chamber systems is their vulnerability to disturbance from ambient air movements. The major concern with the closed-chamber method is its inability to perform continuous measurements as the accumulated water vapor needs to be purged after each reading. A different approach in closed-chamber systems has been introduced by Imhof (2001), where a condenser is used to remove water vapor from the closed measurement chamber, thus enabling continuous flux measurement to be made without purging.

The TEWL measurement devices could be modified with the use of appropriate adapters to suit the contour of the nail plate to measure TOWL. More information on different devices used to measure the TOWL/TEWL is provided in Section 7.4.

7.4 BIOENGINEERING TECHNIQUES TO MEASURE TOWL/TEWL

7.4.1 Open-Chamber System

Three commercially available instruments are Tewameter (Figure 7.1) (Courage + Khazaka electronic GmbH, Köln, Germany), Servo Med EP1 evaporimeter (Servo Med, Varberg, Sweden), and DermaLab System (Cortex Technologies, Hadsund, Denmark) (Primavera et al. 2005). They share the same principle of measurement. They are both open cylinder systems that measure water evaporation rate based on diffusion principles. The vapor density gradient is measured indirectly by two pairs of sensors (temperature and RH) inside a hollow cylinder, and the resulting data are analyzed by a microprocessor. Measurement values are given in g/m^2. One major concern with the open-chamber systems is that they require a special environment in which air turbulence does not interfere with their measurements (Hannon and Maibach 2005; Tagami et al. 2002).

In a study conducted by Kronauer et al. (2001), the investigators determined the TOWL in both healthy and diseased nails and compared it to TEWL. They used the Servo Med EP1 evaporimeter (Servo Med) coupled with a probe protection cover to avoid air convection, and they used Plasticine to connect the probe closely with the nail plate so as to avoid escape of water vapor from under the probe. The investigators have found that TOWL behaved contrary to TEWL, where affected nails showed significantly lower TOWL than did nails of healthy test subjects. This decrease in TOWL is attributed to the formation of a stratum granulosum, which is normally absent in healthy nails. Water loss from dystrophic nails is less than that from healthy nails, whereas affected skin loses more water than healthy skin (Kronauer et al. 2001)

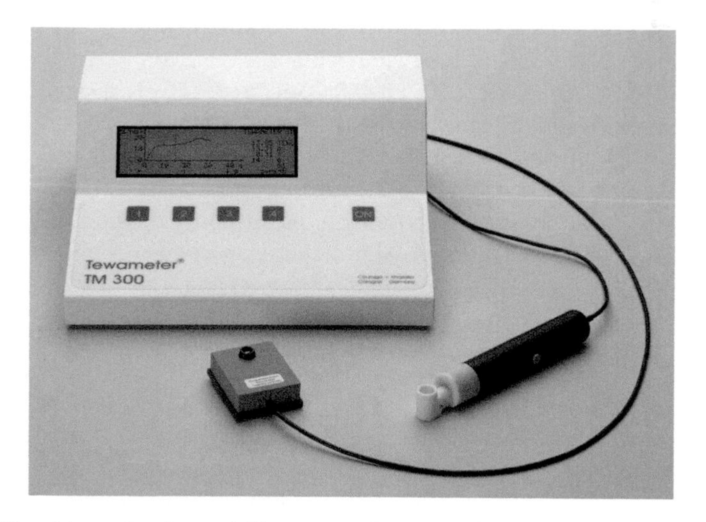

FIGURE 7.1 **(See color insert.)** Tewameter for measuring transepidermal water loss in skin. (Courtesy of Courage + Khazaka electronic GmbH, Köln, Germany.)

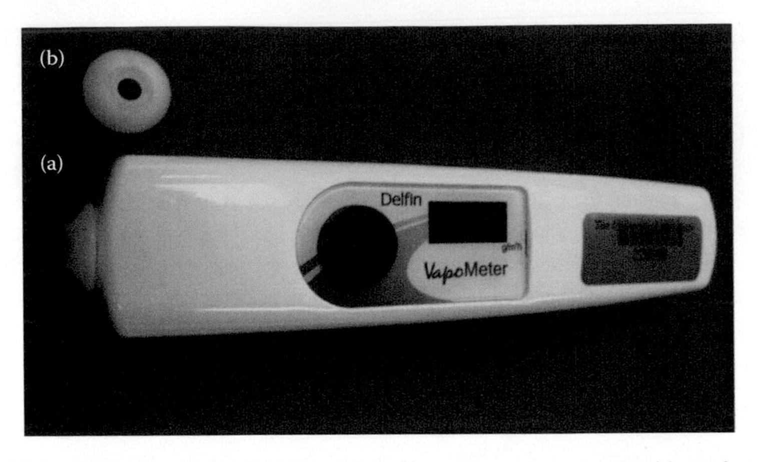

FIGURE 7.2 **(See color insert.)** (a) VapoMeter™ used to measure transepidermal water loss or transonychial water loss. (b) Nail adapter. (Courtesy of Delfin Technologies Ltd, Kuopio, Finland.)

7.4.2 Closed-Chamber System

The VapoMeter™ (Delfin Technologies Ltd, Kuopio, Finland) is a small portable closed-chamber system that calculates water evaporation rates from the increase in RH in the measurement chamber; the instrument compares the ambient humidity with the humidity within the chamber after a filling time of 10 sec (Fluhr et al. 2006) (Figure 7.2). Compared with open-chamber techniques, measurement times are standardized and relatively short (10 sec). However, continuous measurements are not possible because the water vapor captured during measurements needs to be allowed to evaporate before starting the next measurement. The values of the different measurements are expressed in g/m^2/h, with a maximum obtainable value of 300 g/m^2/h (Distante and Berardesca 1995). Nuutinen et al. (2003) evaluated the VapoMeter's performance, a new device at the time, on difficult sites such as nail, lip, and skin of the eyelid *in vivo*. TOWL and TEWL measurements were performed, and the authors concluded that the VapoMeter™ was able to measure evaporation rates efficiently at technically difficult anatomical sites. In another study, the VapoMeter™ was used to measure TOWL of human nails in a study that assessed whether protein quantitative tape stripping method was suitable for human nails. TOWL was found to significantly increase in tape-stripped nails (Tudela et al. 2008).

7.4.3 Condenser–Chamber System

AquaFlux AF200 (Biox Systems Ltd, London, UK), uses the patented condenser–chamber method to measure water vapor flux density (Imhof 2001) (Figure 7.3a). The measurement chamber is a hollow cylinder whose lower end acts as a measurement orifice that is placed in contact with the test surface. Its upper end is closed with an aluminum condenser that is maintained below the freezing temperature of water by means of an electronic Peltier cooler. When in contact with the test surface,

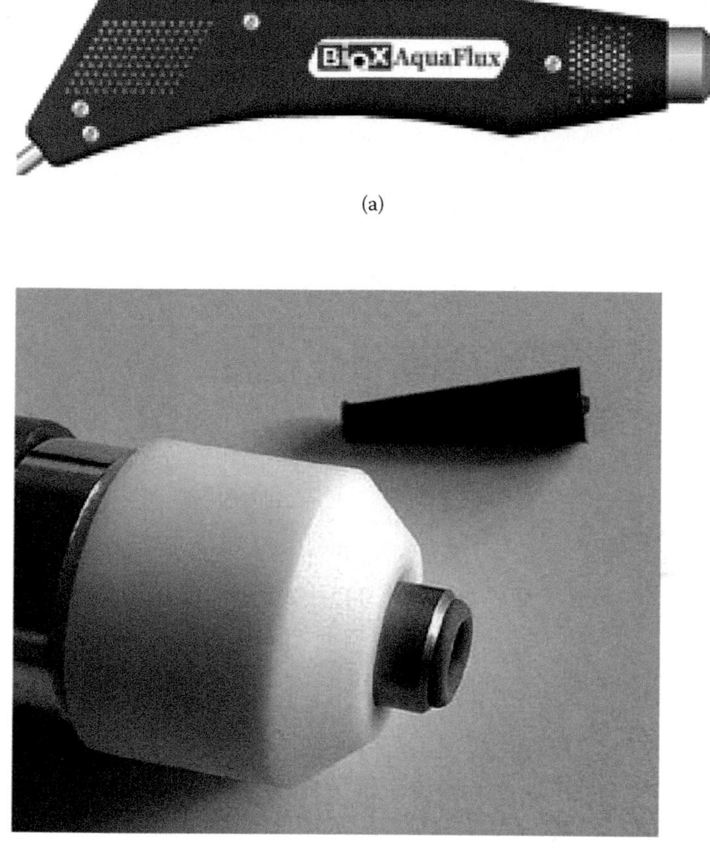

FIGURE 7.3 **(See color insert.)** (a) AquaFlux AF200; (b) TOWL measurement adaptor. (Courtesy of Biox Systems Ltd, London, UK.)

the chamber is closed and the air within it is protected from disturbance from ambient air movements. The condenser controls the humidity in the chamber independently of ambient conditions. It acts as a vapor sink by forming ice on its surface, thus creating a zone of low humidity in its immediate vicinity. By contrast, the test surface acts as a vapor source, creating a zone of higher humidity in its immediate vicinity. This humidity difference causes water vapor to migrate from the source to the sink by passive diffusion, leading to a linear distribution of humidity parallel to the axis of the chamber under steady conditions. Water vapor flux is calculated from measurements of this humidity gradient and Fick's first law of diffusion (Berg et al. 2002). AquaFlux is capable of measuring TOWL when a special nail adaptor is used (Figure 7.3b). An advantage over the conventional closed-chamber system is that continuous flux versus time measurements can be recorded for many hours because the water vapor entering the measurement chamber is continuously removed by the condenser.

Recently, Xiao et al. (2012) have developed a new method in studying both hair and nail water-holding capacity by the use of the condenser chamber method; they have also developed a mathematical model for modeling the hair and nail desorption process.

7.5 MEDICAL IMAGING

Different techniques are being used for noninvasive nail imaging to improve our knowledge of nail morphology and physiology *in vivo* and to reduce the number of nail biopsies. Imaging systems such as ultrasound, optical coherence tomography (OCT), and confocal laser scanning microscopy (CLSM) make use of different kinds of rays, waves, and physical phenomena (light, ultrasound, electromagnetic waves, etc.) to image general tissue structures.

7.5.1 Ultrasound

Sonography (ultrasound imaging) utilizes acoustical waves. It has many benefits over other imaging techniques as it is easy to apply, widely available, and inexpensive. Ultrasound imaging is based on the time delay that occurs when the ultrasound beams are reflected from different tissues (Richert et al. 2009). The structure of the tissue, its echogenicity, and several technical parameters reflect on the clarity of those images. New generations of high-resolution ultrasound machines, reaching up to 15 MHz, can process large amounts of data. New techniques have become widely available, which allow instantaneous integration of several overlapping ultrasound scans taken at different angles to produce a compound image with better information content. This technique, called real-time spatial compound imaging, reduces artifacts and provides a sharper image. Ultrasound can supply information on the anatomy and pathological processes in real time with the possibility of measuring and calculating parameters of thickness with an accuracy of approximately 100 μm (depending on the device used) (Richert et al. 2009; Wortsman and Jemec 2006).

7.5.2 Optical Coherence Tomography

OCT works by emitting infrared light, its reflection is measured, and the intensity is imaged as a function of the position of the material reflecting the wave. OCT has greater resolution than high-frequency ultrasounds, thus having the ability to discriminate between subtle differences that are not detected by ultrasound, and providing more information about the nail unit (Mogensen et al. 2007). It provides real-time images with a possible penetration depth of about 2 mm and a resolution of 10 μm. OCT is well-established in diagnosis in human skin, but it is just evolving for the study of nail disorders (Sattler et al. 2011).

7.5.3 Confocal Laser Scanning Microscopy

CLSM is another noninvasive diagnostic imaging technology tool based on light reflection. Similar to OCT, different layers can be differentiated by CLSM, according to the intensity of reflection. Yet the main advantage of using CLSM technique in healthy nails is the depiction of detailed information on the single cells, thus providing better information on the microscopic structures of the nail plate (Sattler et al. 2011). While few data are available on OCT and nails, no reports exist so far on CLSM and healthy nails. Only one report is available on CLSM and diseased nails, in which CLSM was able to confirm the diagnosis of onychomycosis in one patient. It provides real-time images with a possible penetration depth of about 2 mm and a resolution of 10 μm. OCT is well established in diagnosis in human skin, but it is just evolving for the study of nail disorders (Sattler et al. 2011).

7.6 CONCLUSION

The nail is clearly a very important part of the human body. Nail function has recently been explored after a delay of so many years. Nail bioengineering is an expanding field in investigative dermatology; thus, the nail is readily available for use of instrumentation, and we hope that the many different techniques available in the field of dermatology may be utilized. Clearly, bioengineering of the nail techniques will play a pivotal role in strategies to enhance understanding of ungual drug delivery.

REFERENCES

Berg, E. P., F. Pascut, L. I. Ciortea, D. O'driscoll, P. Xiao, and R. E. Imhof. 2002. "AquaFlux-A New Instrument for Water Vapor Flux Density Measurements." *Proceedings of the 4th International Symposium on Humidity and Moisture.* Taiwan ROC: Center for Measurements Standards, ITRI.

Distante, F., and E. Berardesca. 1995. "Transepidermal Water Loss." In *Bioengineering of the Skin: Methods and Instrumentation*, edited by E. Beradesca, P. Elsner, K. Wilhelm, and H. I. Maibach, 1–4. Boca Raton: CRC Press.

Elkeeb, R., X. Hui, H. Chan, L. Tian, and H. I. Maibach. 2010. "Correlation of Transepidermal Water Loss with Skin Barrier Properties In Vitro: Comparison of Three Evaporimeters." *Skin Research and Technology* 16: 9–15.

Farahmand, S., L. Tien, X. Hui, and H. I. Maibach. 2009. "Measuring Transepidermal Water Loss: A Comparative In Vivo Study of Condenser-Chamber, Unventilated-Chamber and Open-Chamber Systems." *Skin Research and Technology* 15: 392–8.

Fluhr, J. W., K. R. Feingold, and P. M. Elias. 2006. "Transepidermal Water Loss Reflects Permeability Barrier Status: Validation in Human and Rodent In Vivo and Ex Vivo Models." *Experimental Dermatology* 15: 483–92.

Gunt, H., and G. B. Kasting. 2006. "Hydration Effect on Human Nail Permeability." *Journal of Cosmetic Science* 57: 183–4.

Gupchup, G., and J. L. Zatz. 1999. "Structural Characteristics and Permeability Properties of the Human Nail: A Review." *Journal of Cosmetic Science* 50: 363–85

Hannon, C., and H. I. Maibach. 2005. "Efficacy of Moisturizers Assessed through Bioengineering Techniques." In *Textbook of Cosmetic Dermatology*, 3rd Ed, edited by R. Baran and H. I. Maibach, 573–612. New York: Taylor and Francis.

Imhof, R. E. 2001. Method and Equipment for Measuring Water Vapor Flux from Surfaces. US Patent 6439028.

Kronauer, C., M. Gfesser, J. Ring, and D. Abeck. 2001. "Transonychial Water Loss in Healthy and Diseased Nails." *Acta Dermato-Venereologica* 81: 175–7.

Mogensen, M., J. B. Thomsen, L. T. Skovgaard, and G. B. Jemec. 2007. "Nail Thickness Measurements Using Optical Coherence Tomography and 20-MHz Ultrasonography." *British Journal of Dermatology* 157: 894–900.

Murdan, S., D. Hinsu, and M. Guimier. 2008. "A Few Aspects of Transonychial Water Loss (TOWL): Inter-Individual, and Intra-Individual Inter-Finger, Inter-Hand and Inter-Day Variabilities, and the Influence of Nail Plate Hydration, Filing and Varnish." *European Journal of Pharmaceutics and Biopharmaceutics* 70: 684–9.

Nuutinen, J., I. Harvima, M. R. Lahtinen, and T. Lahtinen. 2003. "Water Loss through the Lip, Nail, Eyelid Skin, Scalp Skin and Axillary Skin Measured with a Closed-Chamber Evaporation Principle." *British Journal of Dermatology* 148: 839–41.

Primavera, G., J. Fluhr, and E. Bersrdesca. 2005. "Standardization of Measurements and Guidelines." In *Bioengineering of the Skin: Water and Stratum Corneum*, edited by J. Fluhr, P. Elsner, E. Berardesca, and H. Maibach, 83–95. Boca Raton: CRC Press.

Richert, B., N. Lateur, A. Theunis, and J. Andre. 2009. "New Tools in Nail Disorders." *Seminars in Cutaneous Medicine and Surgery* 28: 44–8.

Sattler, E., R. Kaestle, G. Rothmund, and J. Welzel. 2012. "Confocal Laser Microscopy, Optical Coherence Tomography and Transonychial Water Loss for in vivo Investigation of Nails." *British Journal of Dermatology* 166(4): 740–6.

Tagami, H., H. Kobayashi, and K. Kikuchi. 2002. "A Portable Device Using a Closed Chamber System for Measuring Transepidermal Water Loss: Comparison with the Conventional Method." *Skin Research and Technology* 8: 7–12.

Tudela, E., A. Lamberbourg, M. Cordoba Diaz, H. Zhai, and H. I. Maibach. 2008. "Tape Stripping on a Human Nail: Quantification of Removal." *Skin Research and Technology* 14: 472–7.

Wortsman, X., and G. B. Jemec. 2006. "Ultrasound Imaging of Nails." *Dermatologic Clinics* 24: 323–8.

Xiao, P., L. I. Ciortea, E. P. Berg, and R. E. Imhof. 2012. "Hair and Nail Water-Holding Capability Measurements by Using Condenser-TEWL Method." *International Journal of Cosmetic Science* 34: 12–6.

8 Trans-Ungual Iontophoresis and Physical Drug Delivery Enhancement

Jinsong Hao and S. Kevin Li

CONTENTS

The nail is highly resistant to the penetration of drugs including antifungal agents. As a result, a number of physical enhancement methods have been investigated to enhance trans-ungual delivery. This chapter covers recent developments in electrically facilitated trans-ungual delivery. Other physical enhancement techniques will be briefly reviewed.

8.1 TRANS-UNGUAL IONTOPHORESIS

Iontophoresis is a method to deliver a compound across a membrane with the assistance of an electric field (Kasting 1992). This technology has been explored to deliver drugs across skin (Viscusi and Witkowski 2005), cornea (Eljarrat-Binstock and

Domb 2006), sclera (Myles et al. 2005), and buccal (Jacobsen 2001), nasal (Lerner et al. 2004), and tympanic membranes (Christodoulou et al. 2003) for systemic and local therapies. Iontophoretic devices that have been marketed for topical drug delivery include Phoresor® and Ocuphor® (Iomed Inc., Salt Lake City, UT), Actyve™ (Vyteris Inc., Fair Lawn, NJ), IONSYS E-TRANS® (Alza Corp., Mountain View, CA), and EyeGate® II Delivery System (EyeGate Pharmaceuticals Inc., Waltham, MA). The wearable electronic disposable delivery (WEDD®) system (Travanti Pharma, Mendota Heights, MN) is a thin, Band-Aid size, low-current iontophoresis system. The first attempt of iontophoretic delivery of prednisolone across human nail *in vivo* was reported in 1986, and the pharmacokinetics of prednisolone in plasma was determined after anodal iontophoresis (i.e., the anode electrode in the donor) at a current density of 0.63 mA/cm^2 for 10 min (James et al. 1986). The feasibility of trans-ungual iontophoresis was further explored in an iontophoresis study of salicylic acid (SA) across human nail *in vitro* (Murthy et al. 2007b). Iontophoretic transport of neutral permeants and the permselective property of human nail were investigated (Dutet and Delgado-Charro 2010a, b; Murthy et al. 2007a). There have been several systematic studies on iontophoretic trans-ungual transport mechanisms (Hao and Li 2008a, b; Hao et al. 2008; Smith et al. 2009, 2010). Enhanced trans-ungual delivery of antifungal agents such as terbinafine (Amichai et al. 2010; Nair et al. 2009a, b,d, 2010, 2011) and ciclopirox (Hao et al. 2009) using iontophoresis has also been demonstrated *in vitro*, ex vivo, and *in vivo*. Moreover, low-voltage direct current was shown to be fungicidal (Kalinowski et al. 2004). Table 8.1 summarizes recent studies of trans-ungual iontophoresis that will be discussed in this chapter.

8.1.1 Theory and Equations

Iontophoretically enhanced transport generally includes three mechanisms: (a) direct interaction of the electric field with the charge of the ionic permeant (i.e., electrophoresis or electromigration); (b) convective solvent flow in the preexisting and/or newly created charged pathways (i.e., electroosmosis); and (c) electric field–induced pore induction (i.e., electroporation or electropermeabilization). For neutral permeants, transport enhancement is mainly due to electroosmosis. For ionic permeants, transport enhancement is primarily a result of both electrophoresis and electroosmosis. Theoretically, iontophoresis increases the steady-state transport flux and reduces the transport lag time (Keister and Kasting 1986). The concentration of a permeant in the membrane can be enhanced by iontophoresis (Kasting 1992). The steady-state iontophoretic flux ($J_{\Delta\psi}$) of a permeant through a homogeneous porous membrane can be described by the modified Nernst–Planck equation (Hao and Li 2008a):

$$J_{\Delta\psi} = \varepsilon_{\mathrm{p}}\left[-\mathrm{HD}\left(\frac{\mathrm{d}C}{\mathrm{d}x}+\frac{CzF}{R_{\mathrm{gas}}T}\frac{\mathrm{d}\psi}{\mathrm{d}x}\right)\pm WvC\right] \tag{8.1}$$

where ψ is the electric potential across the membrane, F is the Faraday constant, R_{gas} is the gas constant, T is the temperature, v is the average velocity of the convective solvent flow, ε_{p} is the combined porosity and tortuosity factor for the membrane,

TABLE 8.1
Permeants Studied in Trans-Ungual Iontophoretic Delivery

Permeants	Experimental Conditions	Iontophoretic Devices	Purposes of Studies	References
Model compounds				
Chloride	Thumbnails of healthy human subjects; 4-(2-hydroxyethyl)-1-piperazineethanesulfonic acid (HEPES) buffer as donor; effective diffusion area 0.4 cm^2	Phoresor constant current device; 0.2 mA anodal iontophoresis; 4 × 30 min; Ag/AgCl electrodes	Study iontophoretic transport *in vivo* by outward trans-ungual extraction of endogenous ions	Dutet and Delgado-Charro (2009)
	Side-by-side diffusion cells; fully hydrated fingernails; $Na^{36}Cl$ in phosphate buffered saline (PBS) as donor, PBS as receptor; effective diffusion area 0.64 cm^2	Phoresor constant current device; cathodal iontophoresis; 0.16 mA/cm^2; Ag and Ag/AgCl electrodes	Study pH effect on iontophoretic transport	Smith et al. (2010)
Glucose	Franz diffusion cells; human nail clippings; 10 mM glucose in phosphate buffered saline (PBS) as donor; PBS as receptor; effective diffusion area 0.25 cm^2	Phoresor constant current device; anodal and cathodal iontophoresis; 0.5 mA/cm^2; Ag/AgCl electrodes	Study electroosmosis enhancement	Murthy et al. (2007a)
Lithium	Side-by-side diffusion cells; human nail clippings; single cation or binary mixture of sodium and lithium with HEPES buffer; effective diffusion area 0.2 cm^2	Kepco APH 1000M power supply; anodal iontophoresis; 0.2 mA; Ag/AgCl electrodes	Study the effects of ion competition and pH on transference numbers and permselectivity	Dutet and Delgado-Charro (2010b)
Mannitol	Side-by-side diffusion cells; fully hydrated fingernails; ^{3}H-mannitol in PBS as donor, PBS as receptor; effective diffusion area 0.64 cm^2	Phoresor constant current device; anodal and cathodal iontophoresis; 0.16 and 0.5 mA/cm^2; Ag and Ag/AgCl electrodes	Study electroosmosis contribution	Hao and Li (2008a)

(*Continued*)

TABLE 8.1 (*Continued*)
Permeants Studied in Trans-Ungual Iontophoretic Delivery

Permeants	Experimental Conditions	Iontophoretic Devices	Purposes of Studies	References
	Side-by-side diffusion cells; fully hydrated fingernails; ^{3}H-mannitol in PBS as donor; PBS as receptors; effective diffusion area 0.64 cm^2	Phoresor constant current device; anodal and cathodal iontophoresis; 0.16 and 0.5 mA/cm^2; Ag and Ag/AgCl electrodes	Study the effects of solution pH and ionic strength on electroosmosis	Hao and Li (2008b)
Salicylic acid	Franz diffusion cells; human nail clippings; salicylic acid in PBS as donor; PBS as receptor; effective diffusion area 0.25 cm^2	Phoresor constant current device; cathodal iontophoresis; 0.16–0.5 mA/cm^2; Ag/AgCl electrodes	Study the effects of pH, ionic strength, and current density on iontophoretic transport	Murthy et al. (2007b)
Sodium	Side-by-side diffusion cells; fully hydrated fingernails; $^{22}NaCl$ in PBS as donor, PBS as receptor; effective diffusion area 0.64 cm^2	Phoresor constant current device; anodal iontophoresis; 0.16 mA/cm^2; Ag and Ag/AgCl electrodes	Study pH effect on iontophoretic transport	Smith et al. (2010)
	Thumbnails of healthy human subjects; HEPES buffer as donor; effective diffusion area 0.4 cm^2	Phoresor constant current device; 0.2 mA cathodal iontophoresis; 4 × 30 min; Ag/AgCl electrodes	Study iontophoretic transport *in vivo* by trans-ungual extraction of endogenous ions	Dutet and Delgado-Charro (2009)
Tetraethyl-ammonium (TEA)	Side-by-side diffusion cells; fully hydrated fingernails; ^{14}C-TEA in PBS as donor, PBS as receptor; effective diffusion area 0.64 cm^2	Phoresor constant current device; anodal and cathodal iontophoresis; 0.16 mA/cm^2; Ag and Ag/AgCl electrodes	Study electrophoresis contribution	Hao and Li (2008a)

TABLE 8.1 (*Continued*)
Permeants Studied in Trans-Ungual Iontophoretic Delivery

Permeants	Experimental Conditions	Iontophoretic Devices	Purposes of Studies	References
	Side-by-side diffusion cells; fully hydrated fingernails; ^{14}C-TEA in TEA as donor, PBS as receptor; effective diffusion area 0.64 cm^2	Phoresor constant current device; anodal and cathodal iontophoresis; 0.16 mA/cm^2; Ag and Ag/AgCl electrodes	Study the effect of penetration enhancer pretreatment on transference number	Hao et al. (2008)
	Side-by-side diffusion cells; fully hydrated fingernails; ^{14}C-TEA in TEA as donor, PBS as receptor; effective diffusion area 0.64 cm^2	Phoresor constant current device; anodal and cathodal iontophoresis; 0.16 mA/cm^2; Ag and Ag/AgCl electrodes	Study ionic strength effect on iontophoretic transport	Smith et al. (2009)
Urea	Side-by-side diffusion cells; fully hydrated fingernails; ^{14}C-urea in PBS as donor, PBS as receptor; effective diffusion area 0.64 cm^2	Phoresor constant current device; anodal and cathodal iontophoresis; 0.16 and 0.5 mA/cm^2; Ag and Ag/AgCl electrodes	Study electroosmosis contribution	Hao and Li (2008a)
	Side-by-side diffusion cells; fully hydrated fingernails; ^{14}C-urea in PBS as donor, PBS as receptors; effective diffusion area 0.64 cm^2	Phoresor constant current device; anodal and cathodal iontophoresis; 0.16 and 0.5 mA/cm^2; Ag and Ag/AgCl electrodes	Study the effects of solution pH and ionic strength on electroosmosis	Hao and Li (2008b)
Drugs				
Ciclopirox	Side-by-side diffusion cells; partially hydrated fingernails; 0.05 M ciclopirox ethanol aqueous solution (pH 8.7) as donor; PBS containing 5% bovine serum albumin as receptor; effective diffusion area 0.64 cm^2	Battery-powered 9-V constant voltage cathodal iontophoresis; Ag and Pt electrodes	Study constant voltage iontophoretic delivery of ciclopirox	Hao et al. (2009)

(*Continued*)

TABLE 8.1 (*Continued*)
Permeants Studied in Trans-Ungual Iontophoretic Delivery

Permeants	Experimental Conditions	Iontophoretic Devices	Purposes of Studies	References
Griseofulvin	Franz diffusion cells; human nail clippings; 0.03 mM griseofulvin in PBS as donor; PBS as receptor; effective diffusion area 0.25 cm^2	Phoresor constant current device; anodal and cathodal iontophoresis; 0.5 mA/cm^2; Ag/AgCl electrodes	Study permselectivity of human nail	Murthy et al. (2007a)
Prednisolone	Thumbnails of human subjects; 1% prednisolone aqueous solution	Constant current 10-min anodal iontophoresis 0.63 mA/cm^2	Explore the possibility of trans-ungual iontophoresis	James et al. (1986)
Terbinafine	Franz diffusion cells; human fingernail; 10 mg/mL terbinafine aqueous solution (pH 3) as donor; water at pH 3 as receptor; effective diffusion area 0.2 cm^2	Phoresor constant current device; alternative 1-h anodal iontophoresis (0.5 mA/cm^2) and 1-h passive delivery for three cycles; Ag/AgCl electrodes	Study iontophoresis of terbinafine	Nair et al. (2009d)
	Franz diffusion cells; human fingernail; 4% terbinafine formulation (pH 3.2) as donor; water at pH 3 as receptor; effective diffusion area 0.2 cm^2	Custom-made constant current device; anodal iontophoresis; 0.1–1 mA/cm^2; 15–60 min; Ag/AgCl electrodes	Develop a topical formulation to enhance terbinafine iontophoretic delivery	Nair et al. (2009a)
	Ex vivo toe model with two applicators; 4% terbinafine formulation at pH 3.2 as donor, effective diffusion exposed areas 1.4 and 6.2 cm^2	Custom-made constant current device; 20-min anodal iontophoresis at 0.5 mA	Develop an ex vivo model to assess iontophoretic delivery of terbinafine	Nair et al. (2009b)
	Patients with onychomycosis; 1% terbinafine gel patch at pH 4.6	Custom-made constant current device; overnight anodal iontophoresis daily, 5 days a week, for 4 weeks; 0.1 mA/cm^2	Clinically test the efficacy of terbinafine delivered by iontophoresis	Amichai et al. (2010)

and C, x, z, and D are the concentration, the position in the membrane, the charge number, and the diffusion coefficient of the permeant, respectively. H and W are hindrance factors for Brownian diffusion and for pressure-induced parabolic convective solvent flow, respectively. These hindrance factors are a function of the ratio of permeant radius to pore radius of the transport pathway (Deen 1987).

Permeability coefficient can be defined as the steady-state flux normalized by the donor concentration. The enhancement factor (E) is the ratio of permeability coefficient of iontophoretic transport to that of passive transport and can be expressed as follows:

$$E = \frac{K \pm \text{Pe}}{1 - \exp(-K \pm \text{Pe})} \tag{8.2}$$

where Pe is the Peclet number that describes the extent of electroosmotic transport, and K describes the extent of electrophoresis transport:

$$K = \frac{zF}{R_{\text{gas}}T}\Delta\psi \tag{8.3}$$

For neutral permeants, the enhancement factor of convective transport due to electroosmosis (E_v) is expressed as

$$E_\text{v} = \frac{\text{Pe}}{1 - \exp(-\text{Pe})} \tag{8.4}$$

The transport lag time for iontophoretic transport ($t_{\text{L},\Delta\psi}$) is related to the transport lag time of passive transport ($t_{\text{L,p}}$) by

$$t_{\text{L},\Delta\psi} = \frac{6t_{\text{L,p}}}{(E)^2}\left(E\coth\frac{E}{2} - 2\right) \tag{8.5}$$

For iontophoretic transport of an ionic permeant, the transference number (t_i) is defined as the fraction of the total current carried by the permeant i and is a measure of iontophoretic delivery efficiency:

$$t_i = \frac{C_i u_i |z_i|}{\sum_j C_j u_j |z_j|} \tag{8.6}$$

where u is the effective electrophoretic mobility and the subscript j represents all ions in the membrane including the permeant i. The transference number can be determined experimentally using the iontophoretic flux ($J_{\Delta\psi,i}$), the charge number (z_i), and the current density applied (I_c):

$$t_i = \frac{|z_i| F J_{\Delta\psi,i}}{I_\text{c}} \tag{8.7}$$

8.1.2 Electroosmosis

Electroosmosis is related to the Debye–Huckel thickness (1/κ) of the electrical double layer in the pores of a charged porous membrane. When an electric field is applied across the charged porous membrane (parallel to the charged surface), forces are exerted on the ions in the double layer in the pores. The ions in the double layer move under the influence of the electric field, carrying the solvent with them by momentum. Electroosmotic flow is formed. The thickness of the electrical double layer in the pores is related to the ionic strength of the solution. The solvent flow velocity in convective transport is zero (in stationary) at the pore wall and increases to a maximum at the pore center. As a result, electroosmosis is a function of the pore charge, pore size, solution ionic strength in the pore, and the applied voltage (Pikal 2001; Sims et al. 1993). An increase in the solution ionic strength in the pore decreases the 1/κ value, and thus, the solvent flow velocity decreases. The surface charge density is related to the fraction of the ionized functional groups on the surface and is a function of pH. The direction of electroosmosis is related to both the polarity of the applied electric field and the net charge of the pores in the membrane. For a negatively charged membrane, the electroosmosis solvent flow is from the anode to the cathode. Transport enhancement of a neutral permeant is thus expected in anodal iontophoresis (i.e., the anode in the donor and the cathode in the receptor). For a positively charged membrane, cathodal iontophoresis (i.e., the cathode in the donor and the anode in the receptor) is anticipated to enhance membrane transport due to the electroosmosis solvent flow from the cathode to the anode.

The keratins of the nail plate were reported to have an isoelectric point (pI) of 4.9–5.4 (Gupchup and Zatz 1999; Marshall 1983; Murdan 2002). Accordingly, electroosmosis can enhance the transport of neutral permeants during iontophoresis when the pH is away from the pI of the nail keratins. The effects of electroosmosis on trans-ungual iontophoretic transport have been studied using neutral permeants such as mannitol, urea, glucose, and griseofulvin (Dutet and Delgado-Charro 2010a; Hao and Li 2008a, b; Murthy et al. 2007a). Under physiological conditions (pH 7.4 and 0.15 M phosphate-buffered saline [PBS]), the permeability coefficients (donor concentration normalized fluxes) of mannitol and urea across fully hydrated cadaver human nail during 0.3 mA anodal iontophoresis (~0.5 mA/cm^2) were 2.8 and 1.5 times of those for passive transport, respectively (Hao and Li 2008a). The Peclet numbers increased proportionally when the electric current was increased from 0.1 to 0.3 mA. In the *in vitro* iontophoretic transport studies under the symmetric conditions (i.e., the same pH in both the donor and the receptor), anodal iontophoretic transport of mannitol across fully hydrated nail was enhanced when the solution pH increased from pH 7.4 to 9 and cathodal iontophoretic transport of mannitol was enhanced at pH 3 from pH 7.4 at a current density of approximately 0.5 mA/cm^2 (Hao and Li 2008b). Similar pH-dependent electroomosis enhancement was also observed in the *in vitro* iontophoretic transport studies under the asymmetric conditions (i.e., varied pH values in the donor and pH 7.1 in the receptor) (Murthy et al. 2007a). In these studies, the iontophoretic transport of two neutral permeants, glucose and griseofulvin, was enhanced by anodal iontophoresis at pH 7 and by cathodal iontophoresis at pH < 5 at a current density of 0.5 mA/cm^2. At pH 5, no significant transport enhancement was shown in both anodal and cathodal iontophoresis.

In electroosmosis dominant transport, the flux of a permeant is directly related to the velocity of the electroosmotic flow and independent of the permeant diffusion coefficient. Therefore, electroosmotic transport enhancement is a function of the molecular size of the permeant. Flux enhancement increases when the diffusion coefficient of the permeant decreases. This is evidenced by the larger Peclet numbers of mannitol (MW 182 Da) than those of urea (MW 60 Da) in anodal iontophoresis (Hao and Li 2008a). In a similar study, the transport of glucose (MW 180 Da) across fully hydrated human nail clippings can be enhanced approximately sixfold by anodal iontophoresis at 0.5 mA/cm^2 over passive transport at pH 7.1 (Murthy et al. 2007a).

According to electrokinetic theory, membrane electroosmosis is a function of the solution ionic strength. The electrical double layer affects the convective solvent flow velocity in electroosmotic transport of neutral permeants. When $1/\kappa \ll$ pore radius, $1/\kappa$ is proportional to the solvent flow velocity and Peclet number. In an *in vitro* anodal iontophoretic transport study at pH 7.4 under the symmetric conditions (i.e., the same ionic strength in both the donor and the receptor), the Peclet numbers normalized by the voltage across the fully hydrated nail increased by approximately four times for both mannitol and urea when the ionic strength decreased from 0.7 to 0.04 M (Hao and Li 2008b). The observed fourfold increase in electroosmosis from 0.7 to 0.04 M ionic strength is consistent with the electrokinetic theory as the $1/\kappa$ value increases from around 0.4 to 2 nm when the solution ionic strength decreases from 0.7 to 0.04 M (Hao and Li 2008b).

8.1.3 Electrophoresis

The cations in a solution move from the anode to the cathode and the anions move from the cathode to the anode under an electric field. Anodal iontophoresis enhances the transport of cationic permeants, and cathodal iontophoresis enhances the transport of anionic permeants. Anodal iontophoretic transport of positively charged tetraethylammonium (TEA) ion and neutral mannitol through fully hydrated human nail under a constant current of 0.1 mA (current density of 0.16 mA/cm^2) was investigated (Hao and Li 2008a). Iontophoretic enhancement of charged TEA was significantly greater than that of mannitol, and the iontophoretic enhancement factors for TEA and mannitol over passive transport were 29 and 1.6, respectively. This shows that the contribution of electroosmosis to iontophoretic transport was generally less than 10% of electrophoresis contribution for TEA across the fully hydrated nail under the condition of pH 7.4 and 0.15 M. In general, electrophoresis is the dominant transport mechanism of positively charged small permeants across the fully hydrated nail during iontophoresis, and the electroosmosis contribution is marginal.

8.1.4 Electropermeabilization

Due to the low lipid content in the nail (<1%), electropermeabilization that often occurs in biomembrane during iontophoresis, such as in transdermal iontophoresis, was not expected as a significant flux-enhancing mechanism in trans-ungual iontophoresis. In a multistage nail transport study over 2 months, it was found that the passive permeability coefficients of mannitol and urea were essentially the same before and after repeated iontophoresis treatments at 0.1 and 0.3 mA (current densities of 0.16 and 0.5 mA/cm^2) (Hao and Li 2008a). The passive permeability coefficient ratios of urea to mannitol were not affected by the applied electric current, indicating no change in the effective pore size of the nails. Together, these results suggest no irreversible electropermeabilization in trans-ungual iontophoretic transport. If the nail properties were altered by the electric field under the studied conditions, such effects were reversible and the nail returned to its normal state after the iontophoresis application. Although a recent study comparing the electrical resistance of hydrated nail (around 4–10 kΩ) during and before 0.1 mA iontophoresis suggests no significant electropermeabilization during trans-ungual iontophoresis (Hao and Li 2008a), the absence of electropermeabilization needs to be confirmed by closely monitoring the nail resistance during iontophoresis using a more precise and accurate measurement method. In addition, electrical resistance is only an indicator of the nail barrier to small electrolytes, and such results may not reveal possible electropermeabilization for large permeants. It has been found that "true" steady-state transport for TEA was not observed even in the 36-h iontophoretic transport experiments for reasons yet to be investigated (Hao and Li 2008a). Possible electropermeabilization of the nail during iontophoresis could not be ruled out.

8.1.5 Transport Efficiency

Transference number measures the efficiency of iontophoretic transport of a permeant. It is related to the concentrations, electromobilities, and charges of the permeant and other ions (coions and counterions) in the membrane according to Equation 8.6. Theoretically, the maximum transference number is obtained under a "single-ion" case in which competing ions are absent. However, endogenous counterions such as Na and Cl are always present in iontophoresis *in vivo*. During iontophoretic delivery, both extraneous and endogenous ions compete with the permeant for the electric current and greatly decrease the transport efficiency of the permeant (Phipps and Gyory 1992). Based on the free aqueous diffusivities of Na (1.3×10^{-5} cm^2/s) and Cl (2.0×10^{-5} cm^2/s) ions, the transference numbers of Na and Cl ions in bulk solution are 0.4 and 0.6, respectively. In the iontophoretic transport studies of Na and Cl ions across fully hydrated nail (Smith et al. 2010), the transference numbers of Na ion were higher than those for Cl ion between pH 7 and 11 due to nail permselectivity. At pH 3–5, the transference numbers followed a similar trend as those in the bulk solution. At pH 1–2, significant ion competition from hydronium was observed. The estimated transference number of hydronium ion across the nail at pH 1 could be as high as 0.8 (Smith et al. 2010). The *in vivo* trans-ungual transference number of Na measured in human subjects was approximately 0.51 (Dutet and Delgado-Charro 2009), which is not significantly different from that determined using the above method (Smith et al. 2010).

A method to enhance iontophoretic efficiency in constant current iontophoresis is to increase the electromobility of the permeant to a greater extent than those of the small coion and counterion such as Na and Cl ions, as predicted by Equation 8.6. This can be achieved by increasing the effective pore size of the nail with an ungual penetration enhancer. The transference number of TEA across the fully hydrated nail under symmetric conditions in terms of pH and ionic strength (pH 7.4 and 0.15 M in the donor and the receptor) was 0.15 (Hao et al. 2008). The transference number was increased by approximately two times to 0.26 after the pretreatment of the fully hydrated nail with 0.05 M thioglycolic acid (TGA) in PBS. The TGA-treated nail also had lower electrical resistance than the untreated nail during 0.1-mA (0.16-mA/cm^2) iontophoresis, implying enlarged effective pore sizes and/or increased porosity in the nail. Another method to enhance the iontophoretic transport efficiency is to increase the concentration of the permeant in the nail by increasing permeant partitioning into the nail associated with changing the solution ionic strength. For a positively charged permeant, permeant partitioning into the nail at the physiological pH increased when the ionic strength was decreased (Smith et al. 2009). However, the effect of ionic strength upon the transference of the positively charged permeant under the asymmetric conditions (i.e., solutions of varied ionic strength in the donor and 0.15 M PBS in the receptor) was not as significant as the full effect under the symmetric conditions (i.e., the same ionic strength at both sides of the fully hydrated human nail). In a study of the effects of solution composition on trans-ungual iontophoresis, the transference numbers of sodium and lithium ions were found to be linearly related to their respective molar fractions in the donor solutions during iontophoresis (Dutet and Delgado-Charro 2010b). Enhanced iontophoretic transport efficiency can therefore be achieved by increasing the molar fraction of the permeant in the donor solution or decreasing the concentration of coions in the solution. In the absence of competing coions, the maximum transference number is related to the effective electrophoretic mobility of the permeant.

8.1.6 Safety Issues

Iontophoresis is well-tolerated in human skin when the electric current applied is not higher than 0.5 mA/cm^2 (Ledger 1992). In constant voltage iontophoresis applications (Hao et al. 2010), no adverse effect or sensation of the human nail were reported when an electrical potential of 1.5 V was applied continuously on the nail for 6 h *in vivo*. This applied potential and current level was found to be safe. However, high electric current (electric current density >0.5 mA/cm^2) across the human nail was observed to cause moderate sensation and silver stains on the nail surface during 9-V iontophoresis for 6 h (Hao et al. 2010). The subjects also felt an electric shock-like sensation when the current was switched on and off during the resistance measurements. It is believed that the changes and unpleasant sensation in the nail were related to the electric current density, which increased to a level higher than the tolerance threshold of human skin to electric current (> 0.5 mA/cm^2). There was no long-term effect on the nail and the adverse effects were reversible. In another study (Dutet and Delgado-Charro 2009), repeated direct constant current iontophoresis at a current density of 0.5 mA/cm^2 was applied on human fingernails for a total of 120 min (4 × 30 min) *in vivo*. While an uncomfortable "tingling effect" at the skin site was reported immediately after the start of the iontophoresis application, the iontophoresis applications were well-tolerated after nail hydration for a short period of time (≤10 min).

8.1.7 Considerations in Trans-Ungual Iontophoretic Delivery

A number of influencing factors should be considered in the development of an effective iontophoretic delivery system for the topical treatment of nail diseases (Table 8.2). Among these factors, nail hydration is one of the most important for effective trans-ungual iontophoretic delivery. Under normal physiological conditions, human nail is partially hydrated and a water content gradient exists from the nail bed to the nail surface. Without hydration, the ions in the nail would not be conductive due to hindered transport in the nail at this partial hydration state. The high initial nail resistance of partially hydrated nail *in vivo* implies that effective trans-ungual iontophoresis without hydration might not be feasible due to safety issues and iontophoresis efficiency. It has been reported that the dry nail *in vivo* started to conduct electricity after 5 min of hydration (James et al. 1986). Therefore, hydration before and/or during iontophoresis treatment would be preferred to lower the nail resistance and increase the effective pore size in the nail in trans-ungual iontophoretic delivery.

The transport pathway in the nail is highly restrictive. Trans-ungual transport is hindered due to the small effective pore size of the pathway. For example, for small molecule TEA with an effective hydration radius of about 0.4 nm, the hindrance factor (H) for diffusion in fully hydrated nail at pH 7.4 was found to be approximately 0.02 estimated from the transport data (Hao and Li 2008a). This hindrance effect will be more significant for larger molecules. Partially hydrated nail is expected to have

TABLE 8.2
Influencing Factors in Trans-Ungual Iontophoretic Delivery

Factors	Effects	References
Nail hydration state	Hydration increases conductivity and effective pore size of nail	Hao et al. (2010), James et al. (1986)
Permeant molecular weight	Increases transport hindrance with increasing permeant molecular weight	Hao and Li (2008a)
Permeant concentration	Increases iontophoretic flux	Murthy et al. (2007b)
Donor pH	Affects permeant ionization, nail charge, electroosmosis enhancement	Hao and Li (2008b), Murthy et al. (2007a, b), Smith et al. (2010)
Donor ionic strength	Affects permeant partitioning and electroosmosis enhancement	Hao and Li (2008b), Smith et al. (2009)
Molar fraction and solution composition	Increase permeant transference with increasing permeant molar fraction or decreasing the concentration of coions	Dutet and Delgado-Charro (2010b)
Enhancer treatment	Affects nail hydration and/or effective pore size; reduces transport hindrance	Hao et al. (2008), Nair et al. (2009c)

smaller effective pore size than fully hydrated nail. The smaller effective pore size of the transport pathway in the partially hydrated nail compared with fully hydrated nail will disproportionally hinder the effective diffusivity of the larger permeant to a greater extent than the small competing ions. Thus, the effect of ion competition can be amplified by hindered transport in the partially hydrated nail and lead to a significant decrease in iontophoretic transport efficiency across the nail. A previous study shows that pretreatment of the nail with penetration enhancers can enhance nail hydration, increase the effective pore size, and hence enhance the iontophoretic transport efficiency (Hao et al. 2008). The method of cotreatment of the nail with enhancers during iontophoresis was not used because the enhancers, which are ionic, may cause significant ion competition to the iontophoretic transport of the permeant. Iontophoretic delivery of an enhancer to the nail prior to trans-ungual iontophoretic delivery was reported to enhance the efficacy of the enhancer by increasing the penetration of the enhancer into the nail (Nair et al. 2009c). Long pretreatment of the nail with the enhancer was thus not required. It might be difficult to iontophoretically deliver permeants or drugs of relatively large molecular sizes through the nail at a sufficient amount due to the intrinsic permeability barrier of the nail.

The donor pH affects not only the ionization of the ionic permeant but also the surface charge density of the transport pathways in the nail. This determines the direction and magnitude of electroosmosis. To take advantage of electroosmosis, anodal iontophoresis at pH above the pI of the nail would be preferred for a positively charged permeant and cathodal iontophoresis at pH below the pI of the nail would be favored for a negatively charged permeant. However, electroosmosis contribution to trans-ungual iontophoresis is generally small compared to that of electrophoresis for ionic permeants of small to moderate sizes in fully hydrated nail (Hao and Li 2008a). In addition, the electroosmosis enhancement of small permeants could be enhanced only to a small extent by changing the pH and ionic strength (Hao and Li 2008b). For trans-ungual transport of large molecules that can better utilize electroosmotic transport enhancement, the transport of these molecules is restricted by size exclusion from the small pores in the nail. Considering all these factors, a trans-ungual iontophoresis system should be tailored to the mechanism of electrophoresis, irrespective of the electroosmosis effect. In the iontophoretic transport of negatively charged SA with a pKa of 3.1 (Murthy et al. 2007b), maximal enhancement was achieved at pH 5 using cathodal iontophoresis. Increasing pH to 7.1 did not enhance iontophoretic transport and decreasing pH to 3 significantly reduced the iontophoretic transport flux enhancement. Ion competition from hydronium was another concern when the donor pH was low (pH <2). At high pH (>10), the barrier integrity of the nail could be altered (Smith et al. 2010).

Aqueous solubility and the concentration of a drug in an iontophoretic delivery system can be critical for effective trans-ungual iontophoretic delivery, particularly for hydrophobic permeants such as antifungal drugs. A study on ionic strength effect suggests that positively charged drugs can effectively partition and transport into the nail at low aqueous drug concentrations at the physiological pH due to charge–charge interactions (Smith et al. 2009). It also suggests that the disadvantages of delivering cationic drugs of low solubilities might not be significant due to enhanced partitioning of the drugs into the nail at low solution ionic strength. For negatively charged

SA, the enhancement of cathodal iontophoretic transport at pH 7.1 was increased at higher concentrations of SA and an optimal ionic strength of buffer solutions existed for effective delivery of SA (Murthy et al. 2007b). Increasing the mole fraction of a drug ion in the system can reduce the ion competition from endogenous small counterions and exogenous coions according to Equation 8.6. However, increasing the concentration of a drug ion might not be possible owing to the limited drug solubility in the system. Cosolvent can be used in the nail formulation to increase the drug solubility, but the incorporation of a cosolvent is likely to dehydrate the nail and decrease the nail permeability and trans-ungual iontophoretic transport. The presence of a cosolvent may also affect the pKa of a permeant and thus the ionization of the permeant for effective iontophoretic delivery. There can also be side effects associated with the cosolvent in the treatment.

8.1.8 Trans-Ungual Iontophoretic Delivery of Antifungal Agents

8.1.8.1 Ciclopirox

Penlac®, a nail lacquer of ciclopirox olamine, is applied daily for 48 weeks, and repeated dosing for 1–2 weeks is usually required to achieve ciclopirox concentrations in the deeper layers of the nail higher than its minimum inhibitory concentration (MIC) (Bohn and Kraemer 2000). Cathodal iontophoretic transport of ciclopirox across partially hydrated human nail has been explored *in vitro* in side-by-side diffusion cells with constant voltage iontophoresis using a 9-V battery (Hao et al. 2009). Iontophoresis increased the flux of ciclopirox across the nail by approximately 10 times compared to passive delivery from the same formulation or from Penlac. A significant amount of ciclopirox was loaded into and released from the nails after iontophoresis. The estimated ciclopirox concentrations in the nail and calculated for the tissues beneath the nail plate were found to be above the MICs of ciclopirox for dermatophytic molds. The apparent transport lag time decreased from 20 h in passive transport to 2 h in iontophoretic transport. The shorter transport lag time, enhanced ciclopirox penetration across the nail, and extended ciclopirox release from the drug depot in the nail suggest the possibility of reducing the treatment time and frequency as well as enhancing treatment efficacy using iontophoresis. Although ciclopirox penetration was enhanced by iontophoresis, the enhancement factor was much lower than that predicted from the Nernst–Planck equation at 9 V. This discrepancy between the experimental data and the Nernst–Planck prediction as well as the low ciclopirox transference number were discussed (Hao et al. 2009).

8.1.8.2 Terbinafine

The water solubility of terbinafine is pH-dependent and significantly increased at lower pH (e.g., pH 3). Anodal iontophoresis of terbinafine in aqueous solution across fully hydrated nail was investigated using Franz diffusion cells at a constant current of 0.5 mA/cm^2 for 1 h followed by 1 h passive delivery, and this iontophoresis and passive cycle was repeated twice for a total treatment of 6 h (Nair et al. 2009d). Iontophoresis increased the amount of drug permeated across the nail and the drug load in the nail by approximately sixfold compared to passive delivery. Iontophoresis resulted in significant drug distribution through the nail and into the deeper layers of the nail while the passive method only delivered the drug onto the superficial layer of the nail. Drug permeation through the nail and the drug loaded in the nail were found to correlate with the applied electrical dose (i.e., current × duration), respectively. This suggests the possibility of individualizing treatment by controlling the applied electrical dose. The slow release profile of terbinafine from the drug depot in the nail further suggests the possibility of an intermittent therapy strategy to improve patient compliance.

An improved formulation of terbinafine was tested in a 1-h single trans-ungual anodal iontophoresis application (Nair et al. 2009a) using the method described previously (Nair et al. 2009d). Similar findings on the drug permeation, loading, and release as those reported previously (Nair et al. 2009d) were obtained. Iontophoresis significantly enhanced terbinafine permeation through the nail compared to passive delivery. The iontophoretic fluxes were 0.29 μg/cm^2/min at a current density of 0.5 mA/cm^2 and 0.44 μg/cm^2/min at 1 mA/cm^2. Iontophoresis also increased the drug loading into the nail both in the active diffusion area (i.e., the area in direct contact with the donor solution) and in the peripheral area (i.e., the area surrounding the active area but not in contact with the donor solution). Microbiological studies showed that an effective level of terbinafine was released from the ventral side of the drug-loaded nails into a release medium in an agar plate with antifungal activity that was observed to last for over 50 days. These results suggest the possibility of delivering an effective amount of the drug into and across the nail using iontophoresis at a safe current density (e.g., $\leq$0.5 mA/cm^2) in a short duration ($\leq$1 h) to achieve antifungal therapy.

The formulation developed by Nair et al. (2009b) was also tested in an ex vivo toe model with two applicators (nail-only and nail/skin applicators) in iontophoretic delivery of terbinafine. In this study, anodal iontophoresis was applied on the cadaver toenail at 0.5 mA for 20 min and the total electrical dose was 10 mA min. With the nail-only applicator and iontophoresis, a sufficient amount of terbinafine (2.8 μg/g tissue) was delivered into the nail bed and no measurable drug was delivered into the surrounding skin area. With the nail/skin applicator, drug permeation into the surrounding skin was observed after iontophoretic delivery. The amounts of terbinafine in the nail delivered by both applicators were estimated to be higher than the MIC of the drug. Effective amounts of terbinafine were released from the drug-loaded nails for over 60 days after iontophoresis. It was suggested that the nail-only applicator could be utilized for drug delivery to the nail plate and the nail bed, and the nail/skin applicator would allow drug delivery into the surrounding skin tissues as well as into the nail to ensure complete eradication of the fungal infection. Nair et al. (2011) further tested the developed terbinafine formulation in onychomycotic nails and found that drug delivery in the onychomycotic nails was not significantly different from that in healthy nails. Nair et al. (2009c, 2010) also combined iontophoresis and other enhancing techniques (e.g., chemical enhancer) to further increase trans-ungual transport of terbinafine.

Recently, the delivery of terbinafine under an electric current of 0.1 mA/cm^2 was tested in patients with onychomycosis (Amichai et al. 2010). The iontophoretic device and setup are shown in Figure 8.1. The patients received the overnight iontophoresis treatment daily, 5 days a week, for 4 weeks. A promising clinical response was reported in patients treated with iontophoresis of terbinafine. At the end of the

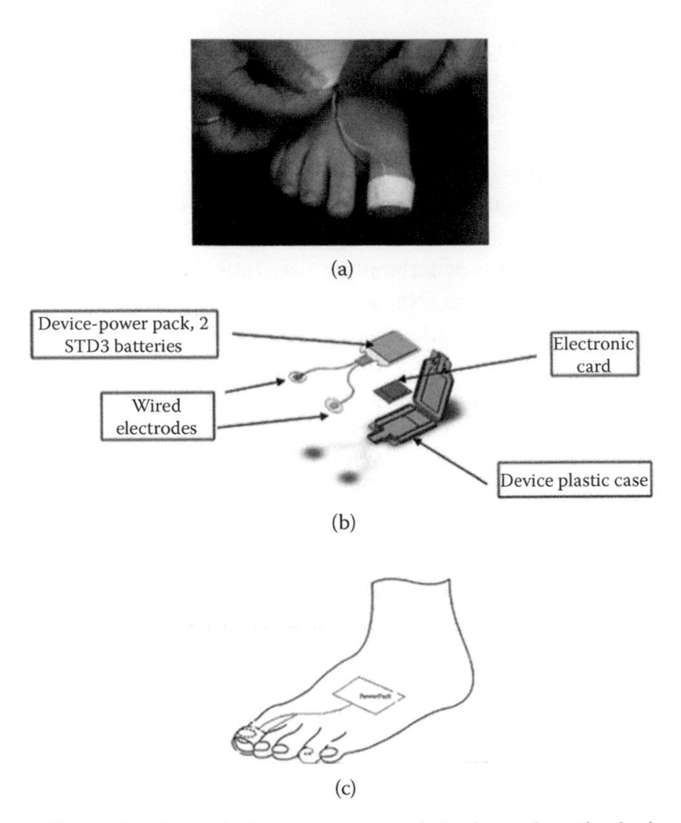

FIGURE 8.1 **(See color insert.)** (a) Appearance of the iontophoretic device. (b) Device components. (c) Device application: the plastic case containing an electronic card and power pack is situated on the upper foot; the treatment patch is placed on the infected nail and two-wired electrodes applied above (active) and beneath (counter) the toe. (Reprinted from Amichai et al., *British Journal of Dermatology*, 162, 46–50, 2010. With permission of Wiley and Blackwell publisher.)

treatment, the percentage of patients showing positive responses was significantly higher in the iontophoresis-treated group than in the passive delivery group. The iontophoresis treatment was concluded to be efficacious, safe, and well-tolerated.

8.2 ELECTROPULSATION-MEDIATED DRUG DELIVERY

Electropulsation technique has been explored to enhance terbinafine delivery across human nail, and protocols of different voltages, length of electrical pulses, and frequencies have been investigated (Nair et al. 2008). In these *in vitro* Franz diffusion cell experiments, the electrical pulses were applied on the human cadaver nail plate for 30 min through an electrode placed in the donor containing the drug solution and an electrode in the receptor containing PBS. Depending on the protocol applied, the permeation of terbinafine was significantly enhanced by electropulsation up to an order of magnitude over the control. The drug loaded into the nail was also enhanced by the electropulsation. The predominant enhancement mechanism of electropulsation was found to be permeabilization of the nail plate. Microscopic studies showed that the electrical pulses caused structural changes in the dorsal layer of the nail plate. These results suggest that electropulsation is a potential method to rapidly deliver drugs across the nail.

8.3 ULTRASOUND-MEDIATED DELIVERY

Low-frequency ultrasound has been explored to enhance drug penetration through bovine hoof membrane using metformin as a model drug (Torkar et al. 2007). The hoof membrane was pretreated with 20 kHz ultrasound for 1 min and a duty cycle of 2 sec on and 2 sec off prior to the transport experiment with modified Franz diffusion cells. The flux of the drug through the ultrasound-treated nail was enhanced by approximately threefold compared to the untreated nail (Torkar et al. 2007). The enhanced drug permeation suggested ultrasound-induced disruption of the bovine hoof membrane, probably due to cavitation mechanism (Murdan 2008). Ultrasound was also found to enhance ciclopirox permeation through human nail from Penlac nail lacquer (Mididoddi et al. 2006). A new ultrasound-mediated drug delivery system was recently developed to increase the permeability of the nail (Abadi and Zderic 2011). The results from the study with canine toenails and a blue dye as a model drug show that the developed system might be a promising approach in trans-ungual drug delivery.

8.4 MICROCONDUIT INDUCTION TECHNIQUE

A microscissioning technique using inert, sharp particles driven by a stream of inert gas at moderate velocity to impinge stochastically on the skin and nail through a mask with an array of microholes to create microconduits in diameter range of 50–200 μm on a small area was developed and applied to ungual drug delivery (Gowrishankar et al. 2009). It was believed that ungual microconduits created by microscissioning could provide the penetration pathways for effective delivery of a drug into the nail bed. In an *in vivo* human study (Gowrishankar et al. 2009), four microconduits were microscissioned in the toenail of a human subject and 1% terbinafine cream was applied to the microconduits daily for 4 months. The treated nail showed no discoloration associated with fungal infection at the end of the treatment. However, the inability to control the depth of the microconduits is one of the disadvantages of this trans-ungual delivery method (Murdan 2008).

PathFormer (Path Scientific LLC, Carlisle, MA) is an electrosurgical handheld medical device that drills holes in the nail. It is a device approved by the U.S. Food and Drug Administration (FDA) for nail trephination and has been reported to successfully relieve patients' pain caused by subungual hematomas (Salter et al. 2006). The PathFormer drills microholes on the nail in a controlled manner by monitoring nail electrical resistance (i.e., mesoscissioning technology) such that the drill penetrates the nail to a specific depth without breaching the nail plate into the nail bed. The electrical resistance decreases from 5 MΩ for undrilled nail to 10–20 kΩ when the nail bed is reached. The procedure was reported to be well tolerated in a pilot study when five holes of diameter 400 μm were drilled in toenails of healthy subjects (Ciocon et al. 2006). The PathFormer was tested with 1% terbinafine cream to deliver terbinafine into toenails of human subjects with onychomycosis in a pilot study (Boker et al. 2007). The cream was applied to the nail on a row of microconduits twice daily for 24 weeks. After the 24-week treatment, 64% of the subjects in the treated group showed improvement in clinical assessment compared to 28% in the placebo group.

8.5 ULTRAVIOLET B THERAPY OF NAIL DISEASES

Ultraviolet B (UVB, 280–320 nm) light is clinically used in phototherapy in psoriasis treatment (Stein et al. 2008). However, exposure of unaffected area to the UVB light may cause irritation and other adverse effects. Among different types of lasers, excimer lasers are a form of UV laser that are commonly used in clinical surgery. The UV light from an excimer laser disrupts the molecular bonds of the surface tissues through ablation rather than heating and burning, and thus leaves the deeper tissues unaffected. Particularly, 308-nm excimer laser emitted from a xenon chloride gas excimer has been shown to be safe and effective in treating psoriasis (Trott et al. 2008). The excimer laser has the benefits of narrowband UVB (NB-UVB) and has been FDA-approved in the treatment of psoriasis. The effects of laser light of different wavelengths and pulse duration on nail plate *in vitro* have been investigated (Neev et al. 1997). Upon exposing the nail to laser light, energy would be absorbed by water or protein in the nail plate and the scattered heat would lead to removal of nail layers. Craters could thus be formed on the laser-treated nail *in vitro*. The reduction of the nail integrity due to the formation of craters was suggested to enhance permeant penetration into the nail. However, very few reports on laser treatment on nail are available. A novel nonlaser 308-nm excimer light therapy was most effective for palmoplantar pustular psoriasis treatment but was ineffective for nail psoriasis treatment (Aubin et al. 2005).

8.6 CONCLUSION

In conclusion, this chapter summarizes recent developments in physical enhancement techniques to improve trans-ungual drug delivery. Although trans-ungual iontophoresis has shown potential to increase drug penetration into and across the nail for the treatment of nail disorders, the clinical efficacy and safety of trans-unugal iontophoresis have not been fully established. Other physical enhancement techniques have not received significant attention as evidenced by the limited research articles published in this area.

REFERENCES

Abadi, D., and V. Zderic. 2011. "Ultrasound-Mediated Nail Drug Delivery System." *Journal of Ultrasound in Medicine* 30: 1723–30.

Amichai, B., B. Nitzan, R. Mosckovitz, and A. Shemer. 2010. "Iontophoretic Delivery of Terbinafine in Onychomycosis: A Preliminary Study." *British Journal of Dermatology* 162: 46–50.

Aubin, F., M. Vigan, E. Puzenat, D. Blanc, C. Drobacheff, P. Deprez, P. Humbert, and R. Laurent. 2005. "Evaluation of a Novel 308-nm Monochromatic Excimer Light Delivery System in Dermatology: A Pilot Study in Different Chronic Localized Dermatoses." *British Journal of Dermatology* 152: 99–103.

Bohn, M., and K. T. Kraemer. 2000. "Dermatopharmacology of Ciclopirox Nail Lacquer Topical Solution 8% in the Treatment of Onychomycosis." *Journal of the American Academy of Dermatology* 43: S57–69.

Boker, A., Y. S. Bae, T. R. Gowrishankar, D. Ciocon, and A. B. Kimball. 2007. "A Double-Blind, Placebo-Controlled, Pilot Study of 1% Terbinafine Cream Delivered via Toenail Microconduits for the Treatment of Subungual Onychomycosis." Paper presented at the 65th Annual Meeting of the American Academy of Dermatology, Washington D.C.

Christodoulou, P., P. G. Doxas, C. E. Papadakis, P. Prassopoulos, T. Maris, and E. S. Helidonis. 2003. "Transtympanic Iontophoresis of Gadopentetate Dimeglumine: Preliminary Results." *Otolaryngology - Head and Neck Surgery* 129: 408–13.

Ciocon, D., T. R. Gowrishankar, T. Herndon, and A. B. Kimball. 2006. "How Low Should You Go: Novel Device for Nail Trephination." *Dermatology Surgery* 32: 828–33.

Deen, W. M. 1987. "Hindered Transport of Large Molecules in Liquid-Filled Pores." *AIChE Journal* 33: 1409–25.

Dutet, J., and M. B. Delgado-Charro. 2009. "In Vivo Transungual Iontophoresis: Effect of DC Current Application on Ionic Transport and on Transonychial Water Loss." *Journal of Controlled Release* 140: 117–25.

Dutet, J., and M. B. Delgado-Charro. 2010a. "Electroosmotic Transport of Mannitol across Human Nail during Constant Current Iontophoresis." *Journal of Pharmacy and Pharmacology* 62: 721–9.

Dutet, J., and M. B. Delgado-Charro. 2010b. "Transungual Iontophoresis of Lithium and Sodium: Effect of pH and Co-Ion Competition on Cationic Transport Numbers." *Journal of Controlled Release* 144: 168–74.

Eljarrat-Binstock, E., and A. J. Domb. 2006. "Iontophoresis: A Non-Invasive Ocular Drug Delivery." *Journal of Controlled Release* 110: 479–89.

Gowrishankar, T. R., T. O. Herndon, and J. C. Weaver. 2009. "Transdermal Drug Delivery by Localized Intervention: Field-Confined Skin Electroporation and Dermal Microscissioning." *IEEE Engineering in Medicine and Biology Magazine* 28: 55–63.

Gupchup, G. V., and J. L. Zatz. 1999. "Structural Characteristics and Permeability Properties of the Human Nail: A Review." *Journal of Cosmetic Science* 50: 363–85.

Hao, J., and S. K. Li. 2008a. "Transungual Iontophoretic Transport of Polar Neutral and Positively Charged Model Permeants: Effects of Electrophoresis and Electroosmosis." *Journal of Pharmaceutical Sciences* 97: 893–905.

Hao, J., and S. K. Li. 2008b. "Mechanistic Study of Electroosmotic Transport across Hydrated Nail Plates: Effects of pH and Ionic Strength." *Journal of Pharmaceutical Sciences* 97: 5186–97.

Hao, J., K. A. Smith, and S. K. Li. 2008. "Chemical Method to Enhance Transungual Transport and Iontophoresis Efficiency." *International Journal of Pharmaceutics* 357: 61–9.

Hao, J., K. A. Smith, and S. K. Li. 2009. "Iontophoretically Enhanced Ciclopirox Delivery into and across Human Nail Plate." *Journal of Pharmaceutical Sciences* 98: 3608–16.

Hao, J., K. A. Smith, and S. K. Li. 2010. "Time-Dependent Electrical Properties of Human Nail upon Hydration In Vivo." *Journal of Pharmaceutical Sciences* 99: 107–18.

Jacobsen, J. 2001. "Buccal Iontophoretic Delivery of Atenolol HCl Employing a New In Vitro Three-Chamber Permeation Cell." *Journal of Controlled Release* 70: 83–95.

James, M. P., R. M. Graham, and J. English. 1986. "Percutaneous Iontophoresis of Prednisolone—A Pharmacokinetic Study." *Clinical and Experimental Dermatology* 11: 54–61.

Kalinowski, D. P., L. E. Edsberg, R. A. Hewson, R. H. Johnson, and M. S. Brogan. 2004. "Low-Voltage Direct Current As a Fungicidal Agent for Treating Onychomycosis." *Journal of the American Podiatric Medical Association* 94: 565–72.

Kasting, G. B. 1992. "Theoretical Models for Iontophoretic Delivery." *Advanced Drug Delivery Reviews* 9: 177–99.

Keister, J. C., and G. B. Kasting, 1986. "Ionic Mass-Transport through a Homogeneous Membrane in the Presence of a Uniform Electric Field." *Journal of Membrane Science* 29: 155–67.

Ledger, P. W. 1992. "Skin Biological Issues in Electrically Enhanced Transdermal Delivery." *Advanced Drug Delivery Reviews* 9: 289–307.

Lerner, E. N., E. H. van Zanten, and G. R. Stewart. 2004. "Enhanced Delivery of Octreotide to the Brain via Transnasal Iontophoretic Administration." *Journal of Drug Targeting* 12: 273–80.

Marshall, R. C. 1983. "Characterization of the Proteins of Human Hair and Nail by Electrophoresis." *Journal of Investigative Dermatology* 80: 519–24.

Mididoddi, P. K., S. Upadhye, J. Raymond, C. Church, S. Davis, and M. A. Repka. 2006. "Influence of Etching and Ultrasound on the Permeability of Ciclopirox through the Human Nail." Paper presented at the Annual Meeting of the American Association of Pharmaceutical Scientists, San Antonio, TX.

Murdan, S. 2002. "Drug Delivery to the Nail Following Topical Application." *International Journal of Pharmaceutics* 236: 1–26.

Murdan, S. 2008. "Enhancing the Nail Permeability of Topically Applied Drugs." *Expert Opinion on Drug Delivery* 5: 1267–82.

Murthy, S. N., D. C. Waddell, H. N. Shivakumar, A. Balaji, and C. P. Bowers. 2007a. "Iontophoretic Permselective Property of Human Nail." *Journal of Dermatology Science* 46: 150–2.

Murthy, S. N., D. E. Wiskirchen, and C. P. Bowers. 2007b. "Iontophoretic Drug Delivery across Human Nail." *Journal of Pharmaceutical Sciences* 96: 305–11.

Myles, M. E., D. M. Neumann, and J. M. Hill. 2005. "Recent Progress in Ocular Drug Delivery for Posterior Segment Disease: Emphasis on Transscleral Iontophoresis." *Advanced Drug Delivery Reviews* 57: 2063–79.

Nair, A. B., B. Chakraborty, and S. N. Murthy. 2010. "Effect of Polyethylene Glycols on the Trans-Ungual Delivery of Terbinafine." *Current Drug Delivery* 7: 407–14.

Nair, A. B., H. D. Kim, B. Chakraborty, J. Singh, M. Zaman, A. Gupta, P. M. Friden, and S. N. Murthy. 2009a. "Ungual and Trans-Ungual Iontophoretic Delivery of Terbinafine for the Treatment of Onychomycosis." *Journal of Pharmaceutical Sciences* 98: 4130–40.

Nair, A. B., H. D. Kim, S. P. Davis, R. Etheredge, M. Barsness, P. M. Friden, and S. N. Murthy. 2009b. "An Ex Vivo Toe Model Used to Assess Applicators for the Iontophoretic Ungual Delivery of Terbinafine." *Pharmaceutical Research* 26: 2194–201.

Nair, A. B., S. M. Sammeta, H. D. Kim, B. Chakraborty, P. M. Friden, and S. N. Murthy. 2009c. "Alteration of the Diffusional Barrier Property of the Nail Leads to Greater Terbinafine Drug Loading and Permeation." *International Journal of Pharmaceutics* 375: 22–7.

Nair, A. B., S. M. Sammeta, and S. N. Murthy. 2008. "Electropulsation for Trans-Ungual Drug Delivery." Paper presented at the Annual Meeting of the American Association of Pharmaceutical Scientists, Altanta, GA.

Nair, A. B., S. R. Vaka, and S. N. Murthy. 2011. "Transungual Delivery of Terbinafine by Iontophoresis in Onychomycotic Nails." *Drug Development and Industrial Pharmacy* 37: 1253–8.

Nair, A. B., S. R. Vaka, S. M. Sammeta, H. D. Kim, P. M. Friden, B. Chakraborty, and S. N. Murthy. 2009d. "Trans-Ungual Iontophoretic Delivery of Terbinafine." *Journal of Pharmaceutical Sciences* 98: 1788–96.

Neev, J., J. S. Nelson, M. Critelli, J. L. McCullough, E. Cheung, W. A. Carrasco, A. M. Rubenchik, L. B. Da Silva, M. D. Perry, and B. C. Stuart. 1997. "Ablation of Human Nail by Pulsed Lasers." *Lasers in Surgery and Medicine* 21: 186–92.

Phipps, J. B., and J. R. Gyory. 1992. "Transdermal Ion Migration." *Advanced Drug Delivery Reviews* 9: 137–76.

Pikal, M. J. 2001. "The Role of Electroosmotic Flow in Transdermal Iontophoresis." *Advanced Drug Delivery Reviews* 46: 281–305.

Salter, S. A., D. H. Ciocon, T. R. Gowrishankar, and A. B. Kimball. 2006. "Controlled Nail Trephination for Subungual Hematoma." *American Journal of Emergency Medicine* 24: 875–7.

Sims, S. M., W. I. Higuchi, V. Srinivasan, and K. D. Peck. 1993. "Ionic Partition Coefficients and Electroosmotic Flow in Cylindrical Pores: Comparison of the Predictions of the Poisson-Boltzmann Equation with Experiment." *Journal of Colloid and Interface Science* 155: 210–20.

Smith, K. A., J. Hao, and S. K. Li. 2009. "Effects of Ionic Strength on Passive and Iontophoretic Transport of Cationic Permeant across Human Nail." *Pharmaceutical Research* 26: 1446–55.

Smith, K. A., J. Hao, and S. K. Li. 2010. "Influence of pH on Transungual Passive and Iontophoretic Transport." *Journal of Pharmaceutical Sciences* 99: 1955–67.

Stein, K. R., D. J. Pearce, and S. R. Feldman. 2008. "Targeted UV Therapy in the Treatment of Psoriasis." *Journal of Dermatological Treatment* 19: 141–5.

Torkar, A., J. Kristl, and S. Murdan. 2007. "Low-Frequency Ultrasound to Enhance Topical Drug Delivery to the Nail." Paper presented at the Annual Meeting of the American Association of Pharmaceutical Scientists, San Diego, CA.

Trott, J., W. Gerber, S. Hammes, and H. M. Ockenfels. 2008. "The Effectiveness of PUVA Treatment in Severe Psoriasis is Significantly Increased by Additional UV 308-nm Excimer Laser Sessions." *European Journal of Dermatology* 18: 55–60.

Viscusi, E. R., and T. A. Witkowski. 2005. "Iontophoresis: The Process Behind Noninvasive Drug Delivery." *Regional Anesthesia and Pain Medicine* 30: 292–4.

9 Pharmacokinetics of Drugs in the Nail Apparatus

Danièle Debruyne and Antoine Coquerel

CONTENTS

9.1 INTRODUCTION

Onychomycosis causes disturbance of the nail morphology and results in thickening, discoloration, splitting, and lifting of the nail from the nail bed (Baker et al. 2007). Old age (poor peripheral circulation, inactivity, inability to cut the toenails, etc.), diabetes, psoriasis or cancer diseases, abnormal nail morphology or repeated nail trauma, immunodeficiency, and genetic factors were identified as risk factors for initial infections (Pierard et al. 2005; Tosti et al. 2005). In onychomycosis, toenails are four to ten times more frequently affected than fingernails, probably because of their slower growth and increased exposure to injury and infecting organisms (Baran and Kaoukhov 2005). Fingernails appear to be involved more often in females whereas toenails are involved more often in males. Clinical manifestations of nail psoriasis (pitting, discoloration, onycholysis and subungual hyperkeratosis, nail plate crumbling, and splinter hemorrhages; Murdan 2008) affect both fingernails and toenails in the majority of patients (Baran 2010).

The goal for antifungal therapy is complete cure, defined as mycological cure (fungi elimination from the nail) and clinical cure (normal appearing nail). The current treatment modalities for onychomycosis include chemical or mechanical debridements, systemic antifungal drug therapies (Finch and Warshaw 2007), and topical agents that may be useful adjuncts. In some cases, topical monotherapy may be preferred due to better drug targeting to the infected site, minimal adverse systemic events and drug interactions, and possibly improved patient compliance (Elkeeb et al. 2010; Vaka et al. 2010). Meanwhile, topical solutions and creams are easily removed by washing or wiping, which further reduces delivery of the drug to the nail plate (Baran and Kaoukhov 2005). Therefore, treatment choice will depend on the severity and nail location of the disease, the thickness of the involved nails, the response to previous therapies, the current comedications, the patient motivation and preference, and the cost of therapy. The current management of nail psoriasis includes topical and oral treatment or biological therapies.

Better drug penetration techniques that would increase the antifungal drug concentration, persistence, and distribution in the deeper layers of the nail might improve the efficacy of topical antifungal treatment (Amichai et al. 2010a). Developing more effective methods for ungual drug delivery is an important objective for the pharmaceutical industry, and newer therapies include topical antifungal formulated to provide delivery of drug to the nail bed and nail plate (Finch and Warshaw 2007). Other emerging therapies centered on altering the nail plate barrier by means of mechanical methods (abrasion and trephination) or chemical treatments (keratolytic enzyme, *N*-acetyl cysteine, etc.) that interact with disulfide bridges in keratin molecules (Vejnovic et al. 2010b). Different techniques are also described in the literature, for example, iontophoretic drug delivery, where the driving force of ions is an electric field. Also ongoing drug development activities have also focused on the discovery of inherently penetrable antifungal such as AN-2690 (Alley et al. 2007).

9.2 DRUGS

Causative agents of onychomycosis are predominantly dermatophytes (*Trichophyton rubrum*, *Trichophyton mentagrophytes*, *Epidermophyton*, and *Microsporum*), occasionally yeasts (*Candida albicans*), and non-dermatophyte molds (*Scopulariopsis* and *Scytalidium*, and phaeoacremonium parasiticum).

Infections range from superficial, causing discoloration of the nail plate, to severe, resulting in separation of the nail plate from the nail bed (onycholysis), accumulation of subungual debris, and nail plate dystrophy (Alley 2007; Hao et al. 2009b). There are three major subtypes of dermatophytic onychomycosis including distal subungual onychomycosis (the most common type), superficial white onychomycosis representing about 10% of all cases, and proximal subungual onychomycosis. Effects from *Candida* infection or colonization occur in four forms affecting the proximal and lateral folds, the nail plate and paronychia, the nail plate with separation from bed, and eventually, total dystrophy (Nunley and Cornelius 2008).

The principal mechanisms of action of antifungal agents involve disruption of spindle and cytoplasmic microtubule function, depletion of or binding to ergosterol (fungistatic activity), and accumulation of intracellular squalene (fungicidal activity) (Elewski 1993). Antifungal medications include imidazoles (e.g., clotrimazole, tioconazole, econazole, ketoconazole, miconazole, sulconazole, and oxiconazole), which have fungistatic properties *in vitro*; allylamines/benzylamines (e.g., naftifine, terbinafine, and butenafine); and polyenes (e.g., nystatin), which have fungistatic and fungicidal properties *in vitro* (Elkeeb et al. 2010). Abafungin is the first member of a novel class of synthetic antifungal compounds, the arylguanidines (Borelli et al. 2008; Pierard et al. 2007). The physical and pharmacological characteristics of some antifungal agents are summarized in Table 9.1.

Currently, the treatment of choice for toenail onychomycosis is oral terbinafine, but oral itraconazole or fluconazole can also be considered. Among the different topical agents are nail lacquers containing ciclopirox olamine 8%, amorolfine 5%, or tioconazole 28%, or a combination of these agents.

TABLE 9.1
Physical and Pharmacological Characteristics of Some Antifungal Agents

Drug	Physicochemical Properties		Pharmacological Properties	Mechanism of Action	Minimum Inhibitory Concentration
	M	Log *P*			
Amorolfine	317.5	5.6	Fungicidal against *Candida* species and dermatophytes (*Trichophyton mentagrophytes*)	Inhibition of D14 reductase and D7-D8 isomerase, which deplete ergosterol	0.01–0.08 μg/mL (dermatophytes); 0.5–16 μg/mL (*Candida parapsilosis*)
Bifonazole	310	4.8	Active against *Malassezia furfur* and *Candida* species	Inhibition of the production of ergosterol by acting on the hydroxymethylglutaryl-CoA and the 24-methylene-dihydrolanosterol pathways	0.04–1.25 μg/mL (dermatophytes, b); 9.05 μg/mL (*Candida* species, c)
Butenafine	317.5	6.6	Active against dermatophytes and *Candida albicans*	Inhibition of ergosterol biosynthesis by blocking squalene epoxidation	0.03–0.25 μg/mL (dermatophytes, d)
Ciclopirox	207.3	2.3	Active against *M. furfur* and *Candida* spp. (a), *Trichophyton rubrum*, *Epidermophyton* spp., and *Scopulariopsis brevicaulis*	Chelation of metal cations that inhibit many enzymes, thus disrupting cellular activities (mitochondrial electron transport and subsequent cellular energy production)	0.016–8.0 μg/mL; 0.05 μg/mL (yeasts, a); 0.04 μg/mL (dermatophytes, a)
Fluconazole	306	1.1	Active against *C. albicans* and *Cryptococcus neoformans*, and *Microsporum* and *Trichophyton* species	Inhibition of the fungal enzyme (CYP450 14-alpha-demethylase) that converts lanosterol to ergosterol	0.063–32.0 μg/mL 16–64 mg/L
Griseofulvin	353	2.2	Fungistatic against dermatophytes: *Microsporum* species, *Epidermophyton floccosum*, and *Trichophyton* species	Inhibition of formation of intracellular microtubules preventing cell division of the fungus	0.43–0.95 μg/mL (dermatophytes, e); 0.25–2 μg/mL (*T. rubrum*, f)

Itraconazole	706	5.66	Fungistatic against dermatophytes (*Trichophyton* species), yeasts (*Candida* species), and other fungi (*Aspergillus*)	Inhibition of the fungal enzyme (CYP450 14-alpha-demethylase) that converts lanosterol to ergosterol	0.016–8.0 μg/mL; 1.0–2.0 mg/L; 0.18 μg/mL (yeasts, a); 2.3 μg/mL (dermatophytes, a); 0.031–1 μg/mL (*T. rubrum*, f)
Ketoconazole	531	4.4	Fungistatic against dermatophytes (*Microsporum* species, *Epidermophyton floccosum*, *Trichophyton* species) and yeasts (*Candida* species)	Inhibition of the fungal enzyme (CYP450 14-alpha-demethylase) that converts lanosterol to ergosterol	0.56 μg/mL (yeasts, a); 0.8 μg/mL (dermatophytes, a)
Sertaconazole	438	6.2	Fungistatic and fungicidal against *Candida* species	Inhibition of the fungal enzyme (CYP450 14-alpha-demethylase) that converts lanosterol to ergosterol	0.25–2 μg/mL (dermatophytes), 0.35–5 μg/mL (yeasts)
Terbinafine	291	5.9	Fungicidal against dermatophytes (*T. rubrum*, *T. mentagrophytes*) and yeasts (*C. parapsilosis* and *C. albicans*); fungistatic against other yeasts and molds	Inhibition of squalene epoxidase, thus blocking the biosynthesis of ergosterol	0.004–2.0 μg/mL; 1.8 μg/mL (yeasts, a); 0.031 μg/mL (*T. rubrum*, f); 0.003–0.006 μg/mL (dermatophytes, h)
Voriconazole	349	2.6	Active against *S. brevicaulis*, *Fusarium* spp., and *Scytalidium dimidiatum*	Inhibition of the fungal enzyme (CYP450 14-alpha-demethylase) that converts lanosterol to ergosterol	0.25–1.0 mg/L
AN-2690	152	1.24	Fungicidal against *T. rubrum*	Inhibition of the cytoplasmic leucyl-tRNA synthetase	0.125–8 μg/mL (g)

Notes: (a) Gupta et al. (2003); (b) Uchida et al. (1996); (c) Carrillo-Munoz and Torres-Rodriguez (1995); (d) Kokjohn et al. (2003); (e) Chadeganipour et al. (2004); (f) Santos and Hamdan (2007) (g) Kaur and Kakkar (2010); (h) Darkes et al. (2003).

The current management of nail psoriasis includes topical treatments such as steroids and vitamin D analogs and systemic or biological therapies.

9.3 CHALLENGES IN ACHIEVING TARGET LEVELS OF DRUG IN THE NAIL APPARATUS IN CASE OF TOPICAL ADMINISTRATION

The challenge for antifungal therapy is to permeate the nail barrier, disseminate throughout the nail unit, and accumulate and maintain effective concentrations into deep nail stratums at amounts above the minimal inhibitory concentration (MIC) to deliver therapeutic levels of active agents to the target site and to kill the pathogen (Alley 2007; Baran and Kaoukhov 2005). The keratinized structure of the human nail plate protects the fingertip and toe-tip but represents a resistive barrier to the permeation of drugs, which limits the accumulation and activity of drugs in the nail on topical application.

The poor permeability, the prolonged transport lag time, and the entrapment of drug in the nail keratin matrix contribute to disappointing topical efficacy in nail diseases (Murdan 2007). The mycological cure rate was reported to be less than 50%, and relapse or reinfection is also common (Finch and Warshaw 2007). Topical monotherapy is considered to be effective only in early distal nail disease when the nail matrix is not involved, that is, in approximately 74% of patients (Baran et al. 2008b). Continuous or intermittent (pulse dose administration) oral treatment is required to reach the nail matrix and the nail bed. Unfortunately, oral therapy is associated with severe side effects, interactions and toxicity, which may require a clinical survey of the hepatic function by means of biological tests. Synergy has been demonstrated among antifungal drugs, and a combination of systemic and topical treatments (ciclopirox olamine and terbinafine or itraconazole) can be proposed to elicit a better therapeutic outcome. Topical antifungal treatment several times a week may also be recommended after the completion of oral therapy to prevent reinfection of the cured nail, although there is little data on the efficacy of this practice (Finch and Warshaw 2007).

9.4 FACTORS INFLUENCING THE BIOAVAILABILITY OF DRUGS FOLLOWING TOPICAL ADMINISTRATION

Different factors influence drug transport into the nail plate: nail properties (thickness, hydration, and integrity), drug–keratin interactions in the nail plate, physicochemical properties of the drug (size, charge, and lipophilicity/hydrophobicity), formulation characteristics (vehicle and drug concentration), and presence of permeation enhancers.

9.4.1 Anatomical and Physiological Factors

9.4.1.1 Nail Growth Rate

Healthy nails grow continuously from the nail matrix. Overall growth of the fingernail or toenail is about 3–4 mm/month or 1.5–2 mm/month, respectively. However, this does vary from person to person. The complete outgrowth takes 4–6 months for fingernails and 12–18 months for toenails. There are many contributing factors such as age, weather, health, and a well-balanced diet that influence nail outgrowth Physiological and environmental factors affect the nail growth: it is greater in children under 14 years and decreases slowly with aging with corneocytes larger in old age; temperature appears to influence it with a faster growth observed in summer compared to winter; growth is accelerated during sleep; malnutrition or poor blood circulation, some illnesses, and certain medications may retard its rate; and in pregnancy, nail growth is faster, but slower in newborns. Each nail grows at a different rate. The index finger generally grows at an accelerated rate compared to the others. The smallest finger and the thumbnail have the slowest nail growth rate. Therefore, nail growth is a factor that plays an important role in the treatment duration.

9.4.1.2 Nail Thickness

The thickness of the nail determined by the nail matrix varies from 0.5 to 0.75 mm for fingernails and can reach 1 mm for toenails. The nail plate gains thickness and density as it grows distally, and the final nail thickness at the distal part of the nail is related to initial thickness of the proximal matrix. Matrix shapes and sizes vary per person. A short matrix produces fewer cells, and as a result, a thinner nail. The longer the matrix, the thicker the nail will be. There is an increase in nail thickness with age, particularly up to 20 years. Factors such as linear rate of nail growth, vascular supply, subungual hyperkeratosis, and drugs may influence thickness. The drug penetration in the deep layers of the nail is more difficult if the nail is thick and can be a factor of therapeutic failure.

9.4.1.3 Nail Barrier Properties

The human nail plate (0.25–0.6 mm) consists of tightly packed dead cells that are highly keratinized and includes three layers depending on the nature of keratin (Chaudari et al. 2006): (i) the ventral layer composed of soft hyponychial is very thin and connects the nail plate to the nail bed; (ii) the intermediate layer represents the majority of the nail thickness (75%) and consists of soft keratin; and (iii) the dorsal layer is constituted of hard keratin (trichocyte) of impermeable nature. Keratin fibers are oriented in a plane parallel to the surface and in the transverse axis; small amounts of lipids mainly represented by cholesterol separate the keratin layers; and proteins rich in cystine, a sulfur-bearing amino acid, act as glue.

Human nail plate consists of 80% protein (mainly α-keratins), water, 1–2% lipids, and trace amounts of minerals, metals, and electrolytes (Fleckman 2005; Hao et al. 2009a).

The thickness of the nail creates a long diffusion pathway for drug delivery. The stable disulfide bonds are responsible for the hardness of the nail. Binding of drug to keratin reduces the availability of the active (free) drug, weakens the concentration gradient, and limits deep penetration (Murthy et al. 2007b). Unlike the skin, the nail plate is not a lipophilic barrier and the water permeation rate of the nail remains around 10 times higher than that of the stratum corneum. Therefore, with its dense keratin fabric network and a high capacity for flux of water, the nail plate is described as a hydrophilic gel (Alley et al. 2007).

Isoelectric point of keratin is thought to be around pH 5 (Murthy et al. 2007b), which means that the proteins in the nail plate are negatively charged, attracting positively charged molecules. At lower pH, keratin is positively charged and at pH 7.4 it is negatively charged (Marshall 1983).

9.4.1.4 Nail Water Content

Water content contributes to the elasticity of the nail (Murdan 2002). Under physiological conditions, the nail plate is partially hydrated and a water gradient exists from the nail bed to the nail surface in the nail plate (Gunt and Kasting 2007). It has been suggested that nail plate could be divided into a superficial dry compartment and a deep humid one (de Berker et al. 2007). It has also been demonstrated that the average water content of the nail plate is significantly lower in winter than in summer (Egawa et al. 2006).

The normal water content in the nail is about 18–20% depending on the relative humidity to which it is exposed (Gunt et al. 2007). The nails approached 90% of complete hydration in half an hour after the nails were soaked in water (Hao et al. 2008). Water hydrates the nail plate, increases distance between the keratin fibers, and enlarges pores that increase molecule permeation. Water diffusivity in hard keratins ranges from about 1.10^{-10} cm^2/s in very dry samples to 1.10^{-7} cm^2/s in hydrated samples (Gunt et al. 2007). Formulations or treatments, which increase nail hydration, have the potential to improve the permeability of topical agents (Gunt and Kasting 2007). For example, increasing the ambient relative humidity from 15% to 100% enhanced permeation of [^{3}H]-ketoconazole by a factor of 3. The effect of nail hydration upon its electric resistance and the importance of nail hydration in trans-ungual iontophoretic delivery have also been demonstrated (Hao et al. 2009b). The *in vivo* nail electric resistance decreases significantly upon 2-h nail hydration and also demonstrates the significance of nail hydration in trans-ungual iontophoretic delivery (Hao et al. 2009b).

A minor lipid pathway also permits the passage of very hydrophobic substances.

9.4.1.5 Peripheral Circulation

The rich blood supply of the nail derives from paired digital arteries giving small collaterals from dorsal and ventral anastomosis. The vascularization of the nail matrix is ensured by the superficial rami and that of the nail bed by the deep rami; the germinal part of the nail bed shows poor vascularization. In the nail bed, numerous capillary loops are observed, with the longest capillary loops being at the hyponychium. In old age, the whole subungual area may show thickening of vessel walls with vascular elastic tissue fragmentation. After oral therapy and considering the mean nail growth rate of 0.4 mm/week, the drug was detected in distal nail clippings earlier than would be expected if incorporation occurred only through the nail matrix by uptake into newly formed nail. To explain this rapid detection at the distal edge of the nail, diffusion through the nail plate from the nail bed has been suggested to be a primary route of entry into this tissue (Dykes et al. 1990; Finlay 1992; Mathieu et al. 1991). After topical administration, antifungal concentrations are commonly observed in blood.

9.4.2 Aging and Environment

Aging can slow down the nail growth and cause brittle, dull, opaque, or yellowish nails. Toenails may become hard and thick, and the tips of the fingernails may fragment.

Other significant causes of nail abnormalities include environmental factors such as exposure to chemicals or harsh detergents, prolonged water exposure, use of certain medications, and injury or trauma, which may affect the nail morphology. It is often a challenge to achieve significant drug levels in the severely abnormal nails; hence, the end results may vary significantly in the same individual and between individuals.

9.4.3 Disease Factors

The signs and symptoms of onychomycosis depend on the subtype form of fungus and its density in the tissue. In distal lateral subungual onychomycosis, the nail begins to become cloudy and shows subungual hyperkeratosis and onycholysis, which is usually yellow-white in color. In time, the nail bed becomes hard and thick and eventually separates from the nail bed. In proximal subungual onychomycosis, white spotting, streaking, or discoloration begin at the nail fold and gradually permeate the nail. With white superficial onychomycosis, the nail roughens and crumbles easily. The nail surface will also have small white patches that are speckled or powdery. Ungual psoriasis often causes ungual hyperkeratosis, onycholysis, onychodystrophy, and paronychia with fine pitting of the nail surface. Besides these specific nail diseases, shape or growth changes of the nail may result from systemic diseases. For example, clubbing is associated with pulmonary malignancy or congenital heart disease, onycholysis with hyperthyroidism or sarcoidosis, koilonychia and Beau's lines with Raynaud's disease or trauma, and yellow nail with nephritic syndrome (Fawcett et al. 2004). In yellow nail syndrome, the nail that grows half as fast grows twice as thick (Moffitt and De Berker 2000).

Regarding data collected from several studies, Debruyne and Coquerel (2001) concluded that the concentrations of oral itraconazole and terbinafine in distal clippings of unaffected nails were similar to those in affected nails after oral administration. Besides, the concentrations of fluconazole in normal nails appeared to be higher than in infected nails. After topical application, Kobayashi et al. (2004) suggested that the permeability through healthy and fungal nail plates was not significantly different even if the flux of drug through very heavily infected fungal nail plate might be higher than through a healthy nail plate. Thus, it is widely accepted that the fungal nail permeability can be estimated from healthy nail permeability data.

9.4.4 Formulation Factors

9.4.4.1 Drug Properties

Topical therapy can be optimized by the use of drugs with correct physicochemical properties for permeation into the nail plate. Nail penetration is dependent on molecular weight, octanol/water partition coefficient, and water solubility. Most existing antifungal drugs are originally designed for oral or skin applications and are consequently fairly lipophilic molecules, only sparingly soluble in water, and have a molecular weight of more than 300 Da (Alley et al. 2007). For optimal ungual permeation and uptake, drug molecules should have a small size and a high degree of water solubility. The pH of aqueous formulations affects the ionization of weakly acidic/base drug with polar drugs permeating to a relatively greater extent compared to nonpolar ones, whereas some unionized compounds may be more permeable than ionized forms due to the repulsion of ions by the charge of the keratin (Murthy et al. 2007a). Nevertheless, the trans-nail flux of drug currently increases with the concentration of drug in contact with the nail plate (Kobayashi et al. 2004).

9.4.4.2 Type of Formulation

In theory, aqueous vehicles are less suitable than lipophilic vehicles, but in practice, these formulations do not adhere to the plate and can be easily cleared. The most convenient preparations are the nail lacquers. Volatile vehicles, used to deliver the drug, evaporate, leaving a water-insoluble and occlusive film on top of the nail plate that slowly releases the drug into the nail plate. The drug concentration in the polymer film is higher than the concentration in the applied nail lacquer. As examples, the ciclopirox concentration is 8% in the nail lacquer, increasing to 34.8% in the film; the amorolfine concentration in the film-forming solution is 5% and progresses to 27% with solvent evaporation (Marty 1995). The water-insoluble film also reduces water loss from the nail upper layer. The polymer film may be consequently regarded as a matrix-type controlled-release device that acts as a drug depot and, by increasing hydration of the nail, enhances drug diffusion (Baran and Kaoukhov 2005). The other benefit results from the barrier formed at the nail surface by the dried vehicle impeding the sustained recontamination of the nail.

By virtue of their positive occlusive influence on lipid and water content in dystrophic nails, nail patches could have beneficial therapeutic effects in onychomycosis. Although the difference was not significant, a light improvement in treatment outcome was observed in patients treated (after avulsion of the involved nail) with ketoconazole 2% cream or oxiconazole 1% cream with occlusion compared to controls (without occlusion). Also, recurrence occurred only in patients who did not receive occlusion.

Photodynamic therapy is defined as a medical treatment by which a combination of a sensitizing drug and visible light causes destruction of selected cells (Pong et al. 1997). A bioadhesive patch containing 5-aminolevulinic acid induced accumulation of the photosensitizer protoporphyrin IX and subsequent photodynamic destruction of fungi (Donnelly et al. 2005; Welsh et al. 2010). After 72 h, almost 90% of the drug had penetrated across the nail barrier.

9.4.4.3 Hydration

Inorganic salts such as sodium sulfite, sodium phosphate, potassium phosphate, calcium phosphate, and ammonium carbonate could be used as potent trans-ungual permeation enhancers (Nair et al. 2009b). Previously, Dorn et al. (1980) reported to prefer usage of 50% potassium iodide ointment in anhydrous lanolin instead of 40% urea for keratinolysis. Mechanistic studies revealed that the enhanced permeation by inorganic salts was mainly due to their ability to increase the nail hydration. The *in vitro* intranail permeation of terbinafine was enhanced (from threefold to fivefold) in the presence of these salts (0.5 M concentration) when compared with the control. Increase in salt concentration up to 1 M increased the permeation that decreased with salt concentration greater than 1 M (Nair et al. 2009b).

Hydrophobins that are water-soluble proteins with an ability to change physical properties of the surface in the way that hydrophobic surface becomes hydrophilic and vice versa are suggested for use in the treatment of nail diseases as stabilizers and enhancers (Vejnovic et al. 2010b). The dddition of hydrophobins in two out of three tested formulations improved the permeability coefficient. A remaining terbinafine reservoir in the nail was also observed compared to the reference (Vejnovic et al. 2010a).

9.4.5 Pretreatment Methods

Nail keratin is compact and hard and, as such, is impermeable, thus restricting drug access to the fungi-causing onychomycosis. Ungual permeation can be enhanced by disrupting the nail plate using physical techniques such as abrasion (filing and nail drilling), etching (tartaric acid and phosphoric acid), ablation, microporation of the nail plate by a handheld microcutting device, or chemical agents such as thiols, sodium sulfite, urea hydrogen peroxide, urea, or enzymes (Kumar and Kimball 2009; Murdan 2008). Some of the pretreatment methods are currently in clinical practice, while some are potential techniques under investigation.

9.4.5.1 Mechanical Methods

Physical permeation enhancement may be superior to chemical methods in delivering hydrophilic and macromolecular agents (Murthy et al. 2007a).

In many cases, especially in the estimated 55% of patients who have single nail onychomycosis, surgery may need to precede drug therapy in order to maximize the prospects of clinical and mycological cure. Mechanical onychomycosis treatments comprise debridement and surgical nail avulsion. Chemical nail avulsion was used primarily as an adjunct to topical or oral therapy and was generally ineffective when used alone (Finch and Warshaw 2007).

Various combinations of oral/topical antifungal and surgical techniques have been proposed, and there has been a resurgence of interest for these combination therapies. There are anecdotal reports of the success of nail avulsion followed by topical antifungal therapy, but this procedure was not found to be a very encouraging modality for the treatment of onychomycosis and is rarely used today (Grover et al. 2007).

Nail abrasion using sandpaper or nail files at the beginning of the treatment with an antifungal nail lacquer may decrease the critical fungal mass and aid penetration (Di Chiacchio et al. 2003). An efficient instrument for this procedure is a high-speed sanding handpiece (Baran et al. 2008a).

9.4.5.2 Chemical Methods

Reduction of disulfide linkages in the nail keratin matrix has been considered for enhancement of nail penetration. Wang and Sun (1998) identified the disulfide, peptide, hydrogen, and polar bonds that could potentially be targeted by chemical enhancers. They can be incorporated in the drug formulations to soften the nail plate, and therefore, they may be more convenient than physical enhancers (Kumar and Kimbal 2009; Murdan 2008).

We briefly report the data of studies that explored the benefit of these additive agents. Quintanar-Guerrero et al. (1999) indicated that urea and salicylic acid had a significant effect on nail surface examined with scanning electron microscopy. Nevertheless, the passage of miconazole nitrate, ketoconazole, and itraconazole through healthy human nail was not improved by pretreatment with salicylic acid for 10 days or by the application of the drug in a 40% urea solution. Keratinase produced by *Paecilomyces marquandii* clearly disrupts the nail plate, acting both on the intercellular matrix that holds the cells of the nail plate together and on the dorsal nail corneocytes (Mohorcic et al. 2007). Cysteine protease such as papain, which breaks the physical and chemical bonds responsible for the stability of nail keratin, have also been investigated as potential nail penetration enhancers (Chaudhari et al. 2006; Pierard et al. 2007). Compounds that contain –SH groups such as acetylcysteine, cysteine, and 2-mercaptoethanol can cleave the disulfide bonds in nail protein (Murdan 2002; Von Hoogdalem et al. 1997). After immersion in aqueous solvents containing *N*-acetyl cysteine or 2-mercaptoethanol, the flux of 5-fluorouracil or tolnaftate measured in nail pieces was significantly increased (Kobayashi et al. 1998). Other compounds such as resorcinol, thioglycolic acid, or urea hydrogen peroxide provide more "open" drug transport channels by acting via disruption of keratin disulfide bonds and formation of pores (Khengar et al. 2007; Von Hoogdalem et al. 1997). As example, the sequential application of urea hydrogen peroxide after thioglycolic acid increased terbinafine flux by approximately nineteen-fold (Brown et al. 2009). Dimethyl sulfoxide (DMSO) interacts with keratin by altering the concentration of lipids present in the dorsal layer of the nail plate in the corneocytes in a concentration-dependent manner (Benson 2005; Notman et al. 2007; Vejnovic et al. 2010b). At least, Hui et al. (2003) have demonstrated that econazole penetrates the nail six times more effectively in a lacquer containing 2-n-nonyl-1,3-dioxolane: the amount of econazole after dosing in the inner part of the human nail was 11.1 ± 2.6 μg/g of nail powder with 2-n-nonyl-1,3-dioxolane in the lacquer and 1.78 ± 0.32 μg/g without 2-n-nonyl-1,3-dioxolane.

9.4.5.3 Physical Methods

Physical permeation enhancements including etching and iontophoresis techniques have been developed to improve the transdermal delivery of drugs. The potential of these methods has been found to be higher than the potential of chemical permeation enhancers.

9.4.5.4 Pretreatment Etching Agents

Lactic, tartaric, glycolic, citric, and phosphoric acids have been evaluated to assess the effect of these etching agents on the delivery of terbinafine hydrochloride and 5-fluorouracil into and across the human nail plate (Vaka et al. 2010). Following a pretreatment with 1% and 10% phosphoric acid gel for 60 s, the permeation of terbinafine hydrochloride was enhanced by twofold and fivefold, respectively, over the control. The morphologic changes investigated by optical microscopy showed a percentage decrease in thickness of 27 ± 3% and 61 ± 5% in nail plates treated with 1% and 10% phosphoric acid, respectively. Besides, 10% lactic acid pretreatment for 60 s has no effect on the permeation of these two drugs through the nail plate. Results were quite similar in the case of 5-fluorouracil, and Vaka et al. (2010) conclude that despite the differences in physicochemical properties between terbinafine hydrochloride (quite lipophilic) and 5-fluorouracil (highly hydrophilic), the *in vitro* permeation and drug load of both compounds were enhanced due to pretreatment with phosphoric acid gel. Nail samples "etched" with phosphoric acid gel for 60 s were also significantly more permeable to ketoconazole (0.125% gel) as compared to normal nails (Repka et al. 2004).

9.4.5.5 Electrical Methods

The mechanisms of iontophoresis-enhanced transport include electrorepulsion/electrophoresis (direct field effect), electroosmosis (convective solvent flow in preexisting and newly created charged pathways), and electropermeabilization of the barrier (field-induced membrane alteration) (Dutet and Delgado-Charro 2010; Elkeeb et al. 2010; Hao et al. 2009b). Murdan, in 2007, measured the depth of penetration of fluorescein into the nail plate after iontophoresis. Confocal imaging showed fluorescein penetration into the nail plate at an average depth of 48 ± 10 μm (~10% of the total thickness of the nail). Iontophoretically enhanced trans-ungual delivery of ionic compounds has been demonstrated *in vitro* by Murthy et al. (2007a, b) who examined transport of salicylic across the human nail plate. Iontophoretic trans-nail flux improved with higher salicylic acid concentrations and higher current density. It appears that about 50–100 mM ionic strength is required for optimal conduction of electric current across the nail. The flux enhancement factor (iontophoretic flux/passive flux) also increased with increase in pH due to increased ionization of salicylic acid.

The electric resistance of the nail plate was measured *in vitro* in side-by-side diffusion cells and compared with those *in vivo*. The *in vivo* electrical resistance decreased significantly upon 2-h nail hydration to a constant value, showing the same pattern as that *in vitro* and demonstrating the effect of nail hydration upon its electric resistance (Hao et al. 2009a).

Moreover, Kalinowski et al. (2004) showed that low-voltage direct current electrostimulation might possess bacteriostatic and fungistatic properties.

9.4.5.6 Laser Therapy

By creating an array of partial microholes in the dorsal part of the nail plate, trans-ungual laser therapy facilitated the permeation of terbinafine applied daily as a topical lacquer solution (Kumar and Kimbal 2009; Welsh et al. 2010).

9.5 TISSUE COLLECTION

In the majority of *in vivo* studies, antifungal drug concentration was determined only in distal nail with samples (10–20 mg) usually taken from nails using clippers or scissors at the nail border. Nail clippings can be cut parallel to the surface in subsequent sections of 20 μm that are collected in separate vials representing nail layers at different depths. Nail dust may also be sampled (up to 30 mg) by carefully scraping the surface of the nail plate with a scalpel. This is the only place from which non-distal nail samples can be obtained from living humans. However, relevant data are expected only after oral administration, owing to possible contamination from the topically applied drug formulation. A novel technique developed by Hui et al. (2003) enables the ex vivo determination of drug concentration within the plate, where fungi reside. A micrometer-precision nail sampling instrument is used to sample the nail core without disturbing the surface. It enables finely controlled drilling into the nail with collection of the powder created by the drilling process. Drilling of the nail occurs through the ventral surface. The dorsal surface and ventrally accessed nail core can be assayed separately. This method permits drug measurement in the intermediate nail plate, which was previously impossible (Elkeeb et al. 2010). Samples may be differentiated into fingernails and toenails and diseased or normal nails. Large samples can be obtained by pooling clippings.

The main *in vitro* models include avulsed human cadaver nail plate (Gunt and Kasting 2007), nail clipping from healthy volunteers, and animal hoof membranes (Mohorcic et al. 2007). Whole nail plates are used in *in vitro* permeation tests, but their scarcity, expense, and variability impede their regular use. Nail clippings from healthy volunteers are a good alternative; however, they provide only a small area for drug penetration and are also fairly scarce due to the large amount of time needed to grow sufficiently long nails. Access to intact human nail is thus often limited, and animal material (hoof membrane) can be used as a human model substitute. However, the membrane surface showed indentations, which were variable from one hoof membrane to another, which would partly explain the high variability seen in drug permeation (Murdan 2007).

9.6 PHARMACOKINETICS OF ANTIFUNGAL DRUGS AFTER TOPICAL ADMINISTRATION

9.6.1 Imidazoles

9.6.1.1 Bifonazole

Bifonazole is an imidazole antifungal with a broad spectrum of activity, available as a cream or a solution for the treatment of nail infections. Chemical avulsion with 40% urea paste has been combined with 1% bifonazole and followed by 1% bifonazole cream alone. The mycologic cure rates ranged from 63% (Bonifaz et al. 1995) to 90% (Roberts et al. 1988). The pharmacokinetics of bifonazole has not been evaluated in depth (Baran et al. 2005).

9.6.1.2 Econazole

Aliquots of lacquer formulation containing 0.45 mg of [^{14}C]-econazole with (test group) or without (control group) 18% 2-n-nonyl-1,3-dioxane were applied twice daily for 14 days to human nails. The total econazole absorbed into the supporting bed cotton ball was nearly 200-fold greater in the test group than that in the control group. The concentration of econazole in the inner part of the nail was 11.1 ± 2.6 μg/mg in the test group and 1.78 ± 0.32 μg/mg in the control group (Hui et al. 2003).

9.6.1.3 Itraconazole

Trey et al. (2007) explored itraconazole for topical treatment of onychomycosis by using hydroxypropyl cellulose hot-melt extruded films. The rate of itraconazole release trended directly with the degree of film hydration and inversely to the hydroxypropyl cellulose molecular weight. Owing to the low solubility of the drug, significant concentrations are stored in fat cells, reducing the amount available for the target nail bed.

9.6.1.4 Oxiconazole

Six healthy volunteers applied an oxiconazole lotion with (nails of one hand) and without acetylcysteine (nails of the other hand) twice daily for 6 weeks. Maximum drug levels were determined in the upper 0–100-μm layer of the sectioned nail clippings, varying between 70 and 1200 ng/mg. Mean AUC $^{0\text{-}8w}$ (area under curve from 0 to 8 weeks) were 3110, 1160, and 380 ng/mg per week in the upper 0–50-μm and 51–100-μm layers and in the medium 101–200-μm layer, respectively. Mean AUC $^{0\text{-}8w}$ were 3110, 1160, and 380 ng/mg per week in the upper 0–50-μm and 51–100-μm layers and in the medium 101–200-μm layer, respectively. Total uptake into the nail was less than 0.2% of the topical dose. With *N*-acetyl cysteine, the AUC $^{0\text{-}8w}$ were 6160, 990, and 200 in the three layers, respectively. The upper 0–50-μm layer showed a tendency of increased oxiconazole contents on delivery with acetylcysteine, but this effect was not significant on AUC $^{0\text{-}8w}$ and C_{max} parameters with regard to the large interindividual variability. Nevertheless, codelivery with acetylcysteine statistically significantly prolonged the mean residence time of oxiconazole in the upper 51–100-μm ring fingernail layer from 3.7–4.9 weeks to 4.1–6.4 weeks (Van Hoogdalem et al. 1997). Kobayashi et al. (2004) confirmed that codelivering *N*-acetyl cysteine was able to enhance oxiconazole (1% lotion) uptake and retention in the upper nail layer but had little effect deeper into the nail plate.

9.6.1.5 Sertaconazole

This imidazole antifungal drug has been only investigated in healthy volunteers. Sixteen subjects were treated for 6 weeks with nail patch containing sertaconazole (3.63 mg) replaced every week. The active ingredient that had penetrated into the nail was evaluated to be 16–71% of the applied dose. The mean sertaconazole concentrations in clippings were 105, 135.5, and 118.6 μg/g at weeks 2, 4, and 6, respectively. Plasma concentrations were all below the lower limit of quantification (0.2 ng/mL) (Susilo et al. 2006).

9.6.2 Allylamines/Benzylamines

Terbinafine is the most potent antifungal agent (MIC: 0.003–0.006 mg/L) against dermatophytes, the key pathogens in onychomycosis (Darkes et al. 2003).

A bilayer nail lacquer formulation formed with an underlying drug-loaded hydrophilic layer and overlying hydrophobic vinyl layer has been developed and compared to a hydrophilic lacquer as control and to a monolayer nail lacquer containing polyethylene glycol 400 as penetration enhancer (plasticizer and humectant). *In vitro*, the mean amounts of terbinafine permeated across the nail plate in 6 days were 1.42 (bilayer), 1.12 (monolayer), and 0.32 μg/cm^2 (control), respectively; the mean amounts of terbinafine loaded from nail plate were 0.59, 0.36, and 0.28 μg/mg, respectively. *In vivo*, the terbinafine contents in the nail clippings collected at the end of 2-week treatment with the different lacquers were 1.27 (bilayer), 0.67 (monolayer), and 0.21 (control) μg/mg, respectively. Two weeks posttreatment, there were 0.11, 0.06, and 0.02 μg/mg, respectively (Shivakumar et al. 2010).

According to Amichai et al. (2010a), iontophoresis loads a high amount of the drug in the nail that was subsequently released by a diffusion process. In passive application and under low current densities, the drug is mostly concentrated in the upper layers of the nail (Amichai et al. 2010a); under high current densities of 400 and 500 μA/cm^2, terbinafine is probably penetrating through the hard keratin and then diffusing into the nail bed. These authors have also demonstrated that (a) an optimal electrolyte concentration (1% NaCl or KCl) is required for an elective delivery and that (b) there is a significant increase in drug delivery into the nail and into the receiving compartment in the presence of 3% DMSO (Amichai et al. 2010a).

To transfer the trans-ungual iontophoretic drug delivery technology to clinical use, two prototype applicators for the iontophoretic delivery of terbinafine for the treatment of onychomycosis were developed: a "nail-only" applicator, which delivers drug only into a part of the nail plate, and a "nail/skin" applicator, which delivers drug into both the whole nail plate and the surrounding soft tissues (Nair et al. 2009a). Both applicator types delivered high levels of terbinafine into the nail. The percentage of terbinafine released in the iontophoretic-loaded nails was found to be 60.11 ± 4.03% and 64.10 ± 6.16% for the nail-only and nail/skin applicator, respectively (Nair et al. 2009a).

Another study evaluated an iontophoretic patch in patients with onychomycosis of the toenail (Amichai et al. 2010b). Terbinafine nail content in the active group of patients treated with terbinafine and an iontophoretic patch at a constant current density of 100 μA/cm^2 was 5.69 μg/cm^2 at 8 weeks (overnight treatment, every day, 5 days per week, for 4 weeks) and 2.41 μg/cm^2 at the end of the follow-up period (8 weeks from the end of treatment). In patients from the passive group, terbinafine content was much lower, 1.34 μg/cm^2 at the fourth visit, but only 0.15 μg/cm^2 at the end of the follow-up period (Amichai et al. 2010a).

9.6.3 Phenylmorpholines

Amorolfine is a morpholine derivative type of antifungal, unrelated to the polyenes, azoles, or allylamines. It is the sole representative of the phenylmorpholines in human therapy. It has activity against dermatophytes, yeasts, and filamentous fungi. The mechanism of action of amorolfine involves the inhibition of two important steps in the ergosterol pathway: Erg24p (delta 14 reductase) and Erg2p (delta 8-delta 7 isomerase). In clinical trials, complete cure rates ranged from 38% to 54%; higher cure rates were found with 5% amorolfine lacquer and with twice-weekly treatment (Lauharanta 1992; Reinel and Clark 1992). Local pain at the application site, redness, and chromonychia are the sole adverse effects reported with amorolfine. In an onychomychosis model, the nail minimum fungicidal concentration of amorolfine ranged from 2 to 32 mg/L, and a mean concentration of 12.28 mg/L was well above the MIC for *C. albicans* (404 μg/L) and sufficient to kill all the strains (Schaller et al. 2009).

Amorolfine nail lacquer was developed as an alternative to traditional formulations such as creams and nail solutions. The film-forming polymer and the solvent were optimized for drug release, stability, and convenience of application. A high drug concentration of 11.72 μg/specimen was reached in the hoof horn after 6 h, increasing to 39.5 μg/specimen within 7 days. Amorolfine was found in the moistened gauze that simulated the nail bed indicating that amorolfine crossed the horn barrier. At the end of the investigation (7 days), 1.8% of the applied dose (500 μg) was available under the nail (Pittrof et al. 1992).

A 48-h treatment of 3H-labeled 5% amorolfine lacquer resulted in concentrations of 2.9–1.2 μg/mg (Franz 1992). Application of a 5% concentration in either an ethanol or a methylene chloride lacquer resulted in permeation rates through nail in the range of 20–100 ng/cm^2/h with greater penetration rates with methylene chloride. Active concentrations of amorolfine persisted for up to 2 weeks in subungual debris. Peak rates occurred 5–25 h after a single application, and comparable profiles were obtained after multiple applications over 8 days.

Polak et al. (1993) reported the kinetics of amorolfine penetration into human nails that had been in contact with methylene chloride- or ethanol-containing lacquer for 24 h. For these authors, the data obtained do not suggest that the solvent used in the lacquer plays any role in the penetration. The concentration profile of amorolfine in the nail plate was typically exponential with 2.9 ± 1.9 mg/g (range: 1–6.7 mg/g) in the upper layer much higher than in the median (1.2 ± 1.3 mg/g) or lower (0.7 ± 0.8 mg/g) layer. The drug concentration in the nails correlated well with the drug content of the formulation and the penetration time (56 μg/cm^2 after 6 h; 188 μg/cm^2 after 7 days). After a lag time, 1–8% of the applied dose was delivered to the nail bed within 7 days.

It results that amorolfine formulated as methylene chloride 5% lacquer is absorbed through the human nail plate. It has been evaluated that, as the vehicle dries, the active compound increases in concentration from 5% to around 25%. Concentrations achieved at the nail bed exceeded those needed to prevent the growth of most fungi responsible for onychomycosis. Application only once a week is required because the high concentration of the drug in the nail plate forms a reservoir (Kaur and Kakkar 2010). Amorolfine appears to be more suitable for drug delivery to human nails because it penetrates into the nails by means of the hydrophilic pathway. Furthermore, amorolfine penetrates very well into fungal cells, owing to its pKa value and the pH value of the nail, as well as the lipophilic properties of its base form (Neuberg et al. 2006).

9.6.4 AN-2690

The oxaborole (AN-2690) represents a new class of antifungal protein synthesis inhibitors, which specifically inhibits the cytoplasmic leucyl-tRNA synthetase. AN-2690 is a small molecule (molecular weight <250 Da), polar (log P <2.5), with good water solubility (>0.1 mg/mL) (Kumar and Kimball 2009). *In vitro*, a significant zone of inhibition against the dermatophyte cultured in the flask below the full thickness human nail plates treated by AN-2690 evidenced a penetration of this antifungal agent through the total nail depth (Alley 2007). At the end of a 28-day treatment of 7.5% solution of AN-2690, the mean nail plate level of AN-2690 was 28.8 ± 22 μg/mg of nail plate. These concentrations were largely above the minimum fungicidal concentrations against dermatophyte fungi after only 24 h of contact.

Its nail penetration efficiency coefficient is 50-fold higher than that of topical ciclopirox (Baker et al. 2007). The small, polar, water-soluble nature of AN-2690 that disseminated through a large area in the receiver cells matched the requirements to penetrate the nail plate, providing a new antifungal agent with an unprecedented ability to penetrate human nails.

9.6.5 Hydroxypyridones

For a long time, ciclopirox olamine has been the only hydroxypyridone used clinically. Ciclopirox as a free acid and rilopirox have also been considered. Hydroxypiridones are active against a broad spectrum of medically relevant fungi including dermatophytes, yeasts, and molds *in vitro*. In contrast with most antifungals, ciclopirox does not interfere with sterol biosynthesis but chelates trivalent metal cations (Fe^{3+} and Al^{3+}) necessary for enzymatic activity, thereby inhibiting enzymes involved in mitochondrial electron transport and subsequent cellular energy production (Bohn and Kraemer 2000). It shows a broad spectrum with activity against dermatophytes and some non-dermatophyte molds.

In an onychomycosis model, the nail minimum fungicidal concentration of ciclopirox olamine ranged from 16 to 32 μg/mL and a mean concentration of 24.13 μg/mL was needed to kill all the strains (Schaller et al. 2009).

Interest mainly focuses on ciclopirox lacquer (Gupta et al. 2000b; Korting and Grundmann-Kollmann 1997), the only topical therapy approved by the FDA for onychomycosis. As it dries, the active lacquer 8% formulation increases in concentration to around 35%, creating a concentration gradient that promotes the penetration of the drug through the dorsal nail surface (Cheschin-Roques et al. 1991).

Bohn and Kraemer (2000) reported penetration data of ^{14}C-labeled ciclopirox into excised human toenails after only one application of the lacquer. After 24 h, the superficial layer contained an average 7812 μg/g, the upper layer 1068 μg/g, and the lower layer 34 μg/g. Nevertheless, the concentration of ciclopirox achieved in the nail bed exceeded the minimum fungicidal concentration for most fungal organisms responsible for onychomycosis. The distal portions of the nail were sampled after 7, 14, 30, and 45 days of treatment in five healthy volunteers applying the lacquer daily to the toenails and removing it once each week. The mean ± SD concentration of ciclopirox increased from 1390 ± 740 μg/g after 7 days of application to 5910 ± 2420 μg/g after 30 days. With repeated daily applications for 45 days, the antifungal agent was homogeneously distributed through all the layers of the nail, with the relative distribution percentage of ciclopirox in each layer of nail varying from 21.2% to 27.8%. When lacquer was removed before each application, the mean ± SD concentration for all individuals was only 890 ± 250 μg/g after 7 days of application and 3350 ± 820 μg/g after 30 days. Although ciclopirox readily penetrates nails, very low concentrations of ciclopirox (0.012–0.080 mg/L) are recoverable systemically, even after long-term use (Bohn and Kramer 2000).

In another group of nine healthy volunteers who applied ciclopirox nail lacquer daily for 45 days with the old lacquer removed before each application, the mean concentration peaked at 3.4 μg/mg after 30 days (Baran and Kaoukhov 2005). Fourteen days after the treatment was concluded, the mean concentration of ciclopirox in the nail was reduced significantly but remained detectable in some subjects.

Ciclopirox is metabolized mainly by glucuronidation.

After application of Penlac® nail lacquer, measured serum concentration ranged 12–80 μg/L. Nail ciclopirox levels are undetectable within 2 weeks of discontinuation of application of the lacquer (Baran et al. 2005). Based on urinary data, systemic absorption of the compound was considered to be less than 5% of the applied dose.

The formulation plays an important role in the enhancement of ciclopirox permeation into and through the human nail plate (Kaur and Kakkar 2010). Indeed, in a study done by Hui et al. (2004), the human nail penetration of ciclopirox was significantly greater with a marketed gel containing 0.77% of ciclopirox than an experimental gel containing 2% of ciclopirox and a marketed lacquer containing 8% of ciclopirox.

It must be noted that, for a better efficacy of antifungal therapy, it is recommended that a health care professional debride and trim the affected nails.

9.6.6 Thiocarbamate

The penetration of tolnaftate, a thiocarbamate, through the fingernail plates of healthy volunteers was determined by measuring the concentration on the ventral nail plate side by high-performance liquid chromatography (HPLC) after the dorsal nail plate was filled with drug suspension. Chemical avulsion with 20% urea paste combined with 2% tolnaftate or solvent systems containing *N*-acetyl cysteine or 2-mercaptoethanol enhanced penetration (Ishii et al. 1983; Kobayashi et al. 1998).

9.6.7 NB-002

NB-002 is an oil-in-water emulsion containing high-energy nanometer-sized particles with a cationic agent (quaternary ammonium compound: cetylpyridinium chloride) as the active ingredient. Treatments (0.25% or 0.5% NB-002 twice daily for 28 days) were tested in 20 subjects with advanced distal subungual onychomycosis of the toenails. There were no detectable levels of surfactant in any of the pharmacokinetic samples determined in HPLC (limit of detection: 1 ng/mL) (Jones et al. 2008).

9.7 Pharmacokinetics of Drugs Used in Nail Psoriasis after Topical Administration

The current management of nail psoriasis includes topical, intralesional, and systemic therapies. Topical treatments are represented by corticosteroids alone or with salicylic acid, tazarotene, urea/propylene glycol, vitamin D analogs (calcipotriol),

5-fluorouracil, anthralin, ciclosporin, and clobetasol. Conventional systemic therapies are also indicated in nail psoriasis: cyclosporine, retinoids, and nimesulide (Lawry 2007). Systemic biologic agents that act as T-cell modulators (alefacept and efalizumab) and anti-Tumor Necrosis Factor alpha anti-TNFα agents (infliximab, adalimumab, and etanercept) are beginning to emerge (Zaiac 2010).

Nail psoriasis is difficult to treat with lengthy, tricky, and disappointing processes (Baran 2010; Zaiac 2010). Despite the recognized burden of this disease, there is limited controlled trial data to support the use of these therapies. In consequence, the efficacy is poorly documented and pharmacokinetic data are nonexistent.

Nevertheless, it is clear that the limited efficacy of topical therapies is related to the real difficulties in topical drug delivery to the involved site, especially the nail deep matrix. Occlusion can be occasionally justified and the direct intralesional injection of the treatment can be proposed for a direct access of topical therapy to the focus of pathology. Corticosteroids, such as triamcinolone acetate, are effective when using this administration pathway but are painful, time consuming, and require local anesthesia (Edwards and De Berker 2009).

9.8 PHARMACOKINETICS OF DRUGS IN THE NAIL APPARATUS FOLLOWING ORAL ADMINISTRATION

9.8.1 ITRACONAZOLE

During and after continuous oral therapy of itraconazole (200 mg) once daily for 3 months, the concentration of itraconazole in the nail slowly increased from 350 to 520 ng/g (week 12) in fingernails and from 100 to 350 ng/g (week 12) in toenails, with the concentrations in toenails continuing to increase until week 36 to a maximum of 700 ng/g mean concentration (Havu et al. 1999). The mean peak concentrations reached in patients treated with 100 mg/day for the same period (3 months) were 110 and 150 ng/g in fingernails and toenails, respectively, at week 9 after therapy. Twenty-seven weeks after stopping administration, itraconazole was always detectable: 35 ng/g in fingernails and 150 ng/g in toenails; in patients treated with 200 mg/day for 3 months, fingernail and toenail concentrations were 140 and 670 ng/g, respectively (Willemsen et al. 1992). A constant nail/plasma ratio of 0.23 from the 2nd to the 12th week of a 200 mg/day treatment was found by Matsumoto et al. (1999); a ratio increasing from 0.3 at week 4 to 1 at week 18 was observed by Mathieu et al. (1991). In conclusion, the uptake of itraconazole by toenails was better, but more variable than that by fingernails. Concentrations over MIC (100 ng/g for most dermatophytes and *Candida* species) persist from the first month of treatment to the third month in fingernail to the sixth.

During and after an intermittent itraconazole schedule, 200 mg twice daily for 1 week followed by a 3-week drug-free period for 36 weeks, the mean concentration of itraconazole from week 4 to week 36 was 122 ng/g in fingernails and 179 ng/g in toenails (Havu et al. 1999). In a second study with an identical three-pulse treatment, the mean itraconazole concentration was 70 and 100 ng/g at week 4 and 424 and 470 ng/g at week 36 in the distal ends of the fingernails and toenails, respectively (De Doncker et al. 1996).

The elimination half-life was evaluated to be around 90 days in fingernails and 150 days in toenails after continuous treatment. It was shorter after intermittent therapy, 30 days (fingernails) and 90 days (toenails). The elimination was much longer in nails than in the plasma (30 h). Maximal nail concentrations are generally reached after treatment has been stopped and antifungal drug remains in the nail at therapeutic concentrations for several months (Debruyne and Coquerel 2001).

9.8.2 Terbinafine

After administration of 250 mg orally once daily for 28 days to 12 healthy volunteers, nail terbinafine concentration increased from 120 ng/g at day 7 to 330 ng/g at day 28. Tissue levels were 390 ng/g 1 day after the stop of medication (nail/blood ratio 1.6) and slowly felt around 90 ng/g 55 days after the last tablet (nail/blood ratio 3) (Faergemann et al. 1993). After oral administration of 250 mg once daily for 7 days, the concentration of terbinafine in peripheral nail clippings was 500 and 200 ng/g 1 day and 90 days after stopping medication, respectively (Faergemann et al. 1994).

In patients taking terbinafine 250 mg/day, terbinafine was detected in distal clippings of nails at 3–18 weeks after starting the therapy. The levels of terbinafine reached a steady state and did not increase with time during therapy. The mean terbinafine in affected toenails (48-week treatment) ranged from 250 to 550 ng/g, and the kinetic profile of terbinafine in normal fingernails was not significantly different from that of affected nails (Finlay 1992). The levels of terbinafine achieved in nail samples exceeded the range of MICs for dermatophytes from week 4 and remained stable up to 3 months (Dykes et al. 1990).

In patients receiving 250 mg/day for up to 12 weeks, a mean concentration of 200 ng/g of terbinafine was observed in the distal ends of the nails from the end of the first week. This regularly increased during treatment: 400 ng/g at week 2, 600 ng/g at week 4, and 700 ng/g at week 12. Thirty weeks after the cessation of therapy, the concentration of terbinafine in the nails was 300 ng/g (Kovarik et al. 1995). Considerable interindividual variations in the drug concentrations in patients' nails have been reported: 210–2030 ng/g in toenails at week 24 and 330–1170 ng/g in fingernails at week 18 (Schatz et al. 1995). This treatment (250 mg/day for 12 weeks) gave medium concentrations well above the MIC from week 2 to week 36 in agreement with the very long half-life evaluated around 90 days.

These studies demonstrated that terbinafine is first detectable in the distal nail within few weeks of starting the oral therapy. It moves rapidly into nail plate, and the nail growth rate of 0.41 mm/week is much too slow to account for early measurable concentrations solely by uptake into newly formed nail. Accordingly, Finlay (1992) suggested an additional diffusion through the nail plate.

9.8.3 Fluconazole

The mean concentrations of fluconazole 40 weeks after a 4-week treatment of 150 mg/week were between 4 and 6 μg/g (normal and infected fingernails) and 3 and 7 μg/g (normal and infected toenails). Nail/blood mean ratios were 1.7 (finger distal clippings), 5.0 (finger dorsal scrapings), 3.9 (toenail distal clippings), and 1.6 (toenail dorsal scrapings). They were quite similar when the dose was 300 mg (Laufen et al. 1999). Fluconazole levels fall slowly after the drug is stopped with a half-life around 60 days for healthy fingernails and toenails, 50 days for diseased fingernails, and 90 days for affected toenails. A longer half-life of 175 days has been found by Faergemann and Laufen (1996) in diseased nails. Fluconazole concentrations remained detectable 5 months after the end of treatment and over MIC during a minimal 2-month period after the last administration, justifying weekly doses.

9.9 CONCLUSION

The clinical efficacy of a drug aimed to treat onychomycosis or nail psoriasis depends not only on its spectrum of activity but also to a great extent on the concentration achieved in the target tissue. Antifungal drugs applied to the nail were shown to passively penetrate through the highly keratinized nail plate *in vitro* but not enough to reach the deep keratin layer and matrix. Indeed, the measured concentrations of drug in the different layers of the nail plate regularly decrease from the upper to the deeper layers and generally result in concentrations lower than the MIC in the nail bed and the nail matrix. This permeation depends on several factors, including the molecular mass, the concentration of the drug agent, the duration of contact with the target nail tissue, and the ability of the compound to penetrate the upper and deep layers of the nail plate. These topical treatments also require regular applications for a long period and complete compliance by the patient. The main consequences are an ineffective therapeutic concentration at the target site, a long duration of treatment, and a suboptimal efficacy.

During the last few years, the main objectives of antifungal therapy development were to force the permeation through the nail by mechanical, chemical, or physical enhancers and target new compounds of low mass and hydrophilic status. Decreasing the nail thickness, disrupting the keratin structure, favoring the aqueous pathways through the nail plate, and promoting the drug penetration by way of low currents (iontophoresis) all contribute to diminishing the nail resistance and favor the drug advancement up to the nail bed. So, novel agents, formulations, and physical methods offering the prospect of more successful topical therapies are in the course of development: topical lacquers, iontophoresis, or light-based therapies. In most cases, significant increases in drug concentration are obtained, above MIC. Among the most common therapies, amorolfine, ciclopirox, and terbinafine largely benefited from these improvements.

Iontophoresis delivers compounds across the nail stratums in significant higher quantities than passive transport within short treatment duration, and portable transungual constant voltage iontophoretic devices are actually in development. They comprise a thin, flexible patch with the battery embedded in the patch between the electrodes. After removal of the device that provides drug delivery, the amount of drug continues to be slowly released from the nail plate to the nail bed and the surrounding tissues. These greatly reduce drug administration frequency and thus improve patient compliance.

Topical therapy has the greatest potential as primary therapy in mild infections, as palliative therapy in those unable to take oral therapy, and as a prophylactic agent.

It must be restricted to patients with mild-to-moderate nail involvement and when the number of nails affected is low. As a general rule, topical medications do not attain the systemic circulation and has a favorable side-effect profile limited mainly to irritation of the surrounding skin. They are preferred in elderly patients or patients receiving multiple medications in order to minimize drug interactions.

As largely documented in a previous review (Debruyne and Coquerel 2001), the concentrations of antifungal drugs (itraconazole, terbinafine, and fluconazole) that reach the nail plate from the capillaries of the nail bed and matrix after oral therapies are effective and persist in the nail for a long time even after treatment is stopped. These data justify promoting the combination of oral and topical therapies in cases of severe and widespread disease. Nail permeability is relevant to topical drugs on the dorsal surface and systemic drugs from the nail matrix and ventral surfaces. In consequence, topical therapy is suitable for superficial onychomycosis restricted to the distal two-thirds of the nail. When the matrix or deepest regions are involved or when a risk of failure is likely, combination therapy should be considered.

REFERENCES

Alley, M. R. K., S. J. Baker, K. R. Beutner, and J. Plattner. 2007. "Recent Progress on the Topical Therapy of Onychomycosis." *Expert Opinion on Investigational Drugs* 16 (2): 157–67.

Amichai, B., R. Mosckovitz, H. Trau, O. Sholto, S. Ben-Yaakov, M. Royz, D. Barak, B. Nitzan, and A. Shemer. 2010a. "Iontophoretic Terbinafine HCl 1% Delivery across Porcine and Human Nails." *Mycopathologia* 169: 343–9.

Amichai, B., B. Nitzan, R. Mosckovitz, and A. Shemer. 2010b. "Iontophoretic Delivery of Terbinafine in Onychomycosis: A Preliminary Study." *British Journal of Dermatology* 162: 46–50.

Baker, S., V. Sanders, K. Hold, and J. Plattner. 2007. "In Vitro Nail Penetration of AN2690, Effect of Vehicle, and Coefficient of Efficacy." *Journal of the American Academy of Dermatology* 56 (2): AB124.

Baran, R. 2010. "The Burden of Nail Psoriasis: An Introduction." *Dermatology* 221: 1–5.

Baran, R., A. K. Gupta, and G. E. Pierard. 2005. "Pharmacotherapy of Onychomycosis." *Expert Opinion on Pharmacotherapy* 6 (4): 609–24.

Baran, R., R. J. Hay, and J. I. Garduno. 2008a. "Review of Antifungal Therapy and the Severity Index for Assessing Onychomycosis: Part I." *Journal of Dermatological Treatment* 19 (2): 72–81.

Baran, R., R. J. Hay, and J. I. Garduno. 2008b. "Review of Antifungal Therapy, Part II: Treatment Rationale, Including Specific Patient Populations." *Journal of Dermatological Treatment* 19 (3): 168–75.

Baran, R., and A. Kaoukhov. 2005. "Topical Antifungal Drugs for the Treatment of Onychomycosis: An Overview of Current Strategies for Monotherapy and Combination Therapy." *Journal of the European Academy of Dermatology and Venereology* 19: 21–9.

Benson, H. A. E. 2005. "Transdermal Drug Delivery Penetration Enhancement Techniques." *Current Drug Delivery* 2: 23–33.

Bohn, M., and K. T. Kraemer. 2000. "The Dermatopharmacologic Profile of Ciclopirox 8% Nail Lacquer." *Journal of the American Podiatric Medical Association* 90 (10): 491–4.

Bonifaz, A., A. Guzman, C. Garcia, J. Sosa, and A. Saul. 1995. "Efficacy and Safety of Bifonazole Urea in the Two-Phase Treatment of Onychomycosis." *International Journal of Dermatology* 34: 500–3.

Borelli, C., M. Schaller, M. Niewerth, K. Nocker, B. Baasner, D. Berg, R. Tiemann, K. et al. 2008. "Modes of Action of the New Arylguanidine Abafungin Beyond Interference with Ergosterol Biosynthesis and In Vitro Activity against Medically Important Fungi." *Chemotherapy* 54 (4): 245–59.

Brown, M. B., R. H. Khengar, R. B. Turner, B. Forbes, M. J. Traynor, C. R. Evans, and S. A. Jones. 2009. "Overcoming the Nail Barrier: A Systematic Investigation of Ungual Chemical Penetration Enhancement." *International Journal of Pharmacy* 370: 61–7.

Carrillo-Munoz, A. J., and J. M. Torres-Rodriguez. 1995. "In Vitro Antifungal Activity of Sertaconazole, Econazole, and Bifonazole against *Candida* spp." *Journal of Antimicrobial Chemotherapy* 36 (4): 713–6.

Chaudhari, P. D., S. P. Chaudhari, and P. K. Kolsure. 2006. "Drug Delivery through Nail—A Review." http://www.pharmainfo.net/reviews/drug-delivery-through-nail-review; 2006.

Cheschin-Roques, C., H. Hanel, S. Pruja-Bougaret, J. Luc, J. Vandermander, and G. Michel. 1991. "Ciclopirox Nail Lacquer 8%: In Vivo Penetration into and through Nails and In Vivo Effect." *Skin Pharmacology* 4: 89–94.

Darkes, M. J. M., L. J. Scott, and K. L. Goa. 2003. "Terbinafine. A Review of Its Use in Onychomycosis in Adults." *American Journal of Clinical Dermatology* 4: 39–65.

de Berker, D. A. R., J. Andre, and R. Baran. 2007. "Nail Biology and Nail Science." *International Journal of Cosmetic Science* 29 (4): 241–75.

Debruyne, D., and A. Coquerel. 2001. "Pharmacokinetics of Antifungal Agents in Onychomycosis." *Clinical Pharmacokinetics* 40 (6): 441–72.

De Doncker, P., J. Decroix, and G. E. Pierard. 1996. "Antifungal Pulse Therapy for Onychomycosis. A Pharmacokinetic and Pharmacodynamic Investigation of Monthly Cycles of 1-Week Pulse Therapy with Itraconazole." *Archives of Dermatology* 132: 34–41.

Di Chiacchio, N., B. V. Kadunc, A. R. T. De Almeida, and C. L. Madeira. 2003. "Nail Abrasion." *Journal of Cosmetic Dermatology* 2: 150–2.

Donnelly, R. F., P. A. McCarron, J. M. Lightowler, and A. D. Woolfson. 2005. "Bioadhesive Patch-Based Delivery of 5-Aminolevulinic Acid to the Nail for Photodynamic Therapy of Onychomycosis." *Journal of Controlled Release* 103: 381–92.

Dorn, M., T. Kienitz, and F. Ryckmanns. 1980. "Onychomycosis: Experience with Nontraumatic Nail Avulsion." *Hautarzt* 31: 30–4.

Dutet, J., and M. B. Delgado-Charro. 2010. "Electroosmotic Transport of Mannitol across Human Nail during Constant Current Iontophoresis." *Journal of Pharmacy and Pharmacology* 62: 721–9.

Dykes, P. J., R. Thomas, and A. Y. Finlay. 1990. "Determination of Terbinafine in Nail Samples during Systemic Treatment for Onychomycoses." *British Journal of Dermatology* 123: 481–6.

Edwards, F., and D. De Berker. 2009. "Nail Psoriasis: Clinical Presentation and Best Practice Recommendations." *Drugs* 69 (17): 2351–61.

Egawa, M., Y. Ozaki, and M. Takahashi. 2006. "In Vivo Measurement of Water Content of the Fingernail and Its Seasonal Change." *Skin Research and Technology* 12: 126–32.

Elewski, B. E. 1993. "Mechanisms of Action of Systemic Antifungal Agents." *Journal of the American Academy of Dermatology* 28: S28–34.

Elkeeb, R., A. Alikhan, L. Elkeeb, X. Hui, and H. Maibach. 2010. "Transungual Delivery: Current Status." *International Journal of Pharmacy* 384 (1–2): 1–8.

Faergemann, J., and H. Laufen. 1996. "Levels of Fluconazole in Normal and Diseased Nails during and after Treatment of Onychomycoses in Toe-Nails with Fluconazole 150 mg Once Weekly." *Acta Dermato-Venereologica* 76 (3): 219–21.

Faergemann, J., H. Zehender, J. Denouel, and L. Millerioux. 1993. "Levels of Terbinafine in Plasma, Stratum Corneum, Dermis-Epidermis (without Stratum Corneum), Sebum, Hair and Nails during and after 250 mg Terbinafine Orally Once Per Day for Four Weeks." *Acta Dermato-Venereologica* 73: 305–9.

Faergemann, J., H. Zehender, and L. Millerioux. 1994. "Levels of Terbinafine in Plasma, Stratum Corneum, Dermis-Epidermis (without Stratum Corneum), Sebum, Hair and Nails during and after 250 mg Terbinafine Orally Once Daily for 7 and 14 Days." *Clinical and Experimental Dermatology* 19: 121–6.

Fawcett, R. S., S. Linford, and D. L. Stulberg. 2004. "Nail Abnormalities: Clues to Systemic Disease." *American Family Physician* 69 (6): 1417–24.

Finch, J. J., and E. M. Warshaw. 2007. "Toenail Onychomycosis: Current and Future Treatment Options." *Dermatologic Therapy* 20 (1): 31–46.

Finlay, A. Y. 1992. "Pharmacokinetics of Terbinafine in the Nail." *British Journal of Dermatology* 126 (suppl. 39): 28–32.

Fleckman, P. 2005. "Structure and Function of the Nail Unit." In *Nails: Diagnosis, Therapy, Surgery*, 3rd ed., edited by R. K. Scher, C. R. Daniel, A. Tosti, B. E. Elewski, P. Fleckman, and P. Rich, 13–25. Philadelphia: Elsevier Saunders.

Franz, T. J. 1992. "Absorption of Amorolfine through Human Nail." *Dermatology* 184: 18–20.

Grover, C., S. Bansal, S. Nanda, B. S. N. Reddy, and V. Kumar. 2007. "Combination of Surgical Avulsion and Topical Therapy for Single Nail Onychomycosis: A Randomized Controlled Trial." *British Journal of Dermatology* 157: 364–8.

Gunt, H. B., and G. B. Kasting. 2007. "Effect of Hydration on the Permeation of Ketoconazole through Human Nail Plate In Vitro." *European Journal of Pharmaceutical Sciences* 32: 254–60.

Gunt, H. B., M. A. Miller, and G. B. Kasting. 2007. "Water Diffusivity in Human Nail Plate." *Journal of Pharmacological Sciences* 96 (12): 3352–62.

Gupta, A. K., P. Fleckman, and R. Baran. 2000b. "Ciclopirox Nail Lacquer Topical Solution 8% in the Treatment of Toenail Onychomycosis." *Journal of the American Academy of Dermatology* 43: S70–80.

Hao, J., and S. K. Li. 2008. "Transungual Iontophoretic Transport of Polar Neutral and Positively Charged Model Permeants. Effects of Electrophoresis and Electroosmosis." *Journal of Pharmacological Sciences* 97: 893–905.

Hao, J., K. A. Smith, and K. Li. 2009a. "Time-Dependent Electric Properties of Human Nail upon Hydration In Vivo." *Journal of Pharmacological Sciences* 99 (1): 107–18.

Hao, J., K. A. Smith, and K. Li. 2009b. "Iontophoretically Enhanced Ciclopirox Delivery into and across Human Nail Plate." *Journal of Pharmacological Sciences* 98 (10): 3608–16.

Havu, V., H. Brandt, and H. Heikkilä. 1999. "Continuous and Intermittent Itraconazole Dosing Schedules for the Treatment of Onychomycosis: A Pharmacokinetic Comparison." *British Journal of Dermatology* 140: 96–101.

Hui, X., T. C. K. Chan, S. Barnadillo, C. Lee, H. I. Maibach, and R. C. Wester. 2003. "Enhanced Econazole Penetration into Human Nail by 2-*n*-Nonyl-1-3-Dioxolane." *Journal of Pharmaceutical Sciences* 92 (1): 142–8.

Hui, X, R. C. Wester, S. Barbadillo, C. Lee, B. Patel, M. Wortzmman, E. H. Gans, and H. I. Maibach. 2004. "Ciclopirox Delivery into the Human Nail Plate." *Journal of Pharmaceutical Sciences* 93 (10): 2545–8.

Ishii, M., T. Hamada, and Y. Asai. 1983. "Treatment of Onychomycosis by ODT Therapy with 20% Urea Ointment and 2% Tolfanate Ointment." *Dermatologica* 167: 273–9.

Jones, T., M. Flack, M. Ijzerman, and J. R. Baker. 2008. "Safety, Tolerance, and Pharmacokinetics of Topical Nanoemulsion (NB-00.2) for the Treatment of Onychomycosis." Annual Meeting of the American Academy of Dermatology, San-Antonio, Texas, February 1–5.

Kalinowski, D. P., L. E. Edsberg, R. A. Hewson, R. H. Johnson, and M. S. Brogan. 2004. "Low-Voltage Direct Current as a Fungicidal Agent for Treating Onychomycosis." *Journal of the American Podiatric Medical Association* 94: 565–72.

Kaur, I. P., and S. Kakkar. 2010. "Topical Delivery of Antifungal Agents." *Expert Opinion on Drug Delivery* 7 (11): 1303–27.

Khengar, R. H., S. A. Jones, R. B. Turner, B. Forbes, and M. B. Brown. 2007. "Nail Swelling as a Pre-Formulation Screen for the Selection and Optimisation of Ungual Penetration Enhancers." *Pharmaceutical Research* 24 (12): 2207–12.

Kobayashi, Y., T. Komatsu, and M. Sumi. 2004. "In Vitro Permeation of Several Drugs through the Human Nail Plate: Relationship between Physicochemical Properties and Nail Permeability of Drugs." *European Journal of Pharmaceutical Sciences* 21: 471–7.

Kobayashi, Y., M. Miyamoto, and K. Sugibayashi. 1998. "Enhancing Effect of N-Acetyl-1-Cysteine or 2-Mercaptoethanol on the In Vitro Permeation of 5-Fluoroacil or Tolfanate through the Human Nail Plate." *Chemical and Pharmaceutical Bulletin* 46: 1797–802.

Kokjohn, K., M. Bradley, B. Griffiths, and M. Ghannoum. 2003. "Evaluation of In Vitro Activity of Ciclopirox Olamine, Butenafine HCl and Econazole Nitrate against Dermatophytes, Yeasts and Bacteria." *International Journal of Dermatology* 42 (suppl. 1): 11–7.

Korting, H. C., and M. Grundmann-Kollmann. 1997. "The Hydroxypyridones: A Class of Antimycotics of Its Own." *Mycoses* 40 (7–8): 243–7.

Kovarik, J. M., E. A. Mueller, and H. Zehznder. 1995. "Multiple-Dose Pharmacokinetics and Distribution in Tissue of Terbinafine and Metabolites." *Antimicrobial Agents and Chemotherapy* 39: 2738–41.

Kumar, S., and A. B. Kimball. 2009. "New Antifungal Therapies for the Treatment of Onychomycosis." *Expert Opinion on Investigational Drugs* 18 (6): 727–34.

Laufen, H., T. Zimmerman, and R. A. Yeates. 1999. "The Uptake of Fluconazole in Finger and Toe Nails." *International Journal of Clinical Pharmacology and Therapeutics* 37: 352–60.

Lauharanta, J. 1992. "Comparative Efficacy and Safety of Amorolfine Nail Lacquer 2% versus 5% Once Weekly." *Clinical and Experimental Dermatology* 17: 41–3.

Lawry, M. 2007. "Biological Therapy and Nail Psoriasis." *Dermatologic Therapy* 20: 60–7.

Marshall, R. C. 1983. "Characterisation of the Proteins of Human Hair and Nail by Electrophoresis." *Journal of Investigative Dermatology* 80: 519–24.

Marty, J. P. L. 1995. "Amorolfine Nail Lacquer: A Novel Formulation." *Journal of the European Academy of Dermatologic Venereology* 4: S17–22.

Mathieu, L., P. De Doncker, and J. Willems. 1991. "Itraconazole Penetrates the Nail via the Nail Matrix and the Nail Bed: An Investigation in Onychomycosis." *Clinical and Experimental Dermatology* 16: 374–6.

Matsumoto, T., H. Tanuma, and S. Nishiyama. 1999. "Clinical and Pharmacokinetic Investigations of Oral Itraconazole in the Treatment of Onychomycosis." *Mycoses* 42 (1–2): 79–91.

Moffitt, D. L., and D. A. De Berker. 2000. "Yellow Nail Syndrome: The Nail that Grows Half As Fast Grows Twice As Thick." *Clinical and Experimental Dermatology* 25: 21–3.

Mohorcic, M., A. Torkar, J. Friedrich, J. Kristl, and S. Murdan. 2007. "An Investigation into Keratinolytic Enzymes to Enhance Ungual Drug Delivery." *International Journal of Pharmacology* 332: 196–201.

Murdan, S. 2002. "Drug Delivery to the Nail Following Topical Applications." *International Journal of Pharmacology* 236: 1–26.

Murdan, S. 2007. "1st Meeting on Topical Drug Delivery to the Nail." *Expert Opinion on Drug Delivery* 4: 453–55.

Murdan, S. 2008. "Enhancing the Nail Permeability of Topically Applied Drugs." *Expert Opinion on Drug Delivery* 5: 1267–82.

Murthy, S. N, D. C. Waddell, H. N. Shivakumar, A. Balaji, and C. P. Bowers. 2007a. "Iontophoretic Perselective Property of Human Nail." *Journal of Dermatological Science* 46: 150–2.

Murthy, S. N., D. E. Wiskirchen, and C. P. Bowers. 2007b. "Iontophoretic Drug Delivery across Human Nail." *Journal of Pharmacological Sciences* 96: 305–11.

Nair, A. B., H. D. Kim, S. P. Davis, R. Etheredge, M. Barsness, P. M. Friden, and S. N. Murthy. 2009a. "An Ex Vivo Toe Model Used to Assess Applicators for the Iontophoretic Ungual Delivery of Terbinafine." *Pharmacology Research* 26 (9): 2194–201.

Nair, A. B., S. M. Sammeta, H. D. Kim, B. Chakraborty, P. M. Friden, and S. N. Murthy. 2009b. "Alteration of the Diffusional Barrier Property of the Nail Leads to Greater Terbinafine Drug Loading and Permeation." *International Journal of Pharmacy* 375: 22–7.

Neuberg, R. H. H., C. Gensbugel, A. Jackel, and S. Wartewig. 2006. "Different Physicochemical Properties of Antimycotic Agents are Relevant for Penetration into and through Human Nails." *Die Pharmazie* 61: 604–7.

Notman, R., W. K. Den Otter, M. G. Noro, W. J. Briels, and J. Anwar. 2007. "The Permeability Enhancing Mechanism of DMSO in Ceramide Bilayers Simulated by Molecular Dynamics." *Biophysical Journal* 93: 2056–68.

Nunley, K. S., and L. Cornelius. 2008. "Current Management of Onychomycosis." *Journal of Hand Surgery* 33A: 1211–24.

Pierard, G. E., J. E. Arrese, P. Quatresooz, and C. Pierard-Franchimont. 2007. "Emerging Therapeutic Agents for Onychomycosis." *Expert Opinion on Emerging Drugs* 12 (3): 345–53.

Pierard, G. E., and C. Pierard-Franchimont. 2005. "The Nail under Fungal Siege in Patients with Type II Diabetes Mellitus." *Mycoses* 48: 339–42.

Pittrof, F., J. Gerhards, and W. Erni. 1992. "Loceryl Nail Lacquer: Realization of a New Galenical Approach to Onychomycosis Therapy." *Clinical and Experimental Dermatology* 17: 26–8.

Polak, A. 1993. "Kinetics of Amorolfine in Human Nails." *Mycoses* 36: 101–3.

Pong, Q., T. Warloe, K. Berg, J. Moan, M. Kongshaug, K. E. Giercksky, and J. M. Nestland. 1997. "5-Aminolevulinic Acid Based Photodynamic Therapy. Clinical Research and Future Challenges." *Cancer* 79: 2282–308.

Quintanar-Guerrero, D., A. Ganem-Quintanar, P. Tapia-Olguin, Y. N. Kalia, and P. Buri. 1999. "The Effect of Keratolytic Agents on the Permeability of Three Imidazole Antimycotic Drugs through the Human Nail." *Drug Development and Industrial Pharmacy* 24 (7): 685–90.

Reinel, D., and C. Clark. 1992. "Comparative Efficacy and Tolerability of Amorolfine Nail Lacquer 5% in Onychomycosis, Once-Weekly Versus Twice-Weekly." *Clinical and Experimental Dermatology* 17: 21–4.

Repka, M. A., P. K. Mididoddi, and S. P. Stodghill. 2004. "Influence of Human Nail Etching for the Assessment for Topical Onychomycosis Therapies." *International Journal of Pharmacy* 282: 95–106.

Roberts, D. T., R. J. Hay, V. R. Doherty, M. D. Richardson, and Y. M. Clayton. 1988. "Topical Treatment of Onychomycosis Using Bifonazole 1% Urea/Paste Paste." *Annals of the New York Academy of Sciences* 544: 586–7.

Santos, D. A, and J. S. Hamdan. 2007. "In Vitro Activities of Four Antifungal Drugs against *Trichophyton rubrum* Isolates Exhibiting Resistance to Fluconazole." *Mycoses* 50 (4): 286–9.

Schaller, M., C. Borelli, U. Berger, B. Walker, S. Schmidt, G. Weindl, and A. Jackel. 2009. "Susceptibility Testing of Amorolfine, Bifonazole and Ciclopiroxolamine against *Trichophyton rubrum* in an In Vitro Model of Dermatophyte Nail Infection." *Medical Mycology* 47 (7): 753–8.

Schatz, F., M. Brautigam, and E. Dobrowolsiki, 1995. "Nail Incorporation Kinetics of Terbinafine in Onychomycosis Patients." *Clinical and Experimental Dermatology* 20: 377–83.

Shivakumar, H. N., S. R. K. Vaka, N. V. Satheesh, H. Chandra, and S. N. Murthy. 2010. "Bilayered Nail Lacquer of Terbinafine Hydrochlorhide for Treatment of Onychomycosis." *Journal of Pharmaceutical Sciences* 10: 4267–76.

Susilo, R., H. C. Korting, W. Greb, and U. P. Strauss. 2006. "Nail Penetration of Sertaconazole with a Sertaconazole-Containing Nail Patch Formulation." *American Journal of Clinical Dermatology* 7 (4): 259–62.

Tosti, A., R. Hay, and R. Arenas-Guzman. 2005. "Patients at Risk of Onychomycosis—Risk Factor Identification and Active Prevention." *Journal of the European Academy of Dermatology and Venereology* 19: 13–6.

Trey, S. M., D. A. Wicks, and P. K. Mididoddi. 2007. "Delivery of Itraconazole from Extruded HPC Flims." *Drug Development and Industrial Pharmacy* 33: 727–35.

Uchida, K., T. Itoyama, and H. Yamaguchi. 1996. "In Vitro Antifungal Activity of Omoconazole Nitrate, a Novel Imidazone Antimycotic Drug, against Clinical Isolates from Patients with Cutaneous Mycosis." *The Japanese Journal of Antibiotics* 49 (8): 818–23.

Vaka, S. R. K., S. N. Murthy, and J. H. O'Haver. 2010. "A Platform for Predicting and Enhancing Model Drug Delivery across the Human Nail Plate." *Drug Development and Industrial Pharmacy* 37 (1): 72–9.

Van Hoogdalem, E. J., W. E. Van der Hoven, I. J. Terpstra, J. Van Zijtveld, and J. S. C. Verschoor. 1997. "Nail Penetration of the Antifungal Agent Oxiconazole after Repeated Topical Application in Healthy Volunteers, and the Effect of Acetylcysteine." *European Journal of Pharmaceutical Sciences* 5: 119–27.

Vejnovic, I., C. Huonder, and G. Betz. 2010a. "Permeation Studies of Novel Terbinafine Formulations Containing Hydrophobins through Human Nails In Vitro." *International Journal of Pharmacy* 397: 67–76.

Vejnovic, I., L. Simmler, and G. Betz. 2010b. "Investigation of Different Formulations for Drug Delivery through the Nail Plate." *International Journal of Pharmacy* 386: 185–94.

Wang, J. C. T., and Y. Sun. 1998. "Human Nail and its Topical Treatment: Brief Review of Current Research and Development of Topical Antifungal Drug Delivery for Onychomycosis Treatment." *Journal of Cosmetic Science* 50: 71–6.

Welsh, O., L. Vera-Cabrera, and E. Welsh. 2010. "Onychomycosis." *Clinical Dermatology* 28: 151–59.

Willemsen, M., P. De Doncker, and J. Willems. 1992. "Posttreatment Itraconazole Levels in the Nail. New Implications for Treatment of Onychomycosis." *Journal of the American Academy of Dermatology* 26: 730–5.

Zaiac, M. 2010. "The Role of Biological Agents in the Treatment of Nail Psoriasis." *American Journal of Clinical Dermatology* 11 (1): 27–9.

10 Onychopharmacokinetics
Proposed Model Insight

Rania Elkeeb, Xiaoying Hui, Laila Elkeeb, Ali Alikhan, and Howard I. Maibach

CONTENTS

10.1 INTRODUCTION

Ungual topical drug delivery serves as an attractive alternative to conventional (oral) therapy to avoid problems such as first-pass metabolism, gastrointestinal (GI) side effects, and poor patient compliance. As a result, the pharmaceutical chemistry community is investing increasing resources toward the development of technology enabling the barrier nature of the nail to be overcome.

To develop appropriate formulations for topical ungual application, there is a requirement for robust and validated *in vitro* techniques and models to enable the accurate prediction of the fate of the drug *in vivo*.

Mathematical models are often used by environmental agencies for safety screening and by pharmaceutical chemists to screen and select appropriate potential drugs for transdermal delivery. Similar to transdermal drug delivery, the prediction of nail permeability could be based on the physicochemical properties of the permeant including molecular weight, log K_{oct}, solubility, and hydrogen donor/acceptor capability. In transdermal drug delivery, attempts have been made to develop quantitative structure–permeability relationship (QSPR) models in order to relate statistically the skin permeation of compounds derived from *in vitro* experiments to the physicochemical parameters of the penetrants (Moss et al. 2002).

Recent studies correlated absorption data from three human *in vivo* data sets including agricultural, steroids, and other organic compounds to physicochemical parameters (Alikhan et al. 2009). Farahmand and Maibach (2009a) evaluated a novel QSPR model that correlated *in vivo* human transdermal patch-based skin permeation parameter data such as plasma concentration to its physicochemical properties and further evaluated the correlation of some widely used skin permeability predictive models with the *in vivo* proposed empirical model (Farahmand and Maibach 2009b). Generally, it is assumed that the physicochemical properties of the penetrating molecule have an influence on nail permeation. We believe that a QSPR model could be developed for nail delivery of compounds to relate statistically the physicochemical parameters of a drug to the experimentally derived ungual permeation data.

This study focuses on published and unpublished data using a novel nail sampling technique for drug penetration, with the goal of defining the relationship between the drug's penetration data determined from experimental values and the corresponding physicochemical properties.

We used these data to construct a preliminary database of drug/chemical nail penetration data for our regression models.

Multiple linear regression (MLR), a common and simple method for constructing QSPR models, was used. The advantage of MLR is that it is simple to use and the derived models are easily interpretable; nevertheless, currently available data present only a first step in onychopharmacokinetic modeling.

10.2 METHODS

10.2.1 Data Obtained from *In Vitro* Permeation Study Using Human Nails

Permeation data were obtained from published and unpublished work performed in one laboratory. Eight studies utilized a Teflon one-chamber diffusion cell (Permegear, Inc., Hellertown, PA) to hold each nail. To approximate physiological conditions, a small cotton ball wetted with normal saline was placed in the chamber to serve as a "nail bed" and to provide moisture for the nail plate measured by sampling the solution on the ventral nail plate at successive time points and calculating drug flux into and through the nail. Data were presented as weight normalized equivalent to normalized drug radioactivity as μg eq/mg nail, except for drug Y, where we had to convert normalized drug radioactivity to weight normalized equivalent.

10.2.2 Nail Sampling to Determine Drug Penetration

A novel technique enables the determination of drug concentration within the plate, where fungi reside. This method relies on a drilling system that samples the nail core without disturbing its surface. This is achieved by the use of a micrometer-precision nail sampling instrument that enables finely controlled drilling into the nail with a collection of the powder created by the drilling process. Drilling of the nail occurs through the ventral surface. The dorsal surface and the ventrally accessed nail core can be assayed separately. The dorsal surface sample contains residual drug, while the core from the ventral side provides drug measurement at the site of disease. This method permits drug measurement in the intermediate nail plate, which was previously impossible. Methodologic details are found in Hui et al. (2002).

10.2.3 Physicochemical Properties Data

Physicochemical properties of the eight drugs/chemicals, including the molecular weight, hydrogen bonding, and log K_{oct}, were collected from PubChem Scholar and original product information literature (Table 10.1). Then, the QSPR model was fitted using SigmaPlot 11 (SPSS Science, Chicago, IL). The Multiple Linear Regression (MLR) equation was determined by a best subset selection procedure to determine which independent variables contribute to the prediction of the dependent variables in the model.

10.2.4 QSPR Model Construction

Statistical analysis of data was conducted as follows: Normality test (Shapiro–Wilk) confirmed that the permeability data for each drug exhibited a normal Gaussian distribution. Therefore, multiple regression analysis was conducted to probe a relation between permeation data for each drug to their corresponding physicochemical

TABLE 10.1
Physicochemical and Ungual Absorption Parameters of Drugs

	Physicochemcial Properties				Nail Absorption (μg eq/mg nail)			
Drug	**MW**	**Log K_{oct}**	**HA**	**HD**	**Dorsal**	**Ventral**	**Remainder**	**References**
Ciclopirox A	207.27	2.3	3	1	71.7 ± 16.79	0.55 ± 0.28	51.97 ± 62.44	Hui et al (2004)
Econazole	381.68	5.5	3	0	0.25 ± 0.03	11.15 ± 2.56	0.16 ± 0.14	Hui et al (2003)
Terbinafine	291.43	5.9	1	0	0.81 ± 0.13	0.74 ± 0.38	0.003 ± 0.001[b]	Unpublished
AN2690	152.00	1.2	3	1	25.6 ± 8.80	20.46 ± 4.72	26.06 ± 12.41	Hui et al (2007a)
Ketoconazole	531.43	3.5	8	0	0.38 ± 0.12	0.81 ± 0.39	0.18 ± 0.09[b]	Unpublished
Panthenol	205.25	−1.2	5	4	10.05 ± 3.52	4.97 ± 0.90	0.80 ± 0.33[b]	Hui et al (2007b)
Drug X[a]	350.00	3.9	3	0	187 ± 90	349 ± 158	188 ± 64	Unpublished
Drug Y[a]	200.00	2.5	3	1	8.5 ± 4.71	1.92 ± 0.30	61.16 ± 30.48[b]	Unpublished

[a] Proprietary compound.
[b] Cotton ball (reservoir data).

properties. To determine the usefulness of a model, the following statistical parameters were considered (Pallant 2007):

1. r (correlation coefficient): This indicates the strength of the relationship between variables. A value of r between 0.3 and 0.49 represents a medium correlation; r between 0.5 and 1 demonstrates an increasingly strong relationship.
2. r^2 (coefficient of determination): This represents the amount of the dependent variable (ventral/dorsal) absorption attributable to the values of the independent (predictor) variables and is a measure of how well the regression model describes the data.
3. p: This is the probability of error in concluding that there is an association between the dependent and the independent variables. The smaller the p value, the greater the probability that there is an association. Traditionally, it is concluded that the independent variable can be used to predict the dependent variable's ventral/dorsal absorption when $p < 0.05$.

10.3 RESULTS

It was assumed that a general linear model would apply:

$$y = a + b_1x_1 + b_2x_2 + b_3x_3 + b_nx_n$$

where y is the dependent variable, a represents a constant, and b_1, b_2, ... b_n are fitted coefficients to the predictors (independent variables) x_1, x_2, ... x_n (such as molecular weight, log K_{oct}, and hydrogen bond donor or acceptor groups on the molecule).

MLR of both ventral and dorsal absorption of eight drugs against physicochemical parameters (Table 10.1) yields the following equations as the best-fit model for both dorsal and ventral absorption:

$$\begin{aligned}\text{Dorsal Abs.} = 365.273 + (1.891 \times \text{MW}) - (134.628 \times \log K_{\text{oct}}) \\ - (111.273 \times \text{HA}) - (85.971 \times \text{HD})\end{aligned} \quad (10.1)$$

$$N = 8,\ R = 0.926,\ R^2 = 0.858,\ \text{Adj}\ R^2 = 0.668$$

The dependent variable Dorsal Abs. can be predicted from a linear combination of the independent variables:

	p
MW	0.034
log K_{oct}	0.026
HA	0.029
HD	0.046

All independent variables (predictors) appear to contribute to predicting Dorsal Abs. ($p < 0.05$).

In this equation and elsewhere:

- HA is the total number of hydrogen bond acceptor groups on the molecule.
- MW is molecular weight.
- HD is the total number of hydrogen bond donor groups on the molecule.
- $\log K_{oct}$ is the logarithmically transformed octanol–water partition coefficient.

All independent variable predictors had a significant ($p < 0.05$) partial effect in the full model:

$$\text{Ventral Abs.} = 593.288 + (3.971 \times \text{MW}) - (256.581 \times \log K_{oct}) - (223.626 \times \text{HA}) - (147.371 \times \text{HD}) \quad (10.2)$$

$$N = 8, R = 0.969, R^2 = 0.939, \text{Adj}\,R^2 = 0.858$$

The dependent variable Ventral Abs. can be predicted from a linear combination of the independent variables:

	p
MW	0.008
$\log K_{oct}$	0.008
HA	0.008
HD	0.019

All independent variables (predictors) appear to contribute to predicting Ventral Abs. ($p < 0.05$).

10.4 DISCUSSION

Among the eight drugs, a significant correlation was demonstrated using the multivariate analysis between the physicochemical parameters and both dorsal and ventral ungual absorptions. Using a best subset regression to show significant correlations, all possible combinations of physicochemical parameters (in sets of two, three, and four variables) were modeled against ventral and dorsal absorption.

Using this model, the molecular weight has the largest contribution in predicting *ungual absorption* in both equations, as compared to the other predictor variables in the model. On excluding drug X, an inverse correlation between dorsal and MW was nearly significant ($r^2 = 0.71$; $p = 0.054$). This finding is in accord with previous studies, where molecular weight was determined to be a main parameter in predicting permeability coefficient (Kobayashi et al. 2004; Mertin and Lippold 1997). Diffusivity of a solute through the nail is potentially a function of both its hydrogen bonding and its size.

However, a significant ($r^2 = 0.75$; $p = 0.038$) inverse correlation between dorsal absorption and *K*oct was observed. Nail permeability has been previously reported to be especially sensitive to drug lipophilicity (Kobayashi et al. 1999). It has been previously believed that there is a lipidic pathway that is believed to be critical for the passage of very hydrophobic compounds.

Contrary to this, Brown et al. studied the *in vitro* permeation of compounds that differed considerably in terms of lipophilicity. The *in vitro* penetrant permeation through nails displayed an inverse correlation to molecular weight rather than log K_{oct}. This finding may indicate that the lipophilicity of a compound traversing the nail plate has less influence than its molecular weight in determining the extent of its permeation. The present study therefore supports previous claims that molecular weight is a critical physicochemical property in ungual drug permeation and proposes that the gel membrane description commonly given to the nail is simple and that nail barrier acts more as a complex hydrophilic filter resistant to large and lipophilic penetrants (Brown et al. 2009). Unlike skin, nail plate chemically consists mainly of proteins as well as 10–30% water and small amounts of lipids (0.1–1%; in skin it is 20%) (Smith et al. 2009). Therefore, a criticism of Equation 10.1 could be that log K_{oct} as a predictive variable in the equation may be found to be collinear with other variables such as HA and HD. There were no other meaningful correlations between physicochemical parameters and ventral absorption. Molecular size and other properties not being considered in this chapter, such as surface charge, salt bond formation, and capacity of molecular distortion, might be useful in building future models, as well as in the use of compartmentalized models to model the ungual pharmacokinetics.

A hypothesized QSPR model was developed based on the available *in vitro* data set for the eight drugs studied in our laboratory. This small sample size weakens the predictive power of the model, and a larger sample is needed to improve predictability. As this model has been derived from data in one laboratory where the study design and methodology of the nail study and nail sampling has been the same with similar sensitivity, precision, and robustness of the analytical method using radioactivity, we have eliminated one main factor contributing to a variation in ungual drug delivery *in vitro*. There are other limitations of current ungual drug permeability studies (Elkeeb et al. 2010), which include the following:

1. Use of animal hooves as a model for nail penetration: animal hooves may not necessarily provide a representative model in which diffusion through the human nail can be evaluated.
2. Use of nail clippings as a model of nail penetration: this model is bereft of nail bed and thus is not a complete model of nail permeation.
3. Hydration-controlled method: modified diffusion cells are a commonly used *in vitro* method in ungual drug permeation. Similar to skin penetration studies, permeation is measured by sampling the solution on the ventral nail plate at successive points in time and calculating drug flux through the nail. These assays super hydrate the nail and possibly alter the physical property and nail permeability. Hui et al. (2005) modified this approach by using a cotton ball soaked in saline to provide moisture (but not saturation) and

hydrating the nail throughout the experiment. This provides more hydration control than conventional assays.

Our model eliminated some limitations as we utilized human cadaver whole nail as well as the hydration-controlled method.

4. Correlation of *in vitro* to *in vivo* studies: the *in vitro* data obtained with the use of modified diffusion require comparison to human *in vivo* data with the use of radioisotopes and possibly attenuated mass spectroscopy. *In vitro* human or animal studies assume that penetration is a passive process and that there is no viable component to it, and we believe that *in vitro* studies must be correlated to those *in vivo*. Until these correlations are defined, biological interpretation remains tenuous (Elkeeb 2009). Additional references are found in an overview (Murdan 2002, 2008). A recent study by Murdan et al. (2011) made it possible to measure nail plate pH and provided baseline values; with this, further studies will be needed to understand the role of nail acidic pH in healthy and diseased nails and correlate it to *in vitro* studies.

10.5 CONCLUSION

Such QSPR models are expected to be consistent with the various routes and mechanisms by which permeants traverse the nail. However, the use of those models in predicting nail permeability remains equivocal. In reality, our hypothesized QSPR models in relation to nail delivery are derived from a database from the same investigator and laboratory with similar experimental protocols; therefore, these QSPR models are inherently less subject to experimental error. As this model has been derived from the available small data set, there remains a need to develop a more robust QSPR model using an extensive data set. There is, therefore, a need to develop and reach a standardized protocol so that future data could be analyzed for formation of future QSPR models, which might form the basis of future QSPR models.

The *in vitro* data obtained with the use of modified diffusion cells, together with nail clippings, avulsed human cadaver nail plates, and animal hooves as models for human nail plates, require comparison to *in vivo* data in humans. *In vitro* human or animal studies assume that penetration is a passive process without a viable component. We believe that *in vitro* studies must be correlated to what happens *in vivo*. Until these correlations are defined, biological interpretations are tenuous, but they provide some testable experiments *in vitro* and *in vivo*.

Prediction of the *in vitro* ungual drug permeation has become increasingly important in the development of ungual therapeutic agents. Development and validation of QSPR models that consider the vehicle effect and use physicochemical endpoints of the chemicals would be an economic alternative for costly and time-consuming *in vitro* and *in vivo* skin permeation studies. We have described several unique associations between ungual absorption and physicochemical factors of eight compounds. Large gaps still exist in our knowledge of ungual chemical absorption.

Future *in vitro* studies, involving larger arrays of chemically unrelated compounds, will hopefully fill these voids using predictive modeling techniques.

REFERENCES

Alikhan, A., S. Farahmand, and H. I. Maibach. 2009. "Correlating Percutaneous Absorption with Physicochemical Parameters In Vivo in Man: Agricultural, Steroid, and Other Organic Compounds." *Journal of Applied Toxicology* 29 (7): 590–6.

Brown, M. B., R. H. Khengar, R. B. Turner, B. Forbes, M. J. Traynor, C. R. Evans, and S. A. Jones. 2009. "Overcoming the Nail Barrier: A Systematic Investigation of Ungual Chemical Penetration Enhancement." *International Journal of Pharmaceutics* 370: 61–7.

Elkeeb, R., A. Alikhan, L. Elkeeb, X. Hui, and H. I. Maibach. 2010. "Transungual Drug Delivery: Current Status." *International Journal of Pharmaceutics* 384: 1–8.

Farahmand, S., and H. I. Maibach. 2009a. "Estimating Skin Permeability from Physicochemical Characteristics of Drugs: A Comparison between Conventional Models and an In Vivo-Based Approach." *International Journal of Pharmaceutics* 375: 41–7.

Farahmand, S., and H. I. Maibach. 2009b. "Transdermal Drug Pharmacokinetics in Man: Interindividual Variability and Partial Prediction." *International Journal of Pharmaceutics* 367: 1–15.

Hui, X., S. J. Baker, R. C. Wester, S. Barbadillo, A. K. Cashmore, V. Sanders, K. M. Hold et al. 2007a. "In Vitro Penetration of a Novel Oxaborole Antifungal (AN2690) into the Human Nail Plate." *Journal of Pharmaceutical Sciences* 96: 2622–31.

Hui, X., T. C. Chan, S. Barbadillo, C. Lee, H. I. Maibach, and R. C. Wester. 2003. "Enhanced Econazole Penetration into Human Nail by 2-n-Nonyl-1,3-Dioxolane." *Journal of Pharmaceutical Sciences* 92: 142–8.

Hui, X., S. B. Hornby, R. C. Wester, S. Barbadillo, Y. Appa, and H. Maibach. 2007b. "In Vitro Human Nail Penetration and Kinetics of Panthenol." *International Journal of Cosmetic Science* 29: 277–82.

Hui, X., Z. Shainhouse, H. Tanojo, A. Anigbogu, G. E. Markus, H. I. Maibach, and R. C. Wester. 2002. "Enhanced Human Nail Drug Delivery: Nail Inner Drug Content Assayed by New Unique Method." *Journal of Pharmaceutical Sciences* 91: 189–95.

Hui, X., R. C. Wester, S. Barbadillo, C. Lee, B. Patel, M. Wortzmman, E. H. Gans, and H. I. Maibach. 2004. "Ciclopirox Delivery into the Human Nail Plate." *Journal of Pharmaceutical Sciences* 93: 2545–8.

Hui, X., R. C. Wester, S. Barbadillo, and H. Maibach. 2005. "Nail Penetration: Enhancement of Topical Delivery of Antifungal Drugs by Chemical Modification of the Human Nail." *Textbook of Cosmetic Dermatology*, edited by R. Baran and H. I. Maibach, vol. 6, 57–63. Taylor and Francis, New York, NY.

Kobayashi, Y., T. Komatsu, M. Sumi, S. Numajiri, M. Miyamoto, D. Kobayashi, K. Sugibayashi, and Y. Morimoto. 2004. "In Vitro Permeation of Several Drugs through the Human Nail Plate: Relationship between Physicochemical Properties and Nail Permeability of Drugs." *European Journal of Pharmaceutical Sciences* 21: 471–7.

Kobayashi, Y., M. Miyamoto, K. Sugibayashi, and Y. Morimoto. 1999. "Drug Permeation through the Three Layers of the Human Nail Plate." *Journal of Pharmacy and Pharmacology* 51: 271–8.

Mertin, D., and B. C. Lippold. 1997. "In-Vitro Permeability of the Human Nail and of a Keratin Membrane from Bovine Hooves: Prediction of the Penetration Rate of Antimycotics through the Nail Plate and Their Efficacy." *Journal of Pharmacy and Pharmacology* 49: 866–72.

Moss, G. P., J. C. Dearden, H. Patel, and M. T. Cronin. 2002. "Quantitative Structure-Permeability Relationships (QSPRs) for Percutaneous Absorption." *Toxicology in Vitro* 16: 299–317.

Murdan, S. 2002. "Drug Delivery to the Nail following Topical Application." *International Journal of Pharmaceutics* 236: 1–26.

Murdan, S. 2008. "Enhancing the Nail Permeability of Topically Applied Drugs." *Expert Opinion on Drug Delivery* 5: 1267–82.

Murdan, S., G. Milcovich, and G. S. Goriparthi. 2011. "An Assessment of the Human Nail Plate pH." *Skin Pharmacology and Physiology* 24: 175–81.

Pallant, J. 2007. *SPSS Survival Manual: A Step by Step Guide to Data Analysis Using SPSS For Windows*, 3rd ed. Berkshire: McGraw-Hill.

Smith, K. A., J. Hao, and S. K. Li. 2009. "Effects of Ionic Strength on Passive and Iontophoretic Transport of Cationic Permeant across Human Nail." *Pharmaceutical Research* 26: 1446–55.

11 Photodynamic Therapy of Nail Diseases

Ryan F. Donnelly, Corona M. Cassidy, and Michael M. Tunney

CONTENTS

11.1 INTRODUCTION

Photodynamic therapy (PDT) is a clinical treatment that combines the effects of visible light irradiation with subsequent biochemical events that arise from the presence of a photosensitizing drug (possessing no dark toxicity) to cause destruction of selected cells (Gannon and Brown 1999). The photosensitizer, when introduced into the body, accumulates in the target cells and a measured light dose of appropriate wavelength is then used to irradiate the target tissue (De Rosa and Bentley 2000; Peng et al. 1997). This activates the drug through a series of electronic excitations and elicits a series of cytotoxic reactions, which can be dependent on, or independent of, the generation of reactive oxygen species (Fritsch et al. 1998).

PDT has progressed considerably from the early application of sunlight and hematoporphyrin derivative to the use of Photofrin® and to second-generation preformed photosensitizers and topical (surface) application of the prodrug 5-aminolevulinic acid (ALA), which leads to in situ synthesis of protoporphyrin IX (PpIX) (Daniell and Hill 1991). Topical PDT is now used for a variety of malignant, dysplastic, hyperplastic, and infectious skin disorders (Henderson and Dougherty 1992; Moan and Peng 2003). Clinical acceptance of topical PDT, in particular, has been accredited to the pioneering work of Kennedy et al. (1990b). The results of this first clinical trial of topical PDT exploited the tumor-selective accumulation of the photosensitizer, PpIX, following topical cutaneous application of ALA. A 90% clearance rate was achieved in 80 lesions treated with 20% w/w ALA in an oil-in-water (o/w) cream followed, 3–6 h later, by local illumination from a 500-W lamp equipped with a 600-nm long-wave-pass filter. The popularity of ALA, as the most commonly studied agent for PDT, is clearly evident in the number of published articles on the topic, which has increased markedly from 2 in 1991 to about 2330 in 2010 (ISI Web of Knowledge).

The detailed mechanism of action of PDT has been discussed extensively elsewhere (Dougherty et al. 1998; Kalka et al. 2000). Briefly, it results from the interaction of photons of visible light, of appropriate wavelength, with intracellular concentrations of photosensitizing molecules (Figure 11.1). Photosensitizers have a stable electronic configuration, which is in a singlet state in their lowest or ground energy level (Konan et al. 2002). This means that there are no unpaired electron spins (Isaacs 1995; Kalyanasundaram 1991). Following absorption of a photon of light of specific wavelength, a molecule is promoted to an excited state, which is also a singlet state and is short-lived, with a half-life between 10^{-6} and 10^{-9} sec (Dougherty et al. 1998; Konan et al. 2002). The photosensitizer can return to the ground state by emitting a photon as light energy, or, in other words, by fluorescence, or by internal conversion with energy lost as heat. Alternatively, the molecule may convert to the triplet state. This conversion occurs through intersystem crossing, which involves a change in the spin of an electron (Ochsner 1997). The triplet-state photosensitizer has lower energy than the singlet state but has a longer lifetime.

The singlet-state sensitizer can interact with surrounding molecules through type I reactions, while the triplet-state sensitizer can interact with its surroundings through type II reactions. The former type of reaction leads to the production of free radicals or radical ions, through hydrogen or electron transfer. These reactive species, after interaction with oxygen, can produce highly reactive oxygen species, such as the superoxide and peroxide anions, which then attack cellular targets (Kalka et al. 2000). However, type I reactions do not necessarily require oxygen and can cause cellular damage directly, through the action of free radicals, which may include sensitizer radicals. Type II reactions, by contrast, require an energy

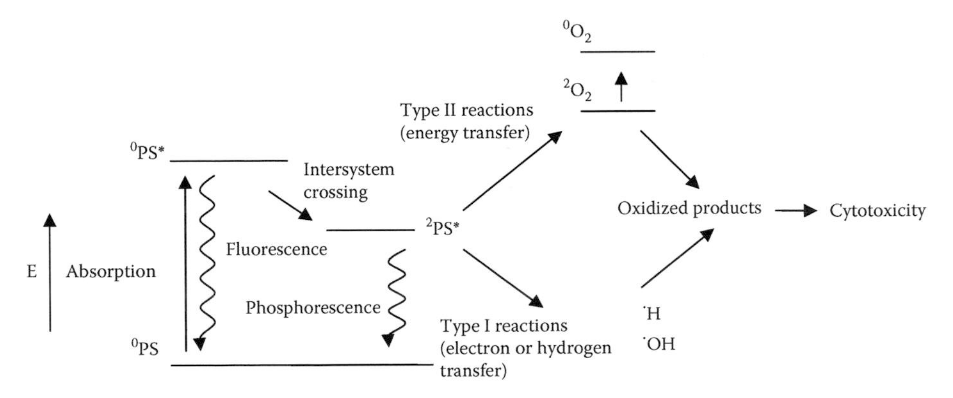

FIGURE 11.1 The mechanism of action of photodynamic therapy. Superscript numbers denote the number of unpaired electron spins in each molecule.

transfer mechanism from the triplet-state sensitizer to molecular oxygen, which itself normally occupies the triplet ground state (De Rosa and Bentley 2000). Although possessing a short lifetime of approximately 10^{-6} sec, a sufficient concentration of highly cytotoxic singlet oxygen is produced to induce irreversible cell damage (Dougherty et al. 1998; Kalka et al. 2000). In addition, the photosensitizer is not necessarily destroyed, but can return to its ground state by phosphorescence without chemical alteration and may be able to repeat the process of energy transfer many times (Ochsner 1997). Alternatively, the sensitizer may return to ground by transferring its energy to molecular oxygen and may even be destroyed by photobleaching due to oxidation (Moan et al. 1997). Evidently, many effects of PDT are oxygen-dependent and rely on the oxygen tension within the target tissue. Type I and type II reactions can occur simultaneously, and the ratio between the two depend on the photosensitizer, substrate, oxygen concentration, and sensitizer to substrate binding (Kalka et al. 2000). Singlet oxygen is, however, widely believed to be the major damaging species in PDT (Gannon and Brown 1999; Konan et al. 2002; Peng et al. 1997). Due to its extreme reactivity, singlet oxygen has a short life span in a cellular environment and limited diffusivity in tissue, allowing it to travel only approximately 0.1 μm (Moan 1990). This, combined with the facts that normal tissue may not contain photosensitizer or may not be perfused by blood vessels damaged by PDT, means that this normal tissue is normally unaffected by exposure to light (Peng et al. 1997).

11.2 PHOTOSENSITIZERS

The efficacy of certain types of dye against microbial species formed the basis of modern chemotherapy over 100 years ago. The selectivity, particularly of cationic dyes, for bacteria over mammalian cells was used by Ehrlich and Browning to develop early synthetic antibacterials. However, much of the impetus for this work was lost at the inception of the antibiotic era, when the action of penicillin was seen as miraculous. The recent renaissance in the use of dyes and their derivatives in cancer treatment (PDT) rely on the fact that the dyes act as photosensitizers.

For a molecule to act as an efficient photosensitizer, it must possess the ability to absorb visible light, becoming excited to the triplet state, and then transfer its energy economically to molecular oxygen. Molecules possessing such characteristics are typically rigid planar structures possessing a high degree of conjugation. The major photosensitizer classes employed to date in photodynamic antimicrobial chemotherapy (PACT) include the porphyrins, the phthalocyanines, and the phenothiaziniums (Figure 11.2). The phenothiaziniums have simple tricyclic planar structures, typically cationic in nature. The most widely used compounds are methylene blue (MB) and toluidine blue (TBO). Both are efficient producers of singlet oxygen, and the maximum absorption wavelength in water is 656 nm for MB and 625 nm for TBO. The porphyrins are heterocyclic macrocycles derived from four pyrrole-like subunits interconnected *via* their α-carbon atoms *via* methine bridges. The absorption spectrum of porphyrins exhibits a maximum in the Soret band in the visible region of the electromagnetic spectrum between 360 and 400 nm, followed by four smaller peaks between 500 and 635 nm (Q bands) (Kalka et al. 2000). The pyrrole groups in phthalocyanines are conjugated to benzene rings and bridges by aza nitrogens rather than

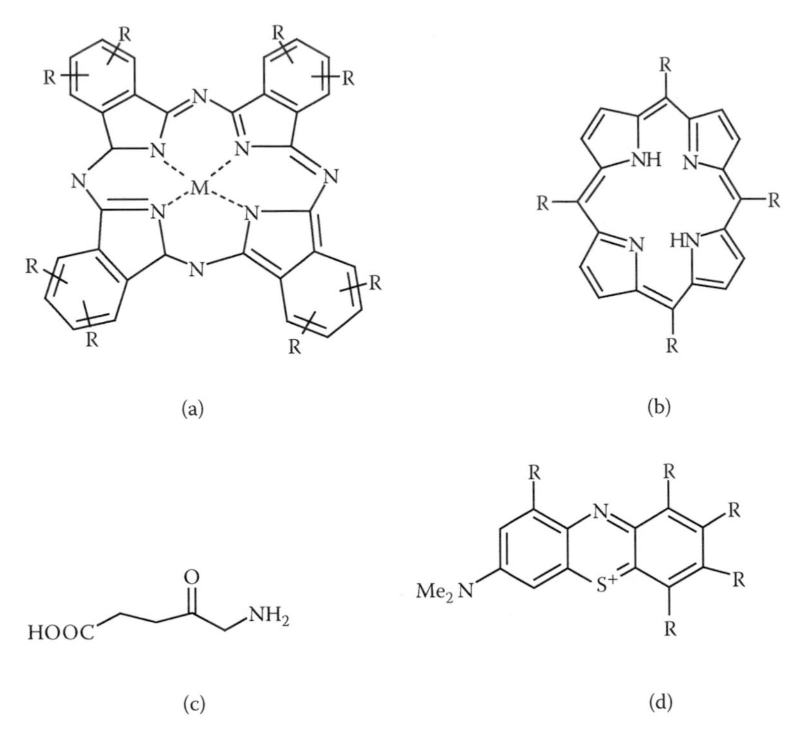

FIGURE 11.2 Basic chemical structures of (a) phthalocyanine, (b) porphyrin, (c) 5-aminolevulinic acid, and (d) phenothiazinium photosensitizers investigated for potential use in photodynamic therapy of nail diseases.

methane carbons. This causes the absorption spectrum to shift to longer wavelengths and the Q bands to become more intense than the Soret peak (Bonnett et al. 2001).

11.3 LOCALIZATION OF PHOTOSENSITIZERS

ALA is a small, water-soluble, prodrug, which is a naturally occurring precursor in the biosynthetic pathway of heme. Administration of excess exogenous ALA avoids the negative feedback control that heme exerts over its biosynthetic pathway. Due to the limited capacity of ferrochelatase to convert PpIX into heme, the presence of excess exogenous ALA in cells induces accumulation of PpIX (Kennedy et al. 1990a, 1996; Meijnders et al. 1996). This effect is pronounced in sebaceous glands and also in neoplastic cells. It has been reported that certain types of neoplastic cells have not only reduced ferrochelatase activity but also enhanced porphobilinogen deaminase (PBGD) activity (De Rosa and Bentley 2000; Kalka et al. 2000; Kennedy and Pottier 1992). The localization of preformed photosensitizers is, however, not completely understood. Various theories exist regarding the preferential uptake by and accumulation of such agents in tumors. New photosensitizers, exhibiting rapid maximal accumulation in tumors, high tumor to normal tissue ratios, and efficient clearance from the body are being actively sought (Gantchev et al. 1996; Kessel 1999).

Preformed, lipophilic sensitizers, such as the porphyrins and phthalocyanines, when administered intravenously, are believed to be transported in the bloodstream bound to lipoproteins such as low-density lipoproteins (LDLs) (Ochsner 1997; Pottier and Kennedy 1990). Tumor cell membranes have been shown to possess disproportionately high numbers of LDL receptors (Jori et al. 1984) leading to large numbers of sensitizer molecules being brought into intimate contact with the tumor cells. Following receptor-mediated endocytosis, the sensitizer molecules may preferentially accumulate in the lipophilic compartments of the cells, including plasma, mitochondrial, endoplasmic reticulum, and nuclear and lysosomal membranes (Ochsner 1997). This rather simplistic view does not provide the whole picture, however. In fact, in *in vitro* tissue culture experiments, tumor cells do not take up any more sensitizer than normal cells (Pottier and Kennedy 1990). The *in vivo* situation is significantly different. Due to the rapidly growing nature of tumors with respect to normal tissue, their microvasculature is substantially altered, meaning they have a disordered blood supply and are less well perfused (Pottier and Kennedy 1990; Vaupel et al. 1981). They exhibit an enhanced vascular permeability to plasma proteins, show poor lymphatic drainage, and have a larger interstitial space. The net result is that sensitizers exhibit enhanced transport to and prolonged residence in tumors. Bound sensitizer, accumulated in the tumor, can then be taken up by the cell (Moan and Sommer 1983).

The uptake and retention mechanisms for free sensitizers located in the interstitial space or tumor microvasculature are distinct from those of bound sensitizers. As a result of the reduced tumor perfusion, tumor cells are forced to undergo anaerobic glycolysis, producing large quantities of lactic acid. Hydrolysis of adenosine triphosphate also occurs (Fuchs et al. 1997). This means that the tumor interstitial pH is significantly lower than that of normal tissue (Wike-Hooley et al. 1984); an average pH value for tumors of pH 6.5 compared to approximately pH 7.5 for normal tissues is not unusual (Pottier and Kennedy 1990; Wike-Hooley et al. 1984). Many photosensitizers are weak acids and, at low pH, will be largely unionized. Therefore, their cellular absorption will be enhanced by the lowered pH in the tumor microenvironment. Indeed, if tumor pH can be further lowered by administration of agents such as glucose (Piot et al. 2001), an increased proportion of tumor cells may be killed directly by PDT (Barrett et al. 1990). Once the sensitizer molecule is within the cell, the higher intracellular pH, which is close to normal intracellular pH of around 6.9, may increase the proportion of ionized sensitizer. This ionized species then becomes temporarily trapped within the cell until such time as the extracellular concentration of sensitizer falls and the complex system of ionic equilibria that exists allows it to diffuse out of the cell as a neutral molecule. This latter principle also applies to sensitizer entering the cell by other means. Hence, there exists a defined timeframe for each lipophilic sensitizer and type of tumor between sensitizer administration and its maximal accumulation within tumor cells (Pottier and Kennedy 1990).

Preformed hydrophilic sensitizers, such as water-soluble phthalocyanines, are largely carried by albumin and other serum proteins after intravenous injection (Ochsner 1997). These sensitizers then accumulate within the interstitial space and the vascular stroma of tumors due to their enhanced vascular permeability to plasma proteins, poor lymphatic drainage, and larger interstitial space (Pottier and Kennedy 1990).

Due to their low lipophilicity, these sensitizers do not readily diffuse across cellular membranes, although a small fraction may be absorbed by pinocytosis or endocytosis (Dougherty et al. 1998; Ochsner 1997). As with lipophilic sensitizers, there exists an optimum time period between administration of hydrophilic sensitizers and their maximal accumulation within tumors. Again, this time period will vary between different sensitizers and tumor types.

Tumors are not the only type of tissue that exhibits accumulation of photosensitizers. For example, the accumulation of certain sensitizers by the rapidly developing retinal neovasculature, which is characteristic of age-related macular degeneration and the plaques of psoriasis, has been used to achieve positive therapeutic outcomes by several workers (Costa et al. 2001; Koh and Ang 2002; Robinson et al. 1999; Schick et al. 1997). In addition, several normal body tissues high in reticuloendothelial components, such as the liver, exhibit accumulation of administered photosensitizers. This is a phenomenon that is not well understood.

It is now well known that cationic photosensitizers are more efficient than their neutral or anionic counterparts in the photodynamic killing of microbial cells. Cationic photosensitizers are more effective, especially as broad-spectrum antibacterials, than their anionic counterparts (Wainwright 1998), as shown by their greater activity against Gram-negative bacteria, which have a more complex structure due to the presence of an outer membrane. The cell envelope of Gram-negative bacteria consists of an inner cytoplasmic membrane and an outer membrane that are separated by the peptidoglycan-containing periplasm. The outer membrane, which is highly negatively charged, forms a physical and functional barrier between the cell and its environment (Nikaido 2003). It has been shown that anionic and neutral photosensitizers can become effective against Gram-negative bacteria when coadministered with a cationic agent such as polymyxin (Malik et al. 1992). However, for simplicity and because, even against more susceptible Gram-positive bacteria, cationic photosensitizers appear to be more effective (Jori and Brown 2004; Wainwright 1998; Wainwright and Crossley 2004), these cationic agents are the predominant type used in PACT.

To date, there have been several reports on the use of photosensitizers and light to kill both yeasts and other fungi. However, there has been much less systematic study on the types of physicochemical properties necessary in a photosensitizer in order to make it effective in mediating photodynamic killing of such microorganisms. Fungi present much more complex targets than bacteria. For example, yeasts, which constitute a large group of rather disparate eukaryotic organisms, are enveloped by a thick external wall composed of a mixture of glucan, mannan, chitin, and lipoproteins and separated from the plasma membrane by a periplasmic space. However, the available evidence suggests that the response of such cells to photodynamic processes is less strictly controlled by structural factors as compared with bacteria (Paardekooper et al. 1995a). Nevertheless, similarities with mammalian cells should be considered and this may indicate the use of cationic photosensitizers, rather than their anionic counterparts, since the latter exhibit facile uptake by mammalian cells (Bonnett 1995).

Uptake of exogenous substances by fungi is generally adversely affected by lipophilicity and positively affected by hydrophilicity and the presence of charged groups. Following uptake, photosensitizers are distributed to subcellular targets.

The pattern of localization is important, as targets adjacent to the photosensitizer have the greatest probability of being involved in photodynamic processes, due to the high reactivity and short lifetime of the singlet oxygen generated. The biochemical and functional effects of photosensitization include inactivation of enzymes and other proteins and peroxidation of lipids, leading to the lysis of cell membranes, lysosomes, and mitochondria (Bertoloni et al. 1992). Thus, singlet oxygen generated by excitation of photosensitizers is a nonspecific oxidizing agent. Consequently, there is no cellular defense against it. Indeed, antioxidant enzymes, such as catalase and superoxide dismutase, are inactivated by it. This means that there should be no difference in susceptibility to PACT between organisms resistant to conventional antifungals and their naïve counterparts. The high reactivity of singlet oxygen has other advantages, because even though the localization of the photosensitizer may be determined by its physiochemical properties, the diffusion of singlet oxygen should be sufficient to be able to inactivate other structures and biomolecules. Therefore, it is unlikely that fungi could readily evolve resistance to singlet oxygen. In addition, photodynamic processes have never been associated with mutagenic effects in microorganisms. Moreover, singlet oxygen is only present during illumination, and fungi are not continuously exposed to it, as they are with conventional antifungals. Furthermore, singlet oxygen cannot travel to other sites in the body, such as the intestinal tract, during treatment. These latter facts make development of resistance even more unlikely.

It has been widely noted that *Candida albicans*, like other yeasts, is slightly more difficult to kill by PACT than Gram-positive bacterial cells, necessitating higher drug and light doses (Zeina et al. 2002). This has been attributed to the presence in the yeasts of a nuclear membrane, the greater cell size and the reduced number of targets for singlet oxygen per unit volume of cell (Codling et al. 2003; Demidova and Hamblin 2005; Zeina et al. 2001). However, it has been shown that the photosensitizer and light doses producing high levels of kill in yeasts *in vitro* do not kill appreciable numbers of human cells under the same conditions and cause no detectable genotoxic or mutagenic effects (Zeina et al. 2003). Should photodynamic killing of fungi be carried out *in vivo*, then the limited diffusion distance of singlet oxygen from its site of generation and the fact that illumination would be limited to the area of infection mean that selectivity for fungi over host cells would be further enhanced.

11.4 LIGHT ADMINISTRATION

By definition, PACT requires a source of light to supply the requisite energy for singlet oxygen production in situ. The energy required is determined by the molecular structure of the photosensitizer, and thus, a different light excitation range is required for the phenothiaziniums (ca. 600–660 nm) than for the phthalocyanines (ca. 630–690 nm). Ideally, light sources should provide a strong output at the requisite wavelength for photoexcitation. Lasers and the less expensive and easier to use filtered incoherent lamps are the most commonly employed sources in PACT today. White or fluorescent light sources may be used. However, for *in vivo* use, emission in the ultraviolet (UV) range should be minimized, due to the risk of mutagenesis.

Similarly, emission in the infrared range is also undesirable, so as to avoid heating of tissue. Typical power outputs for light sources used in antifungal PACT are in the range 10–100 mW cm^{-2}, with typical total light doses being between 10 and 200 J cm^{-2}. In some cases, these may need to be higher than those used in antibacterial PACT to yield comparable rates of kill (Donnelly et al. 2007).

Light fluence through the tissue decreases exponentially with thickness. This decrease is determined by absorption, particularly by hemoglobin, and scattering, parameters that vary between tissue types (Brancaleon and Moseley 2002). Due to the inability of light to penetrate deeply into the tissue, clinical PACT is necessarily limited to areas of the body that can be irradiated from the surface. Thus, antifungal treatment would be restricted to infections of the skin, nails, hair, oral cavity, esophagus, and lower female reproductive tract. In treating such infections, however, some degree of tissue penetration is required to, for example, kill fungi residing below the surface of the skin or in the matrix of the nail. Light in the red region of the spectrum penetrates tissue down to around 3.0 mm, while light in the blue region penetrates down to only around 1.5 mm. Thus, the porphyrins are typically excited by light in the red region of the spectrum, rather than blue light, which they absorb more efficiently (Gannon and Brown 1999). Consequently, much work has been devoted to the phthalocyanines, which absorb more effectively at longer wavelengths.

Endogenous light absorption is important in clinical applications of antifungal PACT. It is essential that photosensitizers used to kill fungi can be photoexcited, and this will not occur if the incident light is absorbed by fungal pigment. Thus, photosensitizers absorbing beyond the range of the pigment are required, with appropriate light sources. As with all proposed protocols, a thorough knowledge of the photoproperties of both target and agent will be essential. It is also vitally important that light can penetrate efficiently through the tissue to reach the site of infection.

11.5 PHOTODYNAMIC THERAPY OF NAIL DISEASES

Given the intractable nature of many nail diseases and the ready accessibility of human nails for application of drug delivery vehicles and illumination with light, it is perhaps unsurprising that a number of research groups and clinicians have shown interest in PDT. *In vitro* studies have, to date, focused on photodynamic killing of the causative pathogens of onychomycosis. The aim has been to determine photosensitizer and light doses necessary for high rates of kill, with a view to clinical application. The practicalities of delivering the photosensitizer precursor ALA across human nail *in vitro* and determination of ALA concentration within porcine hoof matrices, a good model, for human nail, have also been investigated. In addition, successful PDT-based management of plaque psoriasis of the skin (Kim et al. 2007; Robinson et al. 1999; Salah et al. 2009) has prompted use of the technique in treatments of nail psoriasis.

The attractiveness of PDT of nail diseases is not difficult to imagine. If patients suffering from onychomycosis, in particular, could be routinely treated in a single session or even a few sessions rather than having to take drugs with potential for side effects and drug–drug interactions for months, or applying medicated paints for up to a year, then a significant improvement in patient care would be achieved, not to mention considerable economic benefit for health care providers.

11.5.1 *In Vitro* Antimicrobial Studies

Photodynamic destruction of dermatomycetes (*Trichophyton interdigitale*, *Trichophyton rubrum*, *Trichophyton tonsurans*, *Microsporum cookei*, *Microsporum canis*, *Microsporum gypseum*, *Epidermophyton floccosum*, and *Nannizia cajetani*) using two thiophene photosensitizers (2,2′:5′,2″-terthienyl and 5-(4-OH-1-butinyl) 2,2′-bithienyl) was investigated by Romagnoli et al. (1998). A strong and dose-dependent inhibition of growth of all tested strains was observed. However, complete inactivation was never obtained, and the UVA irradiation employed is undesirable for clinical use (Romagnoli et al. 1998).

Upon irradiation with broadband white light, several hydrophilic porphyrinic photosensitizers [deuteroporphyrin, deuteroporphyrin monomethylester, and 5,10,15-tris (4-methylpyridinium)-20-phenyl-(21H, 23H)-porphine trichloride (Sylsens B)] have been shown to exhibit toxic effects toward the dermatophyte *T. rubrum* in suspension culture as well as its isolated microconidia (Smijs and Schuitmaker 2003). A follow-up study using quinolino-(4,5,6,7-efg)-7-de-methyl-8-deethylmesoporphyin dimethylester yielded only a fungistatic effect. However, red light irradiation of Sylsens B produced complete inactivation of fungal spores and destruction of fungal hyphae (Smijs et al. 2004).

Although not a photosensitizer itself, ALA is a naturally occurring precursor in the biosynthetic pathway of heme (Brouillet et al. 1975). However, administration of excess exogenous ALA leads to accumulation of the potent photosensitzser PpIX. ALA is the most commonly used agent in PDT of superficial cutaneous neoplasias, where reduced ferrochelatase and enhanced PBGD activity and a disordered *stratum corneum* barrier lead to selective PpIX accumulation in such lesions. However, in comparison to mammalian cells, fungi display some differences in the enzymatic machinery. Coproporphyrinogen oxidase is found in the cytoplasm, and both ALA synthase and ALA dehydratase are the rate-limiting steps and their activity is controlled by the intracellular free heme pool (Moretti et al. 2000). Therefore, the content of PpIX may be increased by the codelivery of iron chelators, such as ethylenediaminetetraacetic acid (EDTA), hydroxypyridone, and 2,2′-dipyridyl, that may inhibit the conversion of PpIX to heme (Strakhovskaya et al. 1998).

Upon exposure to light, PpIX induces cytotoxic effects through photochemical reactions that damage the plasma membranes and the mitochondria where PpIX is synthesized (Strakhovskaya et al. 1998). Prolonged irradiation causes the additional alteration of other cytoplasmic structures and the inhibition of the synthesis of DNA and RNA (Paardekooper et al. 1995b), but genotoxic and mutagenic effects have never been detected in yeasts (Moretti et al. 2000). The ability of ALA-induced PpIX to kill *C. albicans* upon irradiation has been demonstrated *in vitro* (Donnelly et al. 2005; Monfrecola et al. 2004). It was found that prolonged incubation with ALA was also toxic to *C. albicans* in the absence of irradiation (Donnelly et al. 2005).

Only two studies to date have investigated the potential of ALA-induced PpIX to kill dermatomycetes. *T. rubrum* was shown to synthesize PpIX from exogenously supplied ALA (Kamp et al. 2005). However, 10–14 days of incubation with 1–10 mM ALA were required to induce maximal PpIX concentrations. Increasing the ALA concentration led to reduced growth due to a concomitant decrease in suspension pH. Only incubation with 100 mM ALA for 2 h followed by irradiation (635 nm, 100 J cm^{-2}) caused a significant reduction in viability of *T. interdigitale* microconidia (Donnelly et al. 2005). Incubation with lesser ALA concentrations for 2 h and incubation for 30 min produced no such significant reductions. However, significant kill rates were achieved, with even moderate ALA concentrations when the microorganism was incubated for 6 h followed by irradiation. ALA, in the absence of irradiation, produced no significant kill of *T. interdigitale*. These latter points, perhaps, reflect the slower metabolic rate and innate robustness of dermatomycete microconidia compared with yeast cells.

In the only study published to date on the ability of phthalocyanines to kill dermatomycetes by photodynamic action, lipophilic phthalocyanines were found to exhibit only a fungistatic effect against *T. rubrum* following irradiation. This effect lasted only about 1 week (Smijs and Schuitmaker 2003).

11.5.2 Drug Delivery Studies

Donnelly et al. (2005) studied the *in vitro* penetration of ALA across human nail and into neonate porcine hoof when released from a novel bioadhesive patch containing 50 mg cm^{-2} ALA. The authors proposed that if sufficient concentrations of ALA could be achieved within the nail matrix and at the nail bed, PACT may prove to be a useful treatment for onychomycosis. Patch application for 24 h allowed an ALA concentration of 2.8 mM to be achieved on the ventral side of excised human nail. Application for 48 h induced a concentration of 6.9 mM. Application time had no significant effect on the ALA concentration at mean depths of 2.375 mm in neonate porcine, with application times of 24, 48, and 72 h all producing concentrations of 0.1 mM. Incubation of *C. albicans* and *T. interdigitale* with ALA concentrations of 10.0 mM for 30 min and 6 h, respectively, caused reductions in viability of 87% and 42%, respectively, following irradiation with red light. Incubation with a lesser concentration of 0.1 mM ALA for 30 min and 6 h, respectively, caused reductions in viability of 32% for *C. albicans* and 6% for *T. interdigitale*, following irradiation (Donnelly et al. 2005). These findings led the authors to suggest that ALA penetration across nail may have to be improved using penetration enhancers or by filing of the relatively impenetrable dorsal surface of the nail. Alternatively, iron chelators could be used to increase PpIX production for a given ALA dose. The authors concluded that, with such suitable modifications, ALA-based PACT may prove to be a viable alternative in the treatment of onychomycosis.

Smijs et al. (2007) used an ex vivo human skin model to investigate the ability of porphyrins to kill *T. rubrum*. The photosensitizers, in liquid vehicles, were applied to the skin, which had been previously inoculated with the dermatomycete. It was found that short incubation times (8 h) gave complete kill upon irradiation (108 J cm^{-2}, 580–870 nm), while incubation for longer times (>24 h) prior to irradiation yielded no kill. Water was a more effective delivery vehicle, in terms of kill rate, than the cell culture medium Dulbecco's Modified Eagle's Medium (DMEM) (Smijs et al. 2007).

11.5.3 Clinical Treatment of Onychomycosis

Landsman et al. (2010) investigated use of 870- and 930-nm light exposure, with no photosensitizing agent, in the treatment of mild, moderate, and severe onychomycosis. The light source used was a dual-wavelength near-infrared diode laser operating at physiologic temperatures. The wavelengths used lack the teratogenic danger presented by UV light and the photoablation toxic plume associated with pulsed Nd:YAG lasers (Landsman et al. 2010).

In this randomized controlled study, treatments followed a predefined protocol and laser parameters and occurred on days 1, 14, 42, and 120. Toes were cultured and evaluated, and measurements were taken from standardized photographs obtained periodically during the 180-day follow-up period.

The investigators treated mycologically confirmed onychomycosis in 26 eligible toes (10 mild, 7 moderate, and 9 severe). All of the patients were followed up for 180 days. An independent expert panel, blinded regarding treatment versus control, found that at 180 days, 85% of the eligible treated toenails were improved by clear nail linear extent; 65% showed at least 3 mm, and 26% showed at least 4 mm of clear nail growth. Of the 16 toes with moderate to severe involvement, 10 (63%) improved, as shown by clear nail growth of at least 3 mm. Simultaneous negative culture and periodic acid Schiff was noted in 30% at 180 days. These investigators concluded that there may be a role for this type of laser treatment in the management of onychomycosis, regardless of degree of severity. Ease of delivery and the lack of a need to monitor blood chemistry were claimed to be attractive attributes.

Sotiriou et al. (2010) conducted a single-center open trial of PDT for distal and lateral subungual toenail onychomycosis caused by *T. rubrum*. Thirty Caucasian patients (20 men, 10 women) were enrolled in the study (mean age 59.6 years, age range 41–81). Eligibility requirements for enrollment were (i) clinical features of distal and lateral subungual toenail onychomycosis, (ii) positive direct microscopic examination for fungal elements, (iii) identification of *T. rubrum* in cultures of Sabouraud dextrose agar, and (iv) concomitant conditions that did not allow systemic treatment with antifungal agents. The nail plate was treated for 10 consecutive nights with 20% w/w urea ointment under occlusion and could thus be removed easily with the use of forceps. Patients underwent the following therapeutic procedure: 20% w/w AMA cream was applied topically under an occlusive dressing on the entire nail bed after nail plate and subungual hyperkeratosis removal. After 3 h, irradiation was performed with red light (570–670 nm) from a noncoherent light source at a light dose of 40 J cm^{-2} and a fluence rate of 40 mW cm^{-2}. Prior to PDT, ALA-induced PpIX fluorescence was confirmed each time by UV irradiation, using a Woods lamp. A total of three treatments were performed at biweekly intervals. Only one toenail per subject was treated. Pain during treatment was assessed by the patients using a 10-cm visual analogue scale (VAS) from 0 (no pain) to 10 (unbearable pain).

Clinical and mycological evaluations were performed by the same investigators 12 and 18 months after the last treatment. Criteria for cure were either 100% absence of clinical signs of onychomycosis or subungual hyperkeratosis, leaving less than 10% of the nail plate affected combined with negative mycological laboratory results.

At the end of treatment, and during the follow-up period, no systemic treatment was allowed nor any additional topical treatment on the nails treated with PDT. However, topical treatment was allowed in patients with more than one affected toenail not treated with PDT. This was to try to limit the risk of reinfection of the PDT-treated nail (Sotiriou et al. 2010).

All 30 patients completed the study. Twenty-two patients (73.3%) had onychomycosis of the great toenail and eight (26.6%) had onychomycosis of one toenail other than that of the great toe. Five of the 22 patients with great toenail involvement had involvement of an additional nail. PDT reactions recorded during light treatment were a burning sensation and pain in all patients. Pain management included the use of a fan or cooling sprays in patients who felt intense pain and discomfort. Despite these measures, treatment had to be interrupted for 5–10 min in 12, 8, and 10 patients during the first, second, and third sessions, respectively. The mean VAS score of PDT-associated pain was 6.8 for the first session, 6.4 for the second session, and 6.5 for the third session. Local phototoxic reactions included moderate erythema (30/30), edema (28/30), and blistering (14/30) of the nail bed. These reactions were well tolerated, did not demand additional treatment, and resolved within 7–15 days without further complications.

After 1 year, 13 patients (43.3%) were cured, according to the investigators' criteria. Only five patients (16.6%) showed complete absence of clinical signs, while eight patients (26.6%) showed residual changes affecting less than 10% of the nail plate and negative mycological laboratory results. The remaining patients had changes compatible with dermatophyte infection covering more than 10% of the nail plate as well as positive direct microscopic examination. At month 18, the cure rate dropped to 36.6%. Only 11 of the cured patients at the first follow-up visit had persistent clearance, while clinical and mycological recurrence was seen in two of the patients, with minor residual clinical changes and negative laboratory results at month 12. The remaining patients had a persistent infection that deteriorated as no additional treatment was allowed during the follow-up period (Sotiriou et al. 2010).

Taborda and Taborda (2009) treated three patients with one PDT session using methyl-ALA and 37 J cm^{-2} of red light (633 nm) over a 15-min period. Direct microscopic examination of nail scraping and cultures were taken before and after PDT. Both potassium hydroxide and culture became negative after PDT, and appreciable nail improvement was still seen 4 months after treatment (Taborda and Taborda 2009).

Piraccini et al. (2008) describe a single case of PDT of onychomycosis caused by *T. rubrum*. A 78-year-old woman presented with total onychomycosis of the right big toenail and proximal subungual onychomycosis of the left big toenail. The patient had failed to respond to treatment with topical antifungals (potassium hydroxide and cultures were positive after 18 months of use of amorolfine 5% w/v nail lacquer) and had conditions that contraindicated administration of systemic antifungals (she was affected by hypertension and chronic hepatitis C virus infection and under treatment with warfarin and antihypertensive drugs). The nail plate was first softened with 40% w/w urea ointment under occlusion for 7 days and then clipped off. Nail bed hyperkeratosis was removed and AMA methyl ester 16% w/w cream was applied under an occlusive dressing for a period of 3 h. The irradiation of the nail plate with a Woods lamp showed a strong red fluorescence, confirming PpIX production. Fluorescence was more marked on the exposed nail beds than on the nail plate present on the first left toenail. Irradiation (37 J cm^{-2} for 7 min and 24 sec) was then performed at a distance of 5–8 cm from the nail with a lamp based on red-light-emitting (630 nm) diodes. The entire nail was included in the irradiation field. The patient did not report pain during irradiation, nor did she develop any local side effects afterwards. The treatment was repeated twice at intervals of 15 days, for a total of three PDT sessions during a period of 45 days, and the patient was then followed up every 3 months for 24 months. Treatment was well tolerated with no side effects. Potassium hydroxide (KOH) and cultures were positive after the third PDT session and became negative 3 months after the last treatment. At the month-12 visit, KOH and cultures were still negative and the toenails were considered clinically cured with residual mild traumatic onycholysis. Mycologic and clinical cure persisted until the last follow-up at month 24 (Piraccini et al. 2008).

Watanabe et al. (2008) reported on successful treatment of toenail onychomycosis with PDT in two patients. An 80-year-old Japanese woman presented with a 3-year history of asymptomatic alterations of the toenails of her right foot. She also had Sjögren syndrome, which was well controlled. She had been diagnosed as having onychomycosis and had been treated with topical terbinafine for 2 years, without success. She did not take any oral antifungal agents because of her advanced age. Physical examination revealed white patches on the surface and distal undersurface of her first to third toenails, without other alterations. Direct microscopic examination of superficial nail scrapings with a potassium hydroxide preparation revealed branching hyphae. A diagnosis of distal and lateral subungual onychomycosis was made. A 31-year-old Japanese woman presented with a more than 10-year history of a whitish toenail on her left foot. One year before consultation, she had been treated with oral itraconazole, but the treatment had been discontinued because of gastric pain. Physical examination revealed subungual hyperkeratosis, a yellowish discoloration, and onycholysis in the distal area of her left great toenail. Direct microscopic examination of subungual debris with KOH revealed branching hyphae. A diagnosis of distal and lateral subungual onychomycosis was made (Watanabe et al. 2008).

In treating these patients, a 20% w/w urea ointment was first applied liberally to the diseased nail surface and covered with an occlusive film wrap for 10 h. Then, a 20% w/w solution of ALA methyl ester in aqueous cream was applied to the treated nails, which were sealed with a piece of occlusive film wrap and covered with aluminum foil to protect from light for 5 h. Before PDT, PpIX fluorescence in the nail was confirmed by UV irradiation and spectrophotometer. The fluorescence was observed at the base of the nail and at the periphery of the onychomycosis lesion. Subsequently, the treatment site, including proximal and lateral nail folds, was irradiated both horizontally and vertically with pulsed laser light at a wavelength of 630 nm at 100 J cm^{-2} using an excimer-dye laser. PDT was performed once a week. Both patients had a feeling of some pain during irradiation, but it was tolerable and disappeared within a day. In case 1, the right first toenail was irradiated seven times (total dose, 700 J cm^{-2}). After the irradiation, the lesion substantially improved.

No dermatophyte was detected on KOH or on culture. The second and third toenails, which were used as controls, received only ALA or irradiation, respectively. They did not show clinical improvement after treatment. No recurrence was observed clinically at a 6-month follow-up visit. In case 2, the right first toenail was irradiated six times (total dose, 600 J cm^{-2}). After PDT, the patient's lesion substantially improved, and no dermatophyte was detected by KOH or by culture. No recurrence was observed clinically at a 3-month follow-up visit (Watanabe et al. 2008).

Manevitch et al. (2010) stated that while laser energy can eliminate dermatophytes *in vitro*, direct laser elimination of onychomycosis is not successful, due to difficulties in selectively delivering laser energy to the deeper levels of the nail plate without collateral damage. The authors postulated that femtosecond (fsec) infrared titanium sapphire lasers should circumvent this problem by their nonlinear interactions with biological media. This quality, combined with the deeply penetrating nature of the near-infrared radiation, should allow elimination of deeply seeded nail dermatophytes without associated collateral damage. Nail cuttings obtained from patients with onychomycosis caused by *T. rubrum* underwent fsec laser irradiation using increasing laser intensities, with the focus scanned throughout the whole thickness of the nail specimen. The efficacy of the laser treatment was evaluated by subculture. Scanning electron microscopy was used to determine fsec laser-induced collateral damage. The authors found that an fsec laser fluence of 7×10^{31} photons m^{-2} s^{-1} or above successfully inhibited the growth of the fungus in all samples examined, whereas laser intensities above 1.7×10^{32} photons m^{-2} s^{-1} affected the structure of the nail plate. The authors suggested that these findings indicate that *T. rubrum*-associated onychomycosis may be treated by fsec laser technology without the need for a photosensitizing drug (Manevitch et al. 2010).

11.5.4 Clinical Treatment of Nail Psoriasis

Nail psoriasis is often refractory to traditional treatments, and patients with nail psoriasis usually demand a therapeutic option. Both PDT and pulsed dye laser (PDL) have proved effective for plaque-type psoriasis. Fernandez-Guarino et al. (2009) evaluated both, for the first time, in nail psoriasis.

The authors studied 61 nails treated with PDT and 60 nails treated with PDL in a group of 14 patients. The PDT used PDL as the light source. Sessions were applied monthly, treating one hand with PDT and the other with PDL. The hand treated with PDT was occluded with methyl-AMA cream (16% w/w) for 3 h using a bioadhesive patch. The nails treated were evaluated at baseline and after 3 and 6 months according to the Nail Psoriasis Severity Index (NAPSI) score.

A decrease in NAPSI score was observed with both treatments and in both nail matrix and nail bed involvement. No statistical differences were found between PDT and PDL and between nail matrix and nail bed NAPSI scores. The subjective impression of the patients was good, especially regarding the decrease in pain. The authors concluded that PDL seems to be effective in the treatment of nail psoriasis and improves nail matrix and nail bed involvement (Fernández-Guarino et al. 2009). However, PDT does not seem to play a role in the clinical response.

11.6 CONCLUSIONS

PDT appears to have much promise in treatment of diseases of the nail. The major perceived advantage is that conditions such as onychomycosis that typically require long treatment times may be resolved in one or a few treatments, with ensuing benefits for patients, clinicians, and health care providers. Moreover, the topical nature of PDT, meaning systemic drug administration, is not required; this means that systemic side effects and drug–drug interactions are very unlikely.

Light administration to the nail is straightforward. However, the barrier properties of the nail are such that only relatively small hydrophilic molecules are likely to penetrate effectively in reasonably short application periods. For this reason, ALA has been the most commonly investigated drug. Even in this case, penetration enhancers or nail filing may be required to increase ALA concentrations to useful levels in the nail plate and at the nail bed. Suitable drug delivery systems also warrant further investigation. The ability of ALA to induce production of the potent photosensitizer PpIX in human cells as well as microbial cells means that pain is a strong possibility and will require management.

Following a number of successful clinical trials and pilot studies, industrial investment is now required to fund extensive multisite clinical studies of PDT of common nail disease such as onychomycosis and psoriasis. In this way, drug delivery will be enhanced and treatment regimens will be optimized, with the ultimate aim being to improve therapeutic outcomes for patients.

REFERENCES

Barrett, A. J., J. C. Kennedy, R. A. Jones, P. Nadeau, and R. H. Pottier. 1990. "The Effect of Tissue and Cellular pH on the Selective Biodistribution of Porphyrin-Type Photochemotherapeutic Agents: A Volumetric Titration Study." *Journal of Photochemistry and Photobiology B: Biology* 6: 309–23.

Bertoloni, G., F. Rossi, G. Valduga, G. Jori, H. Ali, and J. E. van Lier. 1992. "Photosensitizing Activity of Water- and Lipid-Soluble Phthalocyanines on Prokaryotic and Eukaryotic Microbial Cells." *Microbios* 71: 33–46.

Bonnett, R. 1995. "Photosensitisers of the Porphyrin and Phthalocyanine Series for Photodynamic Therapy." *Chemical Society Reviews* 24: 19–33.

Bonnett, R., B. D. Djelal, and A. Nguyen. 2001. "Physical and Chemical Studies Related to the Development of m-THPC (Foscan®) for the Photodynamic Therapy (PDT) of Tumours." *Journal of Porphyrins and Phthalocyanines* 5: 652–61, doi:10.1002/jpp.377.

Brancaleon, L., and H. Moseley. 2002. "Laser and Non-Laser Light Sources for Photodynamic Therapy." *Lasers in Medical Science* 17: 173–86, doi:10.1007/s101030200027.

Brouillet, N., G. Arselin-De Chateaubodeau, and C. Volland. 1975. "Studies on Protoporphyrin Biosynthetic Pathway in *Saccharomyces cerevisiae*; characterization of the Tetrapyrrole Intermediates." *Biochimie* 57: 647–55, doi:10.1016/S0300-9084(75)80146-5.

Codling, C. E., J. Y. Maillard, and A. D. Russell. 2003. "Aspects of the Antimicrobial Mechanisms of Action of a Polyquaternium and an Amidoamine." *The Journal of Antimicrobial Chemotherapy* 51: 1153–8, doi:10.1093/jac/dkg228.

Costa, R. A., M. E. Farah, J. A. Cardillo, and R. Belfort Jr. 2001. "Photodynamic Therapy with Indocyanine Green for Occult Subfoveal Choroidal Neovascularization Caused by Age-Related Macular Degeneration." *Current Eye Research* 23: 271–5.

Daniell, M. D., and J. S. Hill. 1991. "A History of Photodynamic Therapy." *Australian and New Zealand Journal of Surgery* 61: 340–8.

De Rosa, F. S., and M. V. Bentley. 2000. "Photodynamic Therapy of Skin Cancers: Sensitizers, Clinical Studies and Future Directives." *Pharmaceutical Research* 17: 1447–55.

Demidova, T. N., and M. R. Hamblin. 2005. "Effect of Cell-Photosensitizer Binding and Cell Density on Microbial Photoinactivation." *Antimicrobial Agents and Chemotherapy* 49: 2329–35, doi:10.1128/AAC.49.6.2329-2335.2005.

Donnelly, R. F., P. A. McCarron, J. M. Lightowler, and A. D. Woolfson. 2005. "Bioadhesive Patch-Based Delivery of 5-Aminolevulinic Acid to the Nail for Photodynamic Therapy of Onychomycosis." *Journal of Controlled Release* 103: 381–92, doi:10.1016/j.jconrel.2004.12.005.

Donnelly, R. F., P. A. McCarron, M. M. Tunney, and D. A. Woolfson. 2007. "Potential of Photodynamic Therapy in Treatment of Fungal Infections of the Mouth. Design and Characterisation of a Mucoadhesive Patch Containing Toluidine Blue O." *Journal of Photochemistry and Photobiology B: Biology* 86: 59–69, doi:10.1016/j.jphotobiol.2006.07.011.

Dougherty, T. J., C. J. Gomer, B. W. Henderson, G. Jori, D. Kessel, M. Korbelik, J. Moan, and Q. Peng. 1998. "Photodynamic Therapy." *Journal of the National Cancer Institute* 90: 889–905.

Fernández-Guarino, M., A. Harto, M. Sánchez-Ronco, I. García-Morales, and P. Jaén. 2009. "Pulsed Dye Laser vs. Photodynamic Therapy in the Treatment of Refractory Nail Psoriasis: A Comparative Pilot Study." *Journal of the European Academy of Dermatology and Venereology* 23: 891–5, doi:10.1111/j.1468-3083.2009.03196.x.

Fritsch, C., K. Lang, W. Neuse, T. Ruzicka, and P. Lehmann. 1998. "Photodynamic Diagnosis and Therapy in Dermatology." *Skin Pharmacology and Applied Skin Physiology* 11: 358–73.

Fuchs, C., R. Riesenberg, J. Siegert, and R. Baumgartner. 1997. "pH-Dependent Formation of 5-Aminolevulinic Acid-Induced Protoporphyrin IX in Fibrosarcoma Cells." *Journal of Photochemistry and Photobiology B: Biology* 40: 49–54.

Gannon, M. J., and S. B. Brown. 1999. "Photodynamic Therapy and Its Applications in Gynaecology." *British Journal of Obstetrics and Gynaecology* 106: 1246–54.

Gantchev, T. G., N. Brasseur, and J. E. van Lier. 1996. "Combination Toxicity of Etoposide (VP-16) and Photosensitisation with a Water-Soluble Aluminium Phthalocyanine in K562 Human Leukaemic Cells." *British Journal of Cancer* 74: 1570–7.

Henderson, B. W., and T. J. Dougherty. 1992. "How Does Photodynamic Therapy Work?" *Photochemistry and Photobiology* 55: 145–57.

Isaacs, N. 1995. *Physical Organic Chemistry*. London: Prentice Hall.

Jori, G., M. Beltramini, E. Reddi, B. Salvato, A. Pagnan, L. Ziron, L. Tomio, and T. Tsanov. 1984. "Evidence for a Major Role of Plasma Lipoproteins as Hematoporphyrin Carriers In Vivo." *Cancer Letters* 24: 291–7.

Jori, G., and S. B. Brown. 2004. "Photosensitized Inactivation of Microorganisms." *Photochemical & Photobiological Sciences* 3: 403–5, doi:10.1039/b311904c.

Kalka, K., H. Merk, and H. Mukhtar. 2000. "Photodynamic Therapy in Dermatology." *Journal of the American Academy of Dermatology* 42: 389, 413. quiz 414–6.

Kalyanasundaram, K. 1991. *Photochemistry of Polypyridine and Porphyrin Complexes*. London: Academic Press.

Kamp, H., H. J. Tietz, M. Lutz, H. Piazena, P. Sowyrda, J. Lademann, and U. Blume-Peytavi. 2005. "Antifungal Effect of 5-Aminolevulinic Acid PDT in *Trichophyton rubrum*." *Mycoses* 48: 101–7, doi: 10.1111/j.1439-0507.2004.01070.x.

Kennedy, J. C., S. L. Marcus, and R. H. Pottier. 1996. "Photodynamic Therapy (PDT) and Photodiagnosis (PD) Using Endogenous Photosensitization Induced by 5-Aminolevulinic Acid (ALA): Mechanisms and Clinical Results." *Journal of Clinical Laser Medicine & Surgery* 14: 289–304.

Kennedy, J. C., and R. H. Pottier. 1992. "Endogenous Protoporphyrin IX, A Clinically Useful Photosensitizer for Photodynamic Therapy." *Journal of Photochemistry and Photobiology B: Biology* 14: 275–92.

Kennedy, J. C., R. H. Pottier, and D. C. Pross. 1990a. "Photodynamic Therapy with Endogenous Protoporphyrin IX: Basic Principles and Present Clinical Experience." *Journal of Photochemistry and Photobiology B: Biology* 6: 143–8.

Kennedy, J. C., R. H. Pottier, and D. C. Pross. 1990b. "Photodynamic Therapy with Endogenous Protoporphyrin IX: Basic Principles and Present Clinical Experience." *Journal of Photochemistry and Photobiology B: Biology* 6: 143–8.

Kessel, D. 1999. "Transport and Localisation of m-THPC In Vitro." *International Journal of Clinical Practice* 53: 263–7.

Kim, J. Y., H. Y. Kang, E. S. Lee, and Y. C. Kim. 2007. "Topical 5-Aminolevulinic Acid Photodynamic Therapy for Intractable Palmoplantar Psoriasis." *The Journal of Dermatology* 34: 37–40, doi:10.1111/j.1346-8138.2007.00213.x.

Koh, A. H., and C. L. Ang. 2002. "Age-Related Macular Degeneration: What's New?" *Annals of the Academy of Medicine, Singapore* 31: 399–404.

Konan, Y. N., R. Gurny, and E. Allemann. 2002. "State of the Art in the Delivery of Photosensitizers for Photodynamic Therapy." *Journal of Photochemistry and Photobiology B: Biology* 66: 89–106.

Landsman, A. S., A. H. Robbins, P. F. Angelini, C. C. Wu, J. Cook, M. Oster, and E. S. Bornstein. 2010. "Treatment of Mild, Moderate, and Severe Onychomycosis Using 870- and 930-nm Light Exposure." *Journal of the American Podiatric Medical Association* 100: 166–77.

Malik, Z., H. Ladan, and Y. Nitzan. 1992. "Photodynamic Inactivation of Gram-Negative Bacteria: Problems and Possible Solutions." *Journal of Photochemistry and Photobiology B: Biology* 14: 262–6.

Manevitch, Z., D. Lev, M. Hochberg, M. Palhan, A. Lewis, and C. D. Enk. 2010. "Direct Antifungal Effect of Femtosecond Laser on *Trichophyton rubrum* Onychomycosis." *Photochemistry and Photobiology* 86: 476–9, doi:10.1111/j.1751-1097.2009.00672.x.

Meijnders, P. J. N., W. M. Star, R. S. De Bruijn, A. D. Treurniet-Donker, M. J. M. Van Mierlo, S. J. M. Wijthoff, B. Naafs, H. Beerman, and P. C. Levendag. 1996. "Clinical Results of Photodynamic Therapy for Superficial Skin Malignancies or Actinic Keratosis Using Topical 5-Aminolevulinic Acid." *Lasers in Medical Science* 11: 123–31.

Moan, J. 1990. "On the Diffusion Length of Singlet Oxygen in Cells and Tissues." *Journal of Photochemistry and Photobiology B: Biology* 6: 343–4, doi: 10.1016/1011-1344(90)85104-5.

Moan, J., and Q. Peng. 2003. "An Outline of the Hundred-Year History of PDT." *Anticancer Research* 23: 3591–600.

Moan, J., and S. Sommer. 1983. "Uptake of the Components of Hematoporphyrin Derivative by Cells and Tumours." *Cancer Letters* 21: 167–74.

Moan, J., G. Streckyte, S. Bagdonas, O. Bech, and K. Berg. 1997. "Photobleaching of Protoporphyrin IX in Cells Incubated with 5-Aminolevulinic Acid." *International Journal of Cancer* 70: 90–7.

Monfrecola, G., E. M. Procaccini, M. Bevilacqua, A. Manco, G. Calabro, and P. Santoianni. 2004. "In Vitro Effect of 5-Aminolevulinic Acid Plus Visible Light on *Candida albicans*." *Photochemical & Photobiological Sciences* 3: 419–22, doi:10.1039/b315629j.

Moretti, M. B., S. C. Garcia, and A. Batlle. 2000. "Porphyrin Biosynthesis Intermediates Are Not Regulating δ-Aminolevulinic Acid Transport in *Saccharomyces cerevisiae*." *Biochemical and Biophysical Research Communications* 272: 946–50, doi:10.1006/bbrc.2000.2874.

Nikaido, H. 2003. "Molecular Basis of Bacterial Outer Membrane Permeability Revisited." *Microbiology and Molecular Biology Reviews* 67: 593–656.

Ochsner, M. 1997. "Photophysical and Photobiological Processes in the Photodynamic Therapy of Tumours." *Journal of Photochemistry and Photobiology B: Biology* 39: 1–18.

Paardekooper, M., A. E. V. Gompel, J. V. Steveninck, and P. J. A. V. D. Broek. 1995a. "The Effect of Photodynamic Treatment of Yeast with the Sensitiser Chloroaluminium Phthalocyanine on Various Cellular Parameters." *Photochemistry and Photobiology* 62: 561–7, doi:10.1111/j.1751-1097.1995.tb02385.x.

Paardekooper, M., A. E. V. Gompel, J. V. Steveninck, and P. J. A. V. D. Broek. 1995b. "Photodynamic Treatment of Yeast with Chloroaluminium-Phthalocyanine: Role of the Monomeric form of the Dye." *Photochemistry and Photobiology* 62: 561–7, doi:10.1111/j.1751-1097.1995.tb02385.x.

Peng, Q., K. Berg, J. Moan, M. Kongshaug, and J. M. Nesland. 1997. "5-Aminolevulinic Acid-Based Photodynamic Therapy: Principles and Experimental Research." *Photochemistry and Photobiology* 65: 235–51.

Piot, B., N. Rousset, P. Lenz, S. Eleouet, J. Carre, V. Vonarx, L. Bourre, and T. Patrice. 2001. "Enhancement of 5-Aminolevulinic Acid-Photodynamic Therapy In Vivo by Decreasing Tumor pH with Glucose and Amiloride." *The Laryngoscope* 111: 2205–13.

Piraccini, B. M., G. Rech, and A. Tosti. 2008. "Photodynamic Therapy of Onychomycosis Caused by *Trichophyton rubrum*." *Journal of the American Academy of Dermatology* 59: S75–6, doi:10.1016/j.jaad.2008.06.015.

Pottier, R., and J. C. Kennedy, 1990. "The Possible Role of Ionic Species in Selective Biodistribution of Photochemotherapeutic Agents toward Neoplastic Tissue." *Journal of Photochemistry and Photobiology B: Biology* 8: 1–16.

Robinson, D. J., P. Collins, M. R. Stringer, D. I. Vernon, G. I. Stables, S. B. Brown, and R. A. Sheehan-Dare. 1999. "Improved Response of Plaque Psoriasis after Multiple Treatments with Topical 5-Aminolevulinic Acid Photodynamic Therapy." *Acta Dermato-Venereologica* 79: 451–5.

Romagnoli, C., D. Mares, G. Sacchetti, and A. Bruni. 1998. "The Photodynamic Effect of 5-(4-Hydroxy-1-Butinyl)-2,2-Bithienyl on Dermatophytes." *Mycological Research* 12: 1519–24.

Salah, M., N. Samy, and M. Fadel. 2009. "Methylene Blue Mediated Photodynamic Therapy for Resistant Plaque Psoriasis." *Journal of Drugs in Dermatology* 8: 42–9.

Schick, E., A. Ruck, W. H. Boehncke, and R. Kaufmann. 1997. "Topical Photodynamic Therapy Using Methylene Blue and 5-Aminolevulinic Acid in Psoriasis." *The Journal of Dermatological Treatment* 8: 17–9.

Smijs, T. G., J. A. Bouwstra, H. J. Schuitmaker, M. Talebi, and S. Pavel. 2007. "A Novel Ex Vivo Skin Model to Study the Susceptibility of the Dermatophyte *Trichophyton rubrum* to Photodynamic Treatment in Different Growth Phases." *The Journal of Antimicrobial Chemotherapy* 59: 433–0, doi:10.1093/jac/dkl490.

Smijs, T. G., R. N. van der Haas, J. Lugtenburg, Y. Liu, R. L. de Jong, and H. J. Schuitmaker. 2004. "Photodynamic Treatment of the Dermatophyte *Trichophyton rubrum* and its Microconidia with Porphyrin Photosensitizers." *Photochemistry and Photobiology* 80: 197–202, doi:10.1562/2004-04-22-RA-146.

Smijs, T. G. M., and H. J. Schuitmaker. 2003. "Photodynamic Inactivation of the Dermatophyte *Trichophyton rubrum*." *Photochemistry and Photobiology* 77: 556–60, doi:10.1562/0031-8655(2003)0770556PIOTDT2.0.CO2.

Sotiriou, E., T. Koussidou-Eremonti, G. Chaidemenos, Z. Apalla, and D. Ioannides. 2010. "Photodynamic Therapy for Distal and Lateral Subungual Toenail Onychomycosis Caused by *Trichophyton rubrum*: Preliminary Results of a Single-Centre Open Trial." *Acta Dermato-Venereologica* 90: 216–7, doi:10.2340/00015555-0811.

Strakhovskaya, M. G., A. O. Shumarina, G. Y. Fraikin, and A. B. Rubin. 1998. "Synthesis of Protoporphyrin IX Induced by 5-Aminolevulinic Acid in Yeast Cells in the Presence of 2,2-Dipyridyl." *Biochemistry (Moscow)* 63: 725–8.

Taborda, V., and P. Taborda. 2009. "Photodynamic Therapy with Methylaminolevulinate for Treatment of Onychomycosis." *Journal of the American Academy of Dermatology* 60.

Vaupel, P. W., S. Frinak, and H. I. Bicker. 1981. "Heterogeneous Oxygen Partial Pressure and pH Distribution in CH3 Mouse Mammary Adenocarcinoma." *Cancer Letters* 41: 2008–13.

Wainwright, M. 1998. "Photodynamic Antimicrobial Chemotherapy (PACT)." *The Journal of Antimicrobial Chemotherapy* 42: 13–28.

Wainwright, M., and K. B. Crossley. 2004. "Photosensitising Agents—Circumventing Resistance and Breaking Down Biofilms: A review." *International Biodeterioration & Biodegradation* 53: 119–126.

Watanabe, D., C. Kawamura, Y. Masuda, Y. Akita, Y. Tamada, and Y. Matsumoto. 2008. "Successful Treatment of Toenail Onychomycosis with Photodynamic Therapy." *Archives of Dermatology* 144: 19–21, doi:10.1001/archdermatol.2007.17.

Wike-Hooley, J. L., J. Haveman, and H. S. Reinhold. 1984. "The Relevance of Tumour pH to the Treatment of Malignant Disease." *Radiotherapy and Oncology* 2: 343–66.

Zeina, B., J. Greenman, D. Corry, and W. M. Purcell. 2002. "Cytotoxic Effects of Antimicrobial Photodynamic Therapy on Keratinocytes In Vitro." *The British Journal of Dermatology* 146: 568–73.

Zeina, B., J. Greenman, D. Corry, and W. M. Purcell. 2003. "Antimicrobial Photodynamic Therapy: Assessment of Genotoxic Effects on Keratinocytes In Vitro". *The British Journal of Dermatology* 148: 229–32.

Zeina, B., J. Greenman, W. M. Purcell, and B. Das. 2001. "Killing of Cutaneous Microbial Species by Photodynamic Therapy." *The British Journal of Dermatology* 144: 274–8.

12 Nail as a Surrogate for Investigating Drug Use

Katarzyna Madej

CONTENTS

12.1 HISTORY OF NAIL ANALYSIS

Historically, nails have been subjected to elemental analysis in a variety of applications, including arsenic in forensic investigations (Lander et al. 1965; Pounds et al. 1979) and more recently in environmental (Wilhelm et al. 1991) and occupational estimation of metals exposure (Gerhardson et al. 1995).

Suzuki et al. (1984) were the first to detect drugs of abuse and their metabolites (amphetamine [AP] and methamphetamine [MP]) in the nail clippings of MP abusers. Since then, Cirimele et al. (1995) also identified other AP analogs such as 3,4-methylenedioxyamphetamine (MDA) and 3,4-methylenedioxymethamphetamine (MDMA) in this material.

In 1994, the Federal Bureau of Investigation's (FBI) Toxicology Laboratory demonstrated the presence of cocaine and its main metabolite benzoylecgonine in nails originating from cadavers (Miller et al. 1994).

In the nineties, nail analysis was also applied for drug monitoring (Schatz et al. 1995; De Doncker et al. 1996). The pharmacokinetics of therapies with terbinafine (Schatz et al. 1995) and itraconazole (De Doncker et al. 1996) in onchomycosis patients were investigated.

12.2 CHARACTERISTIC FEATURES OF NAIL PLATE

Nails, like hair, are epidermal products, and the analytical findings in nail may be complementary for hair results. Both of these keratinized matrices possess some advantages over conventional biological materials (urine or blood samples) in terms of simply and relatively noninvasive sampling, long-term window, and stability of xenobiotics in the tissue. The study of stability of an anticoagulant (ticlopidine) in hair and nail samples before storage, after 14 days, and again after 60 days was conducted (Pufal et al. 2008). The results showed that ticlopidine was a very stable compound and suggested that hair and nails can be stored at room temperature, but preferably should be protected against light.

However, drugs may appear in nail clippings of some subjects within 1 month of ingestion (the rate of fingernail growth has been reported to be approximately 3 mm per month and approximately 0.9–1.2 mm per month for toenails), and this fact may reduce the usefulness of nail samples for estimating the time of drug exposure.

Nevertheless, in the last few years, analysis of nails has received considerable attention. In the past decade, the review articles concerning analysis of drugs of abuse in various biological specimens, including nail samples, have also appeared (de Oliveira et al. 2007; Saito et al. 2011).

12.3 DIRECTIONS OF NAIL DRUG ANALYSIS

Generally, the drugs studied in nails may be divided into two main groups: (1) illicit drugs and (2) medicaments, especially psychotropic drugs.

To date, the presence of illicit drugs such as AP-type stimulants (Kim et al. 2008; 2010), ketamine (KET) and norketamine (NKT) (Kim et al. 2010), cannabinoids (Lemos et al. 1999; Kim et al. 2008), opiates (Lemos et al. 2000b; Cingolani et al. 2004), cocaine (Garside et al. 1998; Cingolani et al. 2004), phencyclidine (Jenkis and Engelhart 2006), and methadone (Lemos et al. 2000a) has been determined in nails.

From among the second group of licit drugs, anticoagulant agent (ticlopidine, Pufal et al. 2008), β-blocker (atenolol, Pufal 2004) and psychotropic drugs like antipsychotics (haloperidol, Pufal and Piotrowski 2006; flupentixol, Pufal et al. 2008), neuroleptic phenothiazines, and tricyclic antidepressants (Skutera et al. 2008) and sedatives (Irving and Dickson 2007) were examined in the nail.

Recently, ciclopirox (antifungal agent) penetration across human nail plate *in vitro* has been studied (Bu et al. 2010), and the presence of citalopram, the new generation antidepressant, with its metabolite desmethylcitalopram (Pufal et al. 2010), were determined in nails.

12.4 METHODOLOGY

Nail samples are usually collected by cutting the excess overhang of the nail plate using cosmetic nail clippers. The samples of each subject examined are pooled and stored (e.g., in sealed plastic bags) at room temperature with limited light exposure until required for analysis.

12.4.1 Sample Preparation Procedures

Nail sample preparation process for drug determination consists of few steps (at least four steps): (1) decontamination; (2) cutting into small segments; (3) alkaline, acidic, or methanolic digestion; and (4) extraction (usually liquid–liquid extraction [LLE]). These steps are presented schematically in Figure 12.1. For surface decontamination, nails are washed in an appropriate reagent or a sequence of reagents (e.g., water, methanol, acetone, and surfactant sodium dodecyl sulfate [SDS]) using ultrasonic bath. In some cases of nail analysis, solid-phase extraction (SPE) after LLE as well as a derivatization process (when gas chromatography [GC] method is used) may be employed.

LLE, also known as solvent extraction and partitioning, is a technique to isolate and separate analytes from a sample, based on their relative solubility in two different immiscible liquids, usually water and an organic solvent. The analyte

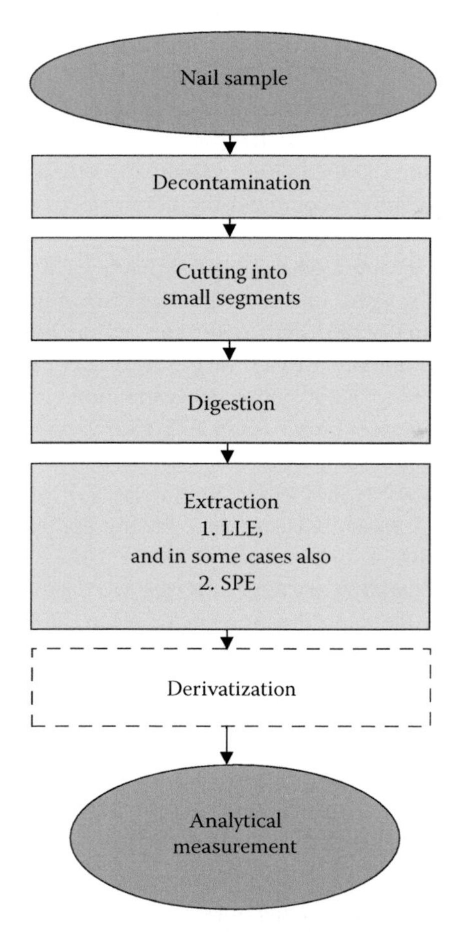

FIGURE 12.1 The steps in nail analysis for drug determination (the dashed box indicates that this step is not the basic one and is usually used when gas chromatography/mass spectrometry analysis is employed).

will partition into the solvent that offers the greatest solubility. Hydrophobic compounds (most drugs are hydrophobic substances) have an affinity for the organic phase, and hydrophilic compounds have an affinity for the aqueous phase. After the partitioning process, two immiscible phases are centrifuged and separated. The organic phase with the isolated analytes is evaporated and then reconstituted in the known volume of an appropriate medium compatible with the analytical method used.

The LLE technique is predominately employed for isolation of drugs from nail samples. The following organic solvents were applied as extraction media: dichloromethane, *n*-hexane, *n*-hexane/chloroform (7:3, v/v), ethyl acetate, dichloromethane/dichloroethane/heptane (19:18:63), and chloroform/2-propyl alcohol. In two cases (Kim et al. 2008; Kim et al. 2010), two-step LLE with ethyl acetate and back extractions with *n*-hexane/ethyl acetate were performed.

The drugs were also isolated from nail clippings using LLE combined with subsequent SPE (Garside et al. 1998; Cingolani et al. 2004). SPE is a sample pretreatment technique, in which an analyte is retained on a small extraction column, while undesirable chemical entities in the sample either pass through or are irreversible bound to a sorbent in the column. The analyte is subsequently eluted from the column and is usually evaporated to dryness and reconstituted in an appropriate medium. SPE technique is known as an effective method for cleanup of biological samples (it is universally recognized as more effective for cleanup than LLE technique).

To increase sensitivity of GC/mass spectrometry (MS) method for measuring relatively nonvolatile substances (most drugs are nonvolatile compounds), a derivatization step is usually required. After extraction and evaporation processes, illicit drugs like cannabinoids, opiates (morphine, 6-acetylmorphine, and cocaine), and AP-type stimulants were subjected to derivatization with such reagents as *N,O-bis*(trimethylsilyl)trifluoroacetamide (BSTFA) catalyzed with 1% trimethychlorosilane (TMCS), propionic anhydride, and *N*-methyl-*N*-trimethylsilyltrifluoroacet amide (MSTFA), respectively. For GC/MS analysis, AP-type stimulants and KET with its metabolite NKT were also derivatized using heptafluorobutyric anhydride (HFBA) (Kim et al. 2010).

The nail samples containing cocaine analytes were prepared for GC/MS analysis by extraction with methanol heating under reflux at 40°C for 16 h and then using SPE (Garside et al. 1998). Before the derivatization step with BSTFA and 1% TMCS, the methanolic extracts including wash residues (after the nail washing step) were cleaned by SPE. Cartridges Clean Screen®, ZSDAU020 (United Chemical Technologies, Horsham, PA), and elution solvent methylene chloride/isopropanol/concentrated ammonium hydroxide (80:20:2, v/v/v) were used. LLE with the successively used SPE were also applied for determination of morphine, 6-acetymorphine, and cocaine (Cingolani et al. 2004). The chloroform/isopropyl alcohol extracts were cleaned using Bond Elut Certify extraction columns. The drugs of abuse were eluted twice with 1 mL of methylene chloride/isopropyl alcohol solution (8:2, v/v) containing 2% ammonium hydroxide and then derivatized with propionic anhydride and analyzed by GC/MS.

12.4.2 Analytical Methods

The drug determinations in nails have been based on several instrumental methods, including GC/MS detection, liquid chromatography with mass spectroscopy detection [LC/MS(MS)], and more rarely, immunoassay or high-performance liquid chromatography with electrochemical detection(HPLC/ED).

GC/MS belongs to commonly exploited and reliable methods, although it usually requires rather complicated specimen preparation including derivatization procedures for nonvolatile compounds. However, due to its specificity, sensitivity, and relatively low costs, the GC/MS method in the selected ion monitoring (SIM) mode has been frequently employed for the detection of drugs of abuse in nails. GC separations of a variety of illicit drugs were performed in such capillary columns as HP-5MS cross-linked 5% diphenyl, 95% dimethylpolysiloxane fused silica, HP-1 cross-linked dimethyl silicone, and DB-5 MS fused silica.

LC/MS(MS) method offers high sensitivity and specificity and reduces sample preparation time compared with GC/MS methods because relatively nonvolatile compounds (like most drugs) can be analyzed and no derivatization is necessary. The main disadvantage of this advanced instrumental technique is its high cost. The LC/MS/(MS) methods developed for determination of drugs in nails operated with such columns as Phenomenex Luna phenyl hexyl (50 mm × 2 mm, 3 μm, 100 Å), Waters Atlantis® T3 (50 mm × 2.1 mm, 5 μm), and RP-C18 (Zorbax Eclipse XDB-C18; 150 mm × 4.6 mm; 5 μm). The mobile phases consisting of acetonitrile and 2 mM ammonium formate or 0.1% formic acid in water (solvent A) and 0.1% formic acid in acetonitrile (solvent B) were applied in an appropriate gradient mode. Drug separations in the RP-C18 column were conducted in an isocratic mode with a mixture of acetonitrile and 0.1% trifluoroacetic acid (TFA), mixed in various proportions. In most cases of LC/MS analyses, electrospray ionization (ESI) mode was used for mass detection and analysis.

In two cases (analyses of cannabis and morphine in fingernail clippings), radioimmunoassay (RIA) methods were employed. In the first case GC/MS, and in the latter HPLC/ED, were applied as confirmation methods for RIA. HPLC separations were made using a PRLP-S polymeric column (150 mm × 4.6 mm) thermostated at 63.8°C.

12.5 APPLICATIONS

12.5.1 Determination of Illicit Drugs in Nails

The drugs of abuse like cocaine, cannabinoids, morphine, AP-type stimulants and KET, and some selected metabolites are among the illicit drugs that have been determined in nails (Table 12.1).

Garside et al. (1998) detected and quantified nine cocaine analytes: anhydroecgonine methyl ester, benzoylecgonine, cocaine, cocaethylene, ecgonine ethyl ester, ecgonine methyl ester, *m*-hydroxybenzoylecgonine, norbenzoylecgonine, and norcocaine in fingernails and toenails taken from 18 suspected cocaine users. Cocaine-related compounds were present in 14 subjects (82.3%), while only 5 subjects (27.7%) had been

Table 12.1
Determination of Illicit Drugs in Nails

Drug	Sample Amount (mg)	Sample Preparation Procedure	Analytical Method	Extraction Recovery (%)	Limit of Detection (LOD) or Limit of Quantification (LOQ)	Precision (%, Relative Standard Deviation)	References
Cocaine analytes	7–50	1. Cutting 2. Washing with methanol 3. Digestion with methanol (heating under reflux at 40°C for 16 h) 4. SPE 5. Derivatization with BSTFA and 1% TMCS	GC/MS	—[a]	LOD = 0.10 ng mg^{-1} (for cocaine) and 0.25 ng mg^{-1} (for metabolites)	—[a]	Garside et al. (1998)
Cannabinoids	2.5–21.5	1. Decontamination by sonication with SDS, water, and methanol 2. Digestion with 1 M NaOH 3. LLE with ethyl acetate 4. Derivatization with BSTFA and 1% TMCS	RIA and GC/MS	>81	LOD < 0.1 ng mg^{-1}	—[a]	Lemos et al. (1999)
Morphine	3.0–96.0	1. Decontamination by sonication with SDS, water, and methanol 2. Digestion with 1 M NaOH 3. LLE with dichloromethane:dichloroethane:heptane (19:18:63)	RIA and HPLC/ED[b]	80.5 (RIA) 86.3 (HPLC)	LOD = 0.05 ng mg^{-1}	—[a]	Lemos et al. (2000)

Morphine, 6-acetylmorphine, and cocaine	50.0	1. Decontamination with methylene chloride 2. Digestion with 37% aq. HCl 3. LLE with chloroform/isopropyl alcohol solution (3:1, v/v) 4. SPE 5. Derivatization with propionic anhydride	GC/MS	—[a]	LOQ = 0.1 ng mg^{-1} (for each drug)	—[a]	Cingolani et al. (2004)
Amphetamine-type stimulants: amphetamine (AP), methamphetamine (MA), 3,4-methylenedioxy-N-amphetamine (MDA), and 3,4-methylenedioxy-N-methamphetamine (MDMA) Cannabinoids: 9-carboxy-11-nor-Δ^9-THC (9-THC) and 11-hydroxy-Δ^9-THC (11-OH-THC)	30.0	1. Decontamination with water and methanol 2. Digestion with 1.0 M NaOH 3. LLE with ethyl acetate and reextracted with n-hexane/ethyl acetate 4. Derivatization with MSTFA	GC/MS	74.0–94.8	LOD < 0.056 ng mg^{-1} LOQ < 0.2 ng mg^{-1}	6.3 (interday)	Kim et al. (2008)
Amphetamine-type stimulants: AP, MA, MDA, MDMA, ketamine, and norketamine	20.0	1. Decontamination with water and methanol 2. Digestion with 1.0 M NaOH (at 95°C for 30 min) 3. LLE with ethyl acetate and reextracted with *n*-hexane/ethyl acetate 4. Derivatization with HFBA	GC/MS	72.3–94.9	LOD < 0.094 ng mg^{-1} LOQ < 0.314 ng mg^{-1}	7.1 (intraday) 10.6 (interday)	Kim et al. (2010)

[a] No data were given.

[b] Electrochemical detection (ED).

found positive in analysis of conventional postmortem materials. Cocaine and its main metabolite benzoylecgonine were the predominant compounds found in all positive nail samples, with the concentrations of cocaine present being approximately 2–10 times greater than benzoylecgonine. Concentrations of cocaine analytes in fingernails were generally greater than in toenails; however, external contamination should be considered.

Fingernail clippings were evaluated as analytical specimens for detection of cannabinoids by RIA and GC/MS methods (Lemos et al. 1999). Cannabinoids were found to be present in all six cases that were examined for this group of illicit drugs by RIA with mean concentration of 1.03 ng mg^{-1}. Using GC/MS method, Δ^9-tetrahydrocannabinol (THC) was detected and quantified in 11 of the 14 examined nails with mean concentration of 1.44 ng mg^{-1}. The main metabolite of 9-THC-11-nor-Δ9-tetrahydrocannabinol-9-carboxylic acid (THCOOH) was detected in two of three fingernail samples with an average concentration of 19.85 ng mg^{-1}. The positive RIA and GC/MS results were achieved 6–9 months after nail collection.

The usefulness of nail as a surrogate for determination of morphine alone (Lemos et al. 2000b) and together with 6-monoacetylmorphine and cocaine (Cingolani et al. 2004) was also investigated. An RIA method was used for morphine screening in heroin users and then positive results were confirmed by the HPLC/ED method (Lemos et al. 2000b). Positive RIA results were obtained with nails from 25 of the 26 subjects with mean morphine concentration of 2.11 ng mg^{-1}. Comparable studies of toenail and hair samples, taken from 18 postmortem bodies of drug abusers, for presence of opiates and cocaine, were carried out (Cingolani et al. 2004). The obtained results revealed that both morphine and cocaine were more concentrated in toenails than in hair. Mean concentrations were 1.27 ng mg^{-1} (in toenails) versus 0.79 ng mg^{-1} (in hair) for morphine and 0.99 ng mg^{-1} (in toenails) versus 0.48 ng mg^{-1} (in hair) for cocaine.

Fingernails were also studied as alternative material for detection and quantification of AP-type stimulants: AP, MA, 3,4-methylenedioxy-*N*-amphetamine (MDA), and 3,4-methylenedioxy-*N*-methamphetamine (MDMA) simultaneously with cannabinoids (Kim et al. 2008) or together with KET and its metabolite NKT (Kim et al. 2010). The GC/MS method was applied for analysis of nine fingernail samples obtained from abusers of illicit drugs (Kim et al. 2008). MA was the most frequently detected drug in association with its major metabolite AP, and Δ^9-tetrahydrocannabinol-11-oic-acid (THC-COOH) was the second most often detected. The concentration ranges of MA, AP, and THCOOH were 0.1–1.41, 0.12– 2.64, and 0.2 ng mg^{-1}, respectively. In contrast to the previous study (Lemos et al. 1999), the presence of the parent drug (9-THC) of THCOOH was not confirmed. The GC/MS method was also used in analysis of seven fingernail samples originated from suspected illicit drug abusers for AP-type stimulants and KET with its metabolite NKT (Kim et al. 2010). MA was the most frequently detected compound and was found in six samples in the concentration range of 0.23–2.09 ng mg^{-1}. AP (metabolite of MA) was detected in four samples at concentrations less than 0.063 ng mg^{-1}. MDMA, MDA, KET, and NKT were detected in only one sample, below the Limit of Detection of the method (for exception of MDMA, 0.46 ng mg^{-1}), and these findings reflected polydrug consumption.

12.5.2 Determination of Medicaments in Nails

Since the present century, nail clippings have also been studied for their usefulness in medicament analysis (Table 12.2).

The LC/ESI/MS (LC/MS by means of ESI) method was applied for determination of atenolol (β-blocker) in nails of patients who had taken this medicine for 6–12 months (Pufal 2004). Mean concentration of atenolol was 0.155 ng mg^{-1} and in comparison with hair (simultaneously sampled from the same patients) results for this drug (1.73 ng mg^{-1}) were lower.

Irving and Dickson (2007) developed a screening LC/MS/MS method for nine sedatives (zopiclone and eight benzodiazepines) and their selected metabolites in human nails and hair samples. The results indicated that the drugs tested were incorporated into nails at concentration levels similar to those in hair. The levels of the drugs in subjects on regular medication suggested that a single dosage of some of these drugs could be detectable.

The possibility of using fingernails and toenails for determination of psychotropic drugs like haloperidol (Pufal and Piotrowski 2006) or flupentixol (Pufal et al. 2008) was also studied by LC/ESI/MS.

Antipsychotic haloperidol was quantified in fingernail and toenail samples collected from subjects who had been administered this medicament at least 6 months prior to nail sampling. The analysis demonstrated the presence of haloperidol in fingernails in a concentration of 67.3 ± 6.49 pg mg^{-1} and in toenails in a concentration of 98.9 ± 9.14 pg mg^{-1}.

Flupentixol (antidepressant and antipsychotic drug) was determined in fingernails and toenails obtained from patients who had been administered this drug for at least 12 months before sample collection. The nails were taken at 4, 6, 7, 8, and 10 months after administration of flupentixol was discontinued, and 10 months after finishing the therapy, this drug was no longer found in nails.

LC/MS method was also applied for analysis of various tricyclic/tetracyclic antidepressants and neuroleptic phenothiazines in nails originating from victims of suicide by hanging (Skutera et al. 2008). The following concentrations of the examined drugs were found: 0.5–65.1 ng mg^{-1} (amitriptyline), 2.8–4.4 ng mg^{-1} (mianserin), 3.3 – 20.6 ng mg^{-1} (chlorpromazine), 3.3 ng mg^{-1} (levomepromazine), and 26.0 ng mg^{-1} (promazine).

Recently, a LC/ESI/MS method for qualitative and quantitative analysis of citalopram (an antidepressant drug of the selective serotonin reuptake inhibitor [SSRI] class) and desmethylcitalopram in nails and hair was developed (Pufal et al. 2010). Determinations of the parent drug and its metabolite were performed in hair and nail samples collected from individuals who had been administered citalopram in therapeutic doses for at least 12 months before sample collection. Hair and nails samples were obtained 4, 6, 9, and 15 months after discontinuing drug administration. The concentrations of citalopram and desmethylcitalopram in nails were 0.40–10.49 and 0.32–3.70 ng mg^{-1}, respectively. For comparison, the concentration ranges for the parent drug and its metabolite in hair were 1.04–8.69 and 0.07–1.27 ng mg^{-1}, respectively.

TABLE 12.2
Determination of Medicaments in Nails

Drug	Sample Amount (mg)	Sample Preparation Procedure	Analytical Method	Extraction Recovery (%)	Detection Limit (LOD) or Quantification Limit (LOQ)	Precision (%, Relative Standard Deviation)	References
Atenolol	100	1. Decontamination by water and then acetone 2. Cutting into 1–2-mm segments 3. Digestion with 0.1 M HCl 4. LLE with dichloromethane	LC/ESI/MS	—[a]	LOD = 0.025 ng mg^{-1}	6.9 (interday)	Pufal (2004)
Alprazolam, clobazam, clonazepam, diazepam, midazolam, oxazepam, temazepam, triazolam, and zopiclone	ca. 50	1. Decontamination by ethanol 2. Cutting into 1–2-mm pieces 3. Digestion with the mixture of trifluoroacetic acid and methanol (1:50) 4. LLE with dichloromethane	LC/MS/MS	—[a]	LOD = 0.01–0.60 pg mg^{-1}	5–59 (for four compounds: diazepam, nordiazepam, zopiclone, and *N*-desmethyl zopiclone)	IIrving and Dickson (2007)
Haloperidol	100	1. Washing with water and then acetone 2. Cutting into 1-mm segments 3. Digestion with 1 M NaOH 4. LLE with *n*-hexane/chloroform (7:3, v/v)	LC/ESI/MS	—[a]	LOD =1.0 pg mg^{-1} LOQ = 2.5 pg mg^{-1}	3.1–16.5	Pufal and Piotrowski (2006)

Flupentixol	≥50	1. Decontamination with water and then with acetone in ultrasonic bath 2. Cutting 3. Digestion with 1 M NaOH 4. LLE with *n*-hexane	LC/ESI/MS	—[a]	LOD = 1.0 pg mg^{-1} LOQ = 2.5 pg mg^{-1}	4.2–9.1	Pufal et al. (2008)
Promazine, chlorpromazine, levomepromazine, thioridazine, amitriptyline, mianserine, and clomipramine	—[a]	1. Digestion with 1 M NaOH 2. LLE with chloroform from pH 8–10 solution	LC/MS	—[a]	—[a]	—[a]	Skutera et al. (2008)
Citalopram and desmethylcitalopram	—[a]	1. Decontamination with water and then with *n*-hexane in ultrasonic bath 2. Cutting 3. Digestion with 1 M NaOH (at 95°C for 10 min) 4. LLE with dichloromethane (pH 8–10)	LC/ESI/MS	—[a]	LOD = 0.01 ng mg^{-1} LOQ = 0.05 ng mg^{-1}	5.2 (intragroup) 5.7 (intergroup)	Pufal et al. (2010)

[a] No data were given.

12.6 CONCLUSION

In comparison to hair, there are a relatively limited number of methods for drug analysis in nails; however, since the nineties, this material has attracted more attention, especially from toxicological analysts.

Studies of drugs (drugs of abuse, as well as medicaments, especially psychotropic drugs) in nail samples revealed that this biological specimen may be used as a complementary or alternative material for hair. Due to its long-term window, nails, similarly to hair, may serve as a potentially useful biological matrix for detection of past use of narcotic drugs in drug abusers. In some cases, morphine and cocaine concentrations were higher in toenails than in hair. Nails also appeared to be useful specimens for detection of medicaments (e.g., psychotropic drugs) after their chronic administration. The study of concentrations of some sedatives from the benzodiazepine group also revealed that even a single dosage of these medicaments could be detected.

Besides the previously mentioned potential of nails as a biological surrogate for drug analysis, advantageous features like noninvasive sampling and stability of drugs in this tissue are also emphasized.

The other direction of nail analysis for drug determination is investigation of pharmacokinetics and pharmacodynamics of some therapeutic agents used for nail therapy. These investigations may improve the efficacy of treatment of nail diseases.

REFERENCES

Bu, W., X. Fan, H. Sexton, and I. Heyman. 2010. "A Direct LC/MS/MS Method for the Determination of Ciclopirox Penetration across Human Nail Plate In Vitro Penetration Studies." *Journal of Pharmaceutical and Biomedical Analysis* 51: 230–5.

Cingolani, M., S. Scavella, R. Mencarelli, D. Mirtella, R. Froldi, and D. Rodriquez. 2004. "Simultaneous Detection and Quantification of Morphine, 6-Acetylmorphine, and Cocaine in Toenails: Comparison with Hair Analysis." *Journal of Analytical Toxicology* 28: 128–31.

Cirimele, V., P. Kintz, and P. Mangin. 1995. "Detection of Amphetamines in Fingernails: An Alternative to Hair Analysis [Letter to the Editors]." *Archives of Toxicology* 70: 68–9.

De Doncker, P., J. Decroix, G. E. Pierard, D. Roelant, R. Woestenborghs, P. Jacqmin et al. 1996. "Antifungal Pulse Therapy for Onchomycosis. A Pharmacokinetic and Pharmacodynamic Investigation of Monthly Cycles of 1-Week pulse Therapy with Itraconazole." *Archives of Dermatology* 132: 34–41.

de Oliveira, C. D. R., M. Rohesing, R. M. de Almeida, W. L. Rocha, and M. Yonamine. 2007. "Recent Advances in Chromatographic Methods to Detect Drugs of Abuse in Alternative Biological Matrices." *Current Pharmaceutical Analysis* 3: 95–109.

Garside, D., J. D. Ropero-Miller, B. A. Goldberger, W. F. Hamilton, and W. R. Maples. 1998. "Identification of Cocaine Analytes in Fingernail and Toenail Specimens." *Journal of Forensic Sciences* 43: 974–9.

Gerhardson, L., V. Englyst, N. G. Lundstrom, G. Nordberg, S. Sandberg, and F. Steinvall. 1995. "Lead in Tissues of Deceased Lead Smelter Workers." *Journal of Trace Elements in Medicine and Biology* 9: 136–43.

Irving, R. C., and S. J. Dickson. 2007. "The Detection of Sedatives in Hair and Nail Samples Using Tandem LC-MS-MS." *Forensic Science International* 166: 58–67.

Jenkis, A. J., and D. A. Engelhart. 2006. "Phencyclidine Detection in Nails." *Journal of Analytical Toxicology* 30: 643–4.

Kim, J. Y., J. C. Cheong, M. K. Kim, J. I. Lee, and M. K. In. 2008. "Simultaneous Determination of Amphetamine-Type Stimulants and Cannabinoids in Fingernails by Gas Chromatography-Mass Spectrometry." *Archives of Pharmacal Research* 31: 805–13.

Kim, J. Y., S. H. Shin, and M. K. In. 2010. "Determination of Amphetamine-Type Stimulants, Ketamine and Metabolites in Fingernails by Gas Chromatography-Mass Spectrometry." *Forensic Science International* 194: 108–14.

Lander, H., P. R. Hodge, and C. S. Crisp. 1965. "Arsenic in the Hair and Nails: Its Significance in Acute Arsenical Poisoning." *Journal of Forensic Medicine* 12: 52–67.

Lemos, N. P., R. A. Anderson, and J. R. Robertson. 1999. "Nail Analysis for Drugs of Abuse: Extraction and Determination of Cannabis in Fingernails by RIA and GC-MS." *Journal of Analytical Toxicology* 23: 147–52.

Lemos, N. P., R. A. Anderson, and J. R. Robertson. 2000a. "The Analysis of Methadone in Nail Clippings from Patients in a Methadone-Maintenance Program." *Journal of Analytical Toxicology* 24: 656–60.

Lemos, N. P., R. A. Anderson, R. Valentini, F. Tagliaro, and R. T. A. Scott. 2000b. "Analysis of Morphine by RIA and HPLC in Fingernail Clippings Obtained from Heroin Users." *Journal of Forensic Sciences* 45: 407–212.

Miller, M., R. Martz, and B. Donnelly. 1994. "Drugs in Keratin Samples from Hair, Fingernails and Toenails." Presented at the Second International Meeting on Clinical and Forensic Aspect of Hair Analysis [Abstract], Genoa, Italy, 6–8 June, 39.

Pounds, C. A., E. F. Pearson, and T. D. Turner. 1979. "Arsenic in Fingernails." *Journal of the Forensic Science Society* 19: 165–73.

Pufal, E. 2004. "Research on the Determination of Xenobiotics in Epidermal Products and Their Usefulness in Forensic Toxicology." *Problems of Forensic Sciences* 59: 38–49.

Pufal, E., and P. Piotrowski. 2006. "Determination of Haloperidol in Fingernail/Toenails by LC-ESI-MS." *Archiwum Medycyny Sądowej i Kryminologii* 56: 187–90.

Pufal, E., P. Piotrowski, G. Rochholz, C. Franzelius, and K. Śliwka. 2008. "Stability of Ticlopidine in Human Biological Samples." *Problems of Forensic Sciences* 74: 168–72.

Pufal, E., M. Sykutera, T. Nowacka, A. Stefanowicz, and K. Śliwka. 2010. "Development of a Method for Estimation of Citalopram and Desmethylcitalopram in Nails and Hairs and Its Usefulness in Forensic Toxicology." *Archiwum Medycyny Sądowej i Kryminologii* 60: 216–22.

Pufal, E., M. Sykutera, and P. Piotrowski. 2008. "Development of a Method for Determining Antidepressant Drugs in Nails and Its Usefulness in Forensic Toxicology." *Archiwum Medycyny Sądowej i Kryminologii* 58: 167–70.

Saito, K., R. Saito, Y. Kikuchi, Y. Iwaszki, R. Ito, and H. Nakazawa. 2011. "Analysis of Drugs of Abuse in Biological Specimens." *Journal of Health Science* 56: 472–87.

Schatz, F., M. Brautigam, E. Dobrowolski, I. Effendy, H. Haberl, H. Mensing et al. 1995. "Nail Incorporation Kinetics of Terbinafine in Onchomycosis Patients." *Clinical and Experimental Dermatology* 20: 377–83.

Skutera, M., E. Pufal, and E. Bloch-Bogusławska. 2008. "Determination of Antidepressants in Biological Material Originating from Victims of Suicide by Hanging." *Archiwum Medycyny Sądowej i Kryminologii* 58: 177–81.

Suzuki, O., H. Hattori, and M. Asano. 1984. "Nails As Useful Materials for Detection of Methamphetamine and Amphetamine Abuse." *Forensic Science International* 24: 9–16.

Wilhelm, M., D. Hafner, I. Lombeck, and F. K. Ohnesorge. 1991. "Monitoring of Cadmium, Copper, Lead and Zinc Status in Young Children Using Toenails: Comparison with Scalp Hair." *The Science of the Total Environment* 103: 199–207.

Index

G

H

I

O

P

Q

R

S

T

U